WEYERHAEUSER ENVIRONMENTAL BOOKS

William Cronon, Editor

WEYERHAEUSER ENVIRONMENTAL BOOKS explore human relationships with natural environments in all their variety and complexity. They seek to cast new light on the ways that natural systems affect human communities, the ways that people affect the environments of which they are a part, and the ways that different cultural conceptions of nature profoundly shape our sense of the world around us. A complete list of the books in the series appears at the end of this book.

9914

HOW TO READ THE AMERICAN WEST

A Field Guide

WILLIAM WYCKOFF

UNIVERSITY OF WASHINGTON PRESS

Seattle and London

How to Read the American West is published with the assistance of a grant from the Weyerhaeuser Environmental Books Endowment, established by the Weyerhaeuser Company Foundation, members of the Weyerhaeuser family, and Janet and Jack Creighton.

Printed and bound in the United States of America
Design by Thomas Eykemans
Composed in Chaparral, typeface design by Carol Twombly
Display type set in Klinic Slab, by Joe Prince
17 16 15 14 5 4 3 2 1

Photographs are by William Wyckoff unless otherwise credited.

UNIVERSITY OF WASHINGTON PRESS
PO Box 50096, Seattle, WA 98145, USA
www.washington.edu/uwpress

LIBRARY OF CONGRESS CATALOGING-IN-PUBLICATION DATA
Wyckoff, William.
How to read the American West : a field guide / William Wyckoff.
pages cm. — (Weyerhaeuser environmental books)
Includes bibliographical references and index.
ISBN 978-0-295-99351-5 (pbk. : alk. paper)
1. West (U.S.)—Historical geography.
2. Regionalism—West (U.S.)
3. Landscapes—West (U.S.)
4. Landscapes—West (U.S.)—Pictorial works.
5. West (U.S.)—Pictorial works.
I. Title.
F590.6.W94 2014
978—dc23 2013031703

The paper used in this publication is acid-free and meets the minimum requirements of American National Standard for Information Sciences—Permanence of Paper for Printed Library Materials, ANSI Z39.48–1984.∞

FRONTISPIECE Sun City, Arizona

CONTENTS

7
RESTAURANT
Cocktail Lounge
urant

FOREWORD

The Many Joys of Landscape Reading

William Cronon

I SHOULD CONFESS HERE AT THE BEGINNING OF WILLIAM WYCKOFF'S wonderful new book that I share with its author a passion for landscape reading—a passion that those who have never experienced this particular pleasure may find eccentric. My students will tell you that I am fond of announcing in class that there is no richer, more complex, more subtly suggestive or revealing historical document in all the world than a landscape. In my large lecture courses, I always include one day when discussion sections leave their classrooms behind and wander across campus to puzzle together over such questions as how the age of buildings can be found in their appearance and the materials from which they are made; why lawns and trees are planted and laid out as they are; and how ideas of nature are expressed in the geographical arrangements of human structures. In smaller classes, one of my greatest delights (which I hope my students wind up sharing) is the full day we spend driving around southern Wisconsin to view everything from roads to houses to parking lots, from quarries to prairies to orchards, from farms to forests to parks, from malls to fast-food restaurants to abandoned industrial sites, and so on and on. Viewed in this multifaceted way, places that we pass every day without much thought suddenly become fascinating and mysterious. Each landscape has endless stories to tell if only we understand the codes that render their details, their surfaces and depths, their peculiarities and contradictions legible.

This fascination for landscape reading is shared among many academic disciplines, though each approaches the exercise with radically different goals. A geologist is perennially on the lookout for places where rocks, usually obscured by vegetation, reveal the underlying stories of deep time. An ecologist focuses on plants and animals to discover the biogeographical relationships that sustain forests, grasslands, marshes, and deserts. A landscape architect or architectural historian looks at human structures, seeing in their layout and design echoes of stylistic and engineering traditions that shape the built environment. I myself leap at the opportunity to accompany these and other specialists who can teach me ways of looking that wouldn't otherwise occur to me, and I urge everyone else to do the same. But if I had to pick one group of colleagues with whom to wander the landscapes of my native land (or any other place, for that matter), I would probably choose geographers more often than not. No one approaches places with more voracious curiosity than they, and no one is more skilled at borrowing from disciplines

Rancho Bar 7 Restaurant, Wickenburg, Arizona

ranging across the sciences and humanities to understand the myriad forces that have shaped a landscape and the stories it has to tell.

Bill Wyckoff, one of the preeminent historical geographers of his generation, is among the best landscape readers I know. Bill combines his disciplinary expertise with a love of travel—almost an obligatory trait among geographers—and a passion for photography, so that his wide-ranging journeys across the American West have given him a deep sense of the best places to illustrate different geographical phenomena. In 2006, he contributed a volume to our Weyerhaeuser Environmental Books series entitled *On the Road Again: Montana's Changing Landscape*, in which he compared old and new photographs to interpret the twentieth-century environmental history of Montana. When that book was finished and we began to talk about whether he might have another project for the series, Bill suggested a photographic field guide to the landscapes of the American West. We could not have been more delighted, and the book you now hold in your hands is the result of those early conversations. *How to Read the American West* combines Bill's knowledge and photographs—all of which he shot afresh for this volume—with the work of editors and designers at the University of Washington Press, with the additional bonus that new printing technologies have made it possible for the first time to illustrate a book in this series entirely in color.

No single volume could ever provide a comprehensive or encyclopedic guide to any landscape, let alone to a region as vast and variegated as the Trans-Mississippi West. Instead, *How to Read the American West* offers something that is arguably better: what Wyckoff calls an "invitation" to join him in visiting some of the most diverse but characteristic landscapes that travelers will likely encounter while exploring the region today. Better still, Bill has designed the book so that his eclectic images are combined with equally eclectic texts that instruct readers not just about particular landscapes but also about the craft of landscape reading itself. As a result, you can enjoy and profit from this guidebook whether you carry it in your glove compartment to consult along the way or read it at your leisure in an easy chair back home. I myself will make sure there is a copy in any vehicle I drive west in the future!

Among the most important lessons this book teaches is that you never want to look at a landscape in just one way. Peruse the chapters and images in chapter 1, on "Nature's Fundament," and you'll learn to recognize the many natural processes and foundations that shape all other landscape features in the West. You'll learn the distinguishing features of everything from lava flows and ash deposits to ancient ocean beds and the tracks of glaciers. Each of these generates characteristic soils, which influence the vegetation that ranges across climate and elevation gradients from deserts to forests. The soils in turn provide the biological contexts for everything from wildfires to wildlife, with natives and exotics jostling each other hither and yon. Above them all is the sky, and Wyckoff encourages us to lift up our eyes to ponder its many meanings as well.

That's just chapter 1. It is followed by chapters on landscapes that reflect different ways of making a living in the West, whether on farms and ranches, in mines and industrial forests—or, for that matter, in tourist resorts and playgrounds. To reach such places, people must travel by foot or car or train or plane, each of which leaves its own distinctive marks on the landscape.

This book will help you understand those as well. Although the West has always been predominantly wild or rural in our national mythology, the vast majority of westerners since far back in the nineteenth century have lived in urban settings, so Wyckoff devotes considerable space to western cityscapes. He also helps us see the many footprints of the federal government, without which the American West would be very different from what it is today. And finally, he shows how groups from different ethnic and religious backgrounds have inscribed their identities on the western terrain.

If it all feels a little scattered and jumbled, that's the way it is with landscapes, especially in the modern era. Opening your mind to the full complexities of a landscape can feel overwhelming, as the seeming cacophony of objects and influences and processes can be hard to untangle. That is precisely why it's so helpful to have a guide like Bill Wyckoff, who knows the ground, has a practiced eye, and can offer suggestions to hold confusion at bay. Take him along as a companion in your western rambles, and there's no telling where you might go or what you might see that you otherwise might have missed.

If you want to get the most out of *How to Read the American West*, I'd encourage you to wander its pages in a spirit of play, much as you would a landscape itself. It's a good idea to spend some time with the table of contents to get a general sense of the book's organization, but from there you can go pretty much anywhere you wish. Read it from cover to cover or dip in and wander around as your fancy takes you. The more familiar you become with its contents, the more easily you'll find your way in the field to the images and interpretations that will remind you how best to understand the places you visit. Once you can identify the features that William Wyckoff puts before you in these pages, you'll be well on your way to reading the western landscape for yourself, with endless stories waiting to be discovered wherever you look. ❋

PREFACE

PREPARATION FOR THIS PROJECT BEGAN EARLY. FOR MANY YEARS, I traveled in the backseat of the family sedan, venturing with parents and grandparents to every corner of the West. We went on camping trips and fishing excursions, attended family reunions, and more. One of my earliest recollections (age six) is traveling east along old U.S. 66 from Los Angeles to Oklahoma City. My grandfather Ralph and Uncle Carl manned the front and I shared the backseat with Leona, my grandmother. We each had jobs: Leona was responsible for "taking care of the boys" up front, which meant occasional reminders to watch their speed and distributing to them judicious doses of sour mash bourbon from a small silver flask. My job was to write down the name, population, and elevation of every town we passed through on a little pad of paper. It was a pretty good system. Among the four of us we made excellent time, and I amassed a long list of towns and a fragmented set of roadside memories that I still retain more than a half-century later.

Perhaps the die was cast. In any event, this book, while it was the focus of my energies only between 2008 and 2013, no doubt bears witness to a lifetime of both travel and academic study across the West. It reflects all the benefits and limitations of my own take on things. It is meant to be a "field guide" in the broadest sense of that term. I wrote it not as a definitive, encyclopedic summation of the West or its landscapes, but rather as an invitation to readers to explore the region with a modest toolbox of ideas, examples, and sensibilities that might enrich their journeys. Look at it as an extended conversation between fellow travelers as they roll down the highway.

How did I come up with one hundred landscape features to explore across the West? Why choose some, and not others? Where to find examples to illustrate them? As you will quickly discover, there is nothing sacrosanct about my list, but I do hope it at least opens the door to your own western journeys. Simply put, the list accumulated slowly over the years, benefited from a basic knowledge of the larger scholarly literature on western American landscapes (see Further Reading) and some valuable feedback from selected colleagues (see the Acknowledgments), and then evolved in a healthy empirical fashion as I put on many miles (more than twenty thousand for this project alone) across the eleven western states and completed visits to all 414 western counties over a five-year period. I was committed to discovery, not verification, and was well rewarded. You really *do* learn from looking, preferably with an open mind.

Attempting to explore individual landscape features in this way also taught me a lot about the interconnectedness of landscape elements, and you will find many cross-references to related features within each description—flagged thus (**36**)—throughout.

Old U.S. 66 east of Grants, New Mexico

I harvested more than fifteen thousand images and talked with hundreds of people in dozens of small-town bars and big-city coffee shops. Many of my best memories involve catching just the right image at the right time of day, often by pure luck. As you wander through the photographs in the pages ahead, know for certain how wonderful it was to be behind the camera in all these places, exploring something in a fresh way and passing on that experience to others.

So what were my most vivid memories? They were a mix of landscapes and personal experiences that included Sunday services at Mission San Xavier del Bac, fall color in Colorado's Collegiate Range, meeting Hopi artist Lawrence Namoki, listening to Latina hip-hop at a bar in Albuquerque, hiking at Point Reyes on a sunny December day, meeting a young migrant worker and her baby in eastern Colorado, scrambling up Angels Landing in Zion National Park, listening to the cottonwoods in Frenchglen, Oregon, surveying Chaco Canyon from a windy mesa top, hanging out at the Venice Beach pier, hiking to the crucifix on the hill above Chimayó in northern New Mexico, being accused by a Wyoming sheriff of being a potential industrial terrorist (be careful what you photograph!), playing pool at the Quartzsite Yacht Club Bar and Grill, and sampling great noodle soup in Southern California's Little Saigon. Each of these experiences and many more are distilled in one form or another throughout this field guide. My hope is that the book offers an incentive for readers to build their own lists. ❋

ACKNOWLEDGMENTS

IN ADDITION TO ALL THE PEOPLE WHO SHARED IDEAS WITH ME INFORmally as I journeyed around the West, special thanks go to Don Meinig and Lary Dilsaver, who spent time sharpening and adjusting my approach and my shifting list of western landscape features to emphasize. Mark Fiege, Paul Starrs, Gary Hausladen, Dale Martin, Mary Murphy, Bill Preston, Matthew Fockler, and Kevin Blake also reviewed elements of the project, and their support and encouragement were vital. Sharing ideas with several graduate classes was also crucial in honing and multiplying examples. More generally, so many of the names that appear in the Further Reading section are friends and professional colleagues. Many of their ideas and insights appear in the book. While I take full responsibility for every word and image, I appreciate their cumulative contributions throughout the text. I also benefited greatly from the comments and observations of anonymous reviewers of the manuscript.

Many other people and institutions also made this project possible. Special thanks go to Montana State University and to the Department of Earth Sciences. Dave Lageson, Steve Custer, and Dave Mogk all offered key personal and departmental support. In particular, a sabbatical leave in 2011–2012 allowed me to complete an initial draft of the manuscript. In addition, visits to the Denver Public Library, History Colorado Center, Wyoming State Archives, Utah Historical Society, Nevada Historical Society, Arizona Historical Society, and Montana Historical Society were instrumental in selecting historical images and in exploring additional historical background on many topics. Thanks to their staffs for their courtesy, insights, and good cheer. Special appreciation goes to Tanya Buckingham at the University of Wisconsin Cartography Laboratory for her advice on graphics and to Lohnes + Wright for designing the book's maps and diagrams.

Additional thanks go to many individuals who gave generously of their time in offering me on-site observations, tours, and guidance. In particular, I enjoyed firsthand insights and tours at Verrado, Arizona (a master-planned community), an open-pit gold mine in Nevada (courtesy of Newmont Mining Corporation), Oregon's Big Muddy Ranch (a former cult community), The Lodge at Ventana Canyon Golf and Racquet Club (near Tucson), the Arvin Farm Labor Center (near Bakersfield), Acoma Pueblo (in western New Mexico), Natalie's Estate Winery (western Oregon), the Tanque Verde Ranch (southern Arizona), Magdalena's old bank building (western New Mexico), El Centro de los Pobres (a help center for migrant workers in Avondale, Colorado), and at many other localities around the region.

Special thanks also go to my extended family at the University of Washington Press. Marianne Keddington-Lang was a terrific editor and sounding board, and I am forever grateful to her keen eye for efficient prose. In addition, her many outstanding colleagues on the staff made every step of this project easier. In particular, I appreciated the sustained and inspirational guidance from both Jacqueline Volin and Tom Eykemans. Don James also offered excellent suggestions and corrections throughout the manuscript. At the Press in Seattle, I especially benefited from the sustained support and guidance of both Pat Soden and Nicole Mitchell. And without Bill Cronon's ceaseless enthusiasm and insights, this book would never have happened. Repeatedly, his wisdom and instincts combined to push this project forward to completion. Once again, it is also a privilege to be a part of the Weyerhaeuser Environmental Books series, and I am grateful to the Weyerhaeuser Company Foundation for its generous assistance in the publication of this book.

Thanks go to my own family for a lifetime of support. My parents and grandparents shared a love of the West, and seeing it through their eyes marked a great beginning to my own understanding of the region. In particular, my dad and I cruised and camped across much of the West in the early 1970s, catching many localities as they were quietly transitioning between "Old West" and "New." Those trips whetted my appetite to continue my explorations in later years, and they gave me a visual baseline of memories and landscapes that still helps me make sense of what I see today. My thanks also go to son Tom and daughter Katie, who were often both willing and captive participants on long family trips. Seeing the West, from the Oregon coast to Zion National Park, through their eyes also taught me a great deal. Finally, special thanks go to my wife, Linda, for her patience, love, and good cheer as I headed out on yet another trip and for her advice and suggestions as the manuscript came together. She was with me on every mile of the journey. ❋

HOW TO READ THE AMERICAN WEST

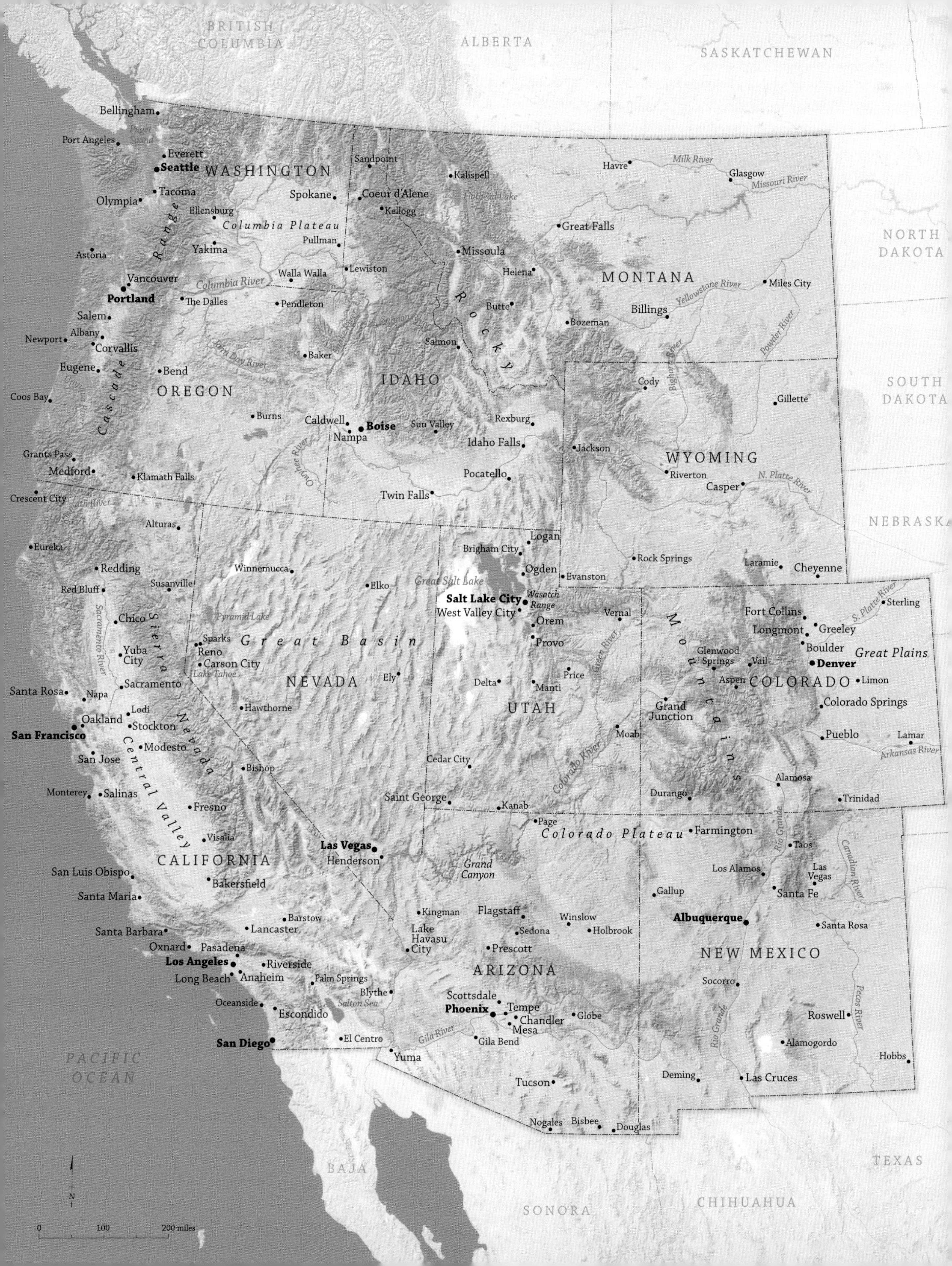

BRITISH COLUMBIA
ALBERTA
SASKATCHEWAN
Bellingham
Puget Sound
Port Angeles
Everett
Seattle
WASHINGTON
Sandpoint
Kalispell
Havre
Milk River
Glasgow
Missouri River
Tacoma
Olympia
Spokane
Coeur d'Alene
Kellogg
Flathead Lake
Ellensburg
Columbia Plateau
Great Falls
Pullman
NORTH DAKOTA
Astoria
Yakima
Missoula
Lewiston
Helena
Walla Walla
Vancouver
Columbia River
MONTANA
Miles City
Yellowstone River
Portland
The Dalles
Pendleton
Butte
Billings
Salem
Bozeman
Albany
Newport
Corvallis
John Day River
Salmon
Powder River
Baker
Bighorn River
Eugene
Bend
IDAHO
Cody
SOUTH DAKOTA
Coos Bay
OREGON
Gillette
Umpqua River
Cascade Range
Rocky Mountains
Burns
Caldwell
Boise
Nampa
Sun Valley
Rexburg
Idaho Falls
Jackson
WYOMING
Grants Pass
Owyhee River
Medford
Klamath Falls
Pocatello
Riverton
N. Platte River
Casper
Crescent City
Twin Falls
Klamath River
NEBRASKA
Alturas
Logan
Eureka
Brigham City
Rock Springs
Laramie
Cheyenne
Redding
Winnemucca
Ogden
Great Salt Lake
Evanston
Red Bluff
Susanville
Elko
Salt Lake City
Wasatch Range
S. Platte River
Sterling
West Valley City
Vernal
Fort Collins
Chico
Sierra Nevada
Pyramid Lake
Orem
Longmont
Greeley
Sacramento River
Sparks
Great Basin
Provo
Green River
Boulder
Great Plains
Yuba City
Reno
Glenwood Springs
Vail
Denver
Carson City
Lake Tahoe
Price
Ely
NEVADA
Delta
Manti
Aspen
COLORADO
Limon
Santa Rosa
Sacramento
Napa
Grand Junction
Colorado Springs
Hawthorne
UTAH
Lodi
Oakland
Stockton
Moab
Pueblo
Lamar
San Francisco
Modesto
Arkansas River
San Jose
Central Valley
Bishop
Cedar City
Colorado River
Alamosa
Monterey
Salinas
Saint George
Durango
Trinidad
Fresno
Kanab
Rio Grande
Page
Colorado Plateau
Farmington
Visalia
Las Vegas
Taos
Canadian River
CALIFORNIA
Henderson
Grand Canyon
Los Alamos
Las Vegas
San Luis Obispo
Bakersfield
Santa Fe
Santa Maria
Gallup
Kingman
Flagstaff
Barstow
Winslow
Albuquerque
Santa Barbara
Lancaster
Sedona
Holbrook
Santa Rosa
Lake Havasu City
Oxnard
Pasadena
Prescott
NEW MEXICO
Los Angeles
Riverside
ARIZONA
Long Beach
Anaheim
Palm Springs
Blythe
Socorro
Oceanside
Salton Sea
Scottsdale
Tempe
Pecos River
Escondido
Phoenix
Chandler
Globe
Roswell
Mesa
Rio Grande
San Diego
El Centro
Gila River
Gila Bend
Alamogordo
PACIFIC OCEAN
Yuma
Hobbs
Deming
Las Cruces
Tucson
Nogales
Bisbee
Douglas
BAJA
TEXAS
N
SONORA
CHIHUAHUA
0
100
200 miles

NAVIGATING WESTERN LANDSCAPES

An Introduction

LANDSCAPES TELL GREAT STORIES. BUT WE NEED TO KNOW WHERE TO look for them and how to make sense of what we find. This field guide to the American West is designed to help you do just that: open your eyes and examine your everyday surroundings—your neighborhood, what you observe on your drive to work, or the scenery you might encounter on your cross-country jaunt—and appreciate in fresh new ways the significance of what you see.

So where to begin such an exploration? The best lessons are learned with intimacy—tactile, terrestrial encounters with your feet planted on the ground. For instance, imagine walking through the quiet streets of Magdalena, a small town in western New Mexico (population nine hundred) that blossomed briefly in the late nineteenth century. Magdalena served as a railhead for livestock being shipped eastward to market. In 1919, when activity there peaked, more than 20,000 cattle and 150,000 sheep passed through. Today, most travelers barely slow down as they cruise through town.

Magdalena's landscape tells many stories. The bank, which was once a symbol of the town's economic promise—the 1906 brick building still stands on the corner of North Main and U.S. 60 (fig. I.1)—went bust during the Great Depression. In its later incarnations, as the accumulation of signs suggests, the building was used as a drugstore and soda fountain, then as Evett's Café. More recently, Laurie Gregg opened the Village Press, an art classroom and print studio. The building keeps being reinvented, chapters added to its century-long story. Outside the bank's front door, "WPA 1938" is chiseled on the sidewalk (fig. I.2). The Works Progress Administration (WPA), part of President Franklin Roosevelt's New Deal, invested in thousands of similar infrastructure projects, giving the West much-needed paved roads, sidewalks, school buildings, county courthouses, and water treatment plants. Many similar signatures of the New Deal endure across the region.

Both the bank building and the faded lettering on the sidewalk outside are reminders that stories are everywhere in the western landscape, often in plain sight. *How to Read the American West* helps you identify such features and suggests how they fit into larger narratives about the history and geography of the West. This field guide also suggests ways to approach looking at landscapes, the rules of engagement you can use to enrich your journey.

The eleven states of the American West

FIG. I.1 Bank building, Magdalena, New Mexico. Many small-town banks across the West met a fate similar to this one's. The building later served as a drugstore, a café, and an art studio and classroom.

FIG. I.2 WPA 1938, Magdalena, New Mexico. The sidewalk in front of the bank was improved with New Deal funds.

The term *landscape* in this book certainly includes the natural world, but our emphasis will be on the material, tangible expression of human settlement on the surface of the earth and how these cultural signatures are everywhere embedded within their larger environmental settings. This notion of landscape also embodies how people—through their experiences—connect these visible elements to place-based meanings and regional identity. It turns out that ordinary sights such as weathered bank buildings and aging sidewalks can tell us a great deal about ourselves, how the West has changed over time, and what the regional landscape may look like tomorrow.

The structure of this book is straightforward. In sections organized around several western settings and themes, it explores a hundred characteristic landscape features in the West. Each feature is described and illustrated. The approach of the text is interpretive, not encyclopedic: representative examples of features are considered, along with useful ways to understand them and

FIG. I.3 Front yards, Fort Collins subdivision. These Colorado homes reflect an emphasis on privacy and security. Small shrubs, trees, and patches of lawn are easy to maintain, but residents spend little time in these tiny front-yard settings. Most social life takes place on the back patio.

see them within the settings where they are found. Whether you are walking through a farm town or a suburb (fig. I.3), this guide will help you make sense of what you see.

Some landscapes in the book are common across the entire United States—commercial strips, golf courses, retirement communities—but in the West they have a particular regional expression. Other landscape elements are unique to the region—dude ranches, Japanese internment camps, hillside letters. All of these features—as extraordinary as Yellowstone National Park or as ordinary as a strip mall—can tell you something about the history of the West and the forces that have shaped it.

TIPS FOR NAVIGATING WESTERN LANDSCAPES

How can you look freshly at everyday western places? As you begin your exploration, consider the following ways of thinking about any western landscape.

Appreciate the role of time.[1] Determining the age of a landscape can be difficult, because the agents that shaped it originated at many different points in time. Wherever you are in the West, you can begin with the deep time of geological making, the long millennia of mountain building and erosion that shaped features as large in scale as Mount Rainier or the Grand Canyon or as small as a rocky hillside. Then you can juxtapose that geological time scale against the newer imprint of Native American peoples, who changed landscapes through their settlement patterns and influences on the distribution of certain plants and animals. Next, Europeans arrived, including seventeenth-century farmers from New Spain and nineteenth-century miners from the Midwest. If you look carefully, you may see evidence of surviving property lines and local roads, perhaps even an old farm or a miner's cabin. Then you can drape across those earlier human signatures the more recent transformations that produced a sprawling landscape of subdi-

FIG. I.4 Adapting to nature, Imperial Valley, California. Plants, animals, and people all find ways to cope with the intense heat characteristic of the Southwest. This parking lot west of El Centro provides a bit of shade from the desert sun.

visions and interstate highways. Every landscape is a layered accumulation from the past, and it takes some practice to make discretely visible these strands of time.

Recognize that landscapes are expressions of the interplay between nature and culture.[2] Every landscape in this guide represents the intertwining of people and environment. Think about how people both transform and adapt to their surroundings (fig. I.4). Even the idea of "wilderness" is a human notion, made possible by strict federal mandates and limitations on activities. On the other hand, a suburban shopping mall is also part of a larger physical setting, the buildings and asphalt reconfiguring local drainage patterns and ecological niches for plants and animals. Simply think about the vegetation you see around the West. It might consist of native plants, imported exotics, or, more likely, a mixture of both. Nature and culture are often impossible to disentangle from each other, and the western landscape contains myriad examples of their conjoined expression.

Follow the path of water.[3] The West's fortunes are rooted in water. Every drop of water is money in the West. Where it is abundant, water has blessed western places with possibilities, but water scarcity can be costly. In both rural and urban settings, much of the western landscape has been sculpted by water's uneven and serendipitous geography and the human desire to control it (fig. I.5)—directing Sierra Nevada snowpacks to Southern California, reworking wetlands in Oregon's deserts, or bringing groundwater to the surface in Phoenix. In the urban and suburban West, consider how water is managed. In Stockton, California, or Albuquerque, New Mexico, vast municipal systems deliver a cool shower to parched summer lawns or a warm bath after dinner. Beyond the city, you can follow river valleys, aqueducts, and irrigation ditches and see windmills and stock ponds. In the recreational West, water in some form is often at center stage, whether at a lakeside marina, a fishing hole, or a ski hill topped with fresh powder.

FIG. I.5 Managing water, Southern California. The concrete-lined creek helps to contain rapid runoff from the snowcapped San Bernardino Mountains near Beaumont, and the nearby storage tank (center) provides a steady supply of drinking water for suburban customers.

Recognize the importance of scale.[4] How large is a landscape? Most simply, landscapes have to be visible, so you can begin small, but not too small. Landscapes are captured in cameras and on sketchpads, not through telescopes or microscopes. They are framed by human perception and experience, by what ordinary vision reveals. Consider suburban lot parcels or the distribution of second homes in a resort community. These are important landscape elements—the basic patterns of landownership, the arrangement of land parcels, and the layout and design of structures upon them. You can also flex your imagination and see larger-scale examples of landscape patterns. Approach Los Angeles from the air and contemplate the vast grid of interlaced city streets, all knitted together by freeways. This grander scale of landscape is defined by topography and traffic flows.

In more abstract ways, you can see how western landscapes are connected to the world beyond. Geographers often argue that "no landscape is local," a phrase evoking the notion that every place is connected to and defined by other places.[5] If you walk through an upscale pedestrian mall in Boise, Tucson, or San Diego, you can find merchandise and luncheon entrées from around the world. They reflect the diverse tastes and preferences of residents and a global economy of resources, labor, and capital that produces everything from Nepalese handbags to Japanese sushi. Workers at the mall may include a mix of immigrants and longtime residents who themselves are also a complex expression of larger patterns of movement. Wherever you are in the West, you can appreciate a landscape at different scales and discover how inevitably it is linked to people, money, and institutions sometimes thousands of miles away.

Pay attention to edges in the landscape. Whether you explore the meeting place of grassland and forest, sample a potpourri of urban neighborhoods, or wander along suburban outskirts, ask why patterns in the landscape are changing.[6] Edges in the landscape are always lessons waiting to be learned. Why do valley grasses and sagebrush exist next to a grove of hillside pine trees?

FIG. I.6 Edges in the landscape, Boise, Idaho. This foothill neighborhood above Boise reveals open space bordered by homes. Residents encounter snakes and coyotes, while native birds find cover in shrubbery imported for shade and landscaping.

FIG. I.7 Low-density living, southwestern Montana. Large view lots, widely scattered homes, and a rural feel appeal to residents of this exurban development west of Bozeman. Contrast the settlement pattern with that of the Boise foothills (fig. I.6).

Similarly, you'll find visual cues as you transect the boundaries in Los Angeles between Koreatown, Latino neighborhoods, and an African American community. Church fronts, graffiti tags, housing types, and other urban sights are suggestive, as are the aromas emanating from neighborhood restaurants. Similarly, transitional landscapes exist at the meeting place of suburbs and open lands, a common circumstance on the periphery of western cities (fig. I.6). What is happening along this ecotone (transition zone) and why? What environmental, political, and economic forces are at work?

Develop an eye for measuring landscape density.[7] Landscape density is how much of something you can see in a particular area. For example, think about two distinctive agricultural settings in the West: In Northern California's Napa Valley, you can count a dozen small vineyards (and tasting rooms) in just a few miles along the highway; farms are small and thickly tucked into the narrow valley. In eastern Wyoming, you can count how many miles there are between isolated ranch houses. Both landscapes are agricultural areas, but the density of operating units varies greatly, which can reveal a good deal about the environments and economics that have shaped these quite different settings.

Similarly, study the different ways westerners have arranged themselves on the suburban fringe. On Boise's north side, upscale homes are carefully

arranged on small lots that face planned open space (fig. I.6). Contrast this pattern with the landscape west of Bozeman, Montana, where similarly upscale houses sit scattered amid a low-density landscape of multiacre properties (fig. I.7). One setting acknowledges the virtues of common open spaces and managed wildlife corridors; the other celebrates independence and a penchant for privacy.

FIG. I.8 Old West landscape, Johannesburg, California. This abandoned mine, with its tall head frame (left, housing the hoist that lifts ore to the surface) and tailings (mine residue) pile (right), sits just south of town.

FIG. I.9 New West landscape, Twin Falls, Idaho. This commercial strip, largely shaped by corporate capital, is a landscape superbly designed for the automobile. Familiar signs and symbols help direct motorists and whet appetites.

Ask who controls the landscape.[8] The West has been shaped by American capitalism, first anchored in the region's abundant natural resources, later complemented by a growing industrial base, and in the twenty-first century reworked to include a service-based economy oriented toward recreational and cultural amenities and the mass consumption of consumer goods. Much of the landscape that results from capitalistic impulses—from an Old West mining operation (fig. I.8) to a New West commercial strip (fig. I.9)—has been organized around the American desire to make money. Viewed through this common lens, landscapes can be seen as sites of investment, places produced to make a profit.

Tracing the nineteenth-century evolution of this economic imperative, you can also discover how much of the West's history is rooted in violence, conquest, and inequality (as defined by race, class, and gender). Powerful interests produced the rationale for subjugating Native peoples, exploiting

Chinese laborers and Latino farmworkers, and paying service workers only a minimum wage. Those same interests made political, economic, and legal arguments that produced enduring changes in the landscape. Indian reservations, railroads, grocery stores, and trailer parks are all tangible manifestations of this legacy. Seeing landscapes as sites of accumulated material value can suggest who the key players were in their formation, why they were produced in the first place, who gained and lost in their creation, and how they function in the larger western economy.

Private ownership and control of western landscapes does have its limits, however. In many western cities and suburbs, zoning regulations and subdivision covenants limit what owners can do with their property—what they can plant in their yards, what color they can paint exterior walls, what they can add on to a house. Such rule making, designed to protect housing values and quality of life, can come from private homeowners' associations, local zoning boards, or city and county planning initiatives.[9] And large portions of the West are part of the public lands system, the majority of which is controlled and shaped by the federal government.[10] In Colorado, 36 percent of the land falls under federal control, and the total is even higher in other western states, such as California (48 percent), Idaho (62 percent), and Nevada (81 percent). Each federal land management agency has rules that guide and restrict development, and the practices that shape public landscapes have changed over time.[11] A national park hotel built in 1904, for example, will likely look very different from a motor lodge built in 1975, and a national forest campground established before World War II will have a different look and feel from one created after 1980.

Make connections between western places and the cultures that shaped them. Places reflect the people who inhabit them.[12] The retirement community of Sun City, Arizona, looks and functions differently from a Latino neighborhood in Yakima, Washington, or an upscale Mormon suburb on the edge of Provo, Utah. The West is home to some of the world's most culturally diverse

FIG. I.10 Variation in the landscape, Los Angeles, California. Users of this book will discover a dizzying array of landscape features across the West. Look for features that offer the unexpected. Storefronts in Koreatown celebrate cultural diversity in Southern California.

populations, and you can trace their signatures and see the way their cultural influences have mingled and adapted to new settings (fig. I.10).

Local cultural identities are often rooted in particular localities: Hopis are deeply attached to the high mesas of northeastern Arizona just as Latino residents in East Los Angeles develop strong connections to their local neighborhoods. Place matters to both groups. Geographer Kent Ryden has written eloquently of the "invisible landscapes" embodied in the stories and experiences that people accumulate in a place. Seeing a landscape through the eyes of residents can teach us more about the connections among landscapes, place, and cultural identity.[13]

Understand how the modern latinization of the West is the region's most extensive recent cultural transformation.[14] The Hispanic legacy in places such as northern New Mexico and coastal California predates the Anglo-American presence, but these traditional enclaves of Hispanic settlement have been almost overwhelmed by more recent arrivals from Mexico and the rest of Latin America. A majority of the nation's contemporary Latino populations live in the West, and their cultural imprint reaches from Southern California to southwestern Montana (fig. I.11). Distinctive Central American communities cluster in neighborhoods in Southern California, while Mexican immigrants in Phoenix retain ties to their Sinaloa, Sonora, or Michoacán roots. The result is a regional landscape enriched and complicated by its Latino character, an imprint that has forever changed the West's cultural geography.

Visit vernacular landscapes.[15] Cultural geographer J. B. Jackson has argued that the "vernacular landscape"—by which he means an ordinary landscape that has taken shape through everyday use—reveals a great deal about the daily lives of people.[16] For example, explore the variety of western residential landscapes (figs. I.12 and I.13), and think about how you and your family organize the space around your home. Some indoor spaces (rooms, garage, storage places) and outdoor ones (yard, patios, play areas) may be gendered, that is,

FIG. I.11 Latinized landscape, Phoenix, Arizona. These west Phoenix businesses serve the area's vibrant Latino population. Many local immigrants come from the nearby Mexican city of Nogales, in the state of Sonora. The site also offers connections with home: the Sigue Corporation (sign, center) offers money transfers across the border.

FIG. I.12 Residential landscape, Manti, Utah. This late-nineteenth-century brick home in the Mormon community of Manti celebrates the Victorian era, complete with decorative gables, elaborate front and side porches, and an eclectic Queen Anne–style roofline, all encircled by a properly tended yard and wrought-iron fence.

FIG. I.13 Residential landscape, Albuquerque, New Mexico. Many westerners live in mobile homes like these on Albuquerque's east side. These settings celebrate the virtues of inexpensive, casual, and convenient living. Exterior landscaping is minimal, and living space is efficiently assembled in affordable and movable single- or double-wide units.

FIG. I.14 Working landscape, El Centro de los Pobres, Avondale, Colorado. Church and community workers in prefab metal buildings provide secondhand goods and health care services for Avondale workers. Most clients are migrant farmworkers from nearby fields.

used or valued differently by men and boys than by women and girls. In the yard, do you like fences, walls, and gates, or do you prefer fewer visible boundaries between neighbors? Do you like lawns, cacti, planters, or casual clutter? How many rooms does your house or apartment have? Where are the doors, windows, and chimneys? How about gables, turrets, and towers? We don't usually think of our home as a "landscape," but how domestic space is organized, produced, and used can tell us a good deal about the residents within.

Also consider landscapes of labor. You might begin with larger landscapes, such as a farm in Colorado's Arkansas Valley or a manufacturing district in Albuquerque or Seattle. Most westerners have jobs in the service sector—in mom-and-pop stores, strip malls, suburban office complexes, and myriad other places. In some of those settings, men and women work together, while in others the work space is gendered. What do these geographies of employment tell you about how workers spend their days?

Some jobs bring workers into direct contact with the public (fig. I.14). How and where do these interactions occur? Some workplaces have an exterior arrangement of windows, merchandise, counter space, and cash registers and an interior arrangement of supply rooms, work spaces, and break rooms. Consider how people navigate through the landscapes of their workday. How do things get built or harvested? What tools and equipment are used? Where do people sit, squat, or stand? Paying close attention to these kinds of details can tell you a great deal about how westerners live and take part in a larger economic system.

Inventory symbolic landscapes and representations of places. Exploring these aspects of the built environment—that is, how a community produces its own history and how places and landscapes are represented in that history (fig. I.15)—is an integral part of learning about the visible scene.[17] Examining publicly designed places, monuments, statues, and museums can suggest how communities think about themselves. Casual displays of vernacular art can also reveal the meanings of local places (fig. I.16).

You can find other representations of place identity in souvenir shops and art galleries, where paintings, photography, postcards, travel brochures, T-shirts, shot glasses, key rings, and decorative plates portray a place (fig. I.17). Landscapes might also be described in historical narratives, essays,

FIG. I.15 Cowboy sculpture near Jal, New Mexico. These metallic monuments to western ranching mark the horizon on a remote hill in far southeastern New Mexico. *The Trail Ahead* is a four-hundred-foot-long metallic sculpture (completed in 2000 by Brian Norwood) that was sponsored by the local chamber of commerce.

FIG. I.16 Landscape as art, Superior, Arizona. Alleyway art in the old copper-mining town of Superior makes visible a place identity defined by history, culture, and labor.

FIG. I.17 Souvenirs of the Southwest, Santa Fe, New Mexico. Dried chiles, cow skulls, and sun ornaments are familiar regional souvenirs, all available in this fashionable shop in Old Town Santa Fe.

FIG. I.18 Panorama from Sentinel Peak Park, Tucson, Arizona. This view includes the downtown skyline (right) and the scenic Santa Catalina Mountains.

FIG. I.19 A different frame of reference: Tucson's west side. A closer look at a portion of the scene highlights a low-income Latino neighborhood just west of the freeway.

novels, or poetry. Virtual exploration can be a part of the itinerary: how does a community's Web site portray and market that community? What parts of town are prominent or invisible?

Remember that what you see depends on the experiences you bring with you, the questions you pose, and the details you emphasize. Scanning the Tucson skyline from Sentinel Peak Park can yield either a chamber of commerce panorama of downtown and the nearby mountains (fig. I.18) or a low-income Latino community (fig. I.19).

We all look at landscapes from our own perspective. Gender, ethnic background, and political leanings shape what we see in the world around us. Put in biographical terms, a fifty-something, relatively affluent white male in Montana will not experience a landscape in the same way that a twenty-something Latina from suburban Phoenix does. The key is to be aware of those differences and to respect them. Many landscape meanings are nested within a single place.[18]

CONSTRUCTING WESTERN MYTHS, INVENTING WESTERN LANDSCAPES

Americans have long shared a fascination with the West and its landscapes.[19] Much of that fascination has its origins in the region's history and mythology—cowboys herding cattle, prospectors panning glitter from the gulch, Native Americans scanning the horizon on a distant ridge. Those myths endure and affect the ways we see the landscape. When examining places such as dude ranches, ghost towns, and Indian reservations, it is important to appreciate how myths have influenced their modern expression and significance.

Historian Howard Lamar reminds us that Thomas Jefferson and others helped invent the idea of the West and that early ideas about the West informed larger national myths regarding economic and political expansion.[20] The West became synonymous with the frontier and with unlimited individual opportunity, particularly for those who were white and male.[21] The West also became the central political prize in the nation's nineteenth-century tale of Manifest Destiny, an imperial vision of continental conquest that marginalized Native peoples, people of color, and many immigrant groups.[22]

Bound up with these myths were notions about the western environment and how the region was a wild land of limitless natural resources, ripe for the taking by an entrepreneurial and acquisitive frontier population. It was America's destiny to domesticate the West, the story went, and to harvest the region's furs, metals, crops, and timber. Before settlement, many saw the West as a land of unlimited natural abundance, and its expansive setting mirrored the ambitions of those who came to subdue it. Once settled, the domesticated landscape of farms and towns carried with it powerful images of national success in conquering the wilderness.

Western landscape images reflected national dreams and ambitions, but this visual vocabulary of elements and meanings did not appear overnight; it had to be created and learned. Cultural historian Anne Hyde makes the point that this encounter with western landscapes required a new language and regional aesthetics.[23] Politicians, travelers, writers, and artists built a new visual lexicon of western landscape meanings, which changed from one generation to the next. Modern ideas about what is valuable, beautiful, or wasteful in the western landscape have grown from these earlier myths and cultural inventions.

So who are the key players in the construction of our notions about the West? In the early nineteenth century (1804–1806), Meriwether Lewis and William Clark carefully described the terrain, animals, and Native peoples, accelerating the process of incorporating the region into our national experience. For Lewis and Clark and for President Jefferson, who initiated their western journey, the landscape came to represent a mix of scientific curiosity, practical challenges, and economic opportunity.[24]

The artists who traveled to the West, particularly during the 1830s and 1840s, focused on depicting Native people and their ways of life. The wide distribution of prints by painters such as Karl (Charles) Bodmer, Alfred Jacob Miller, and George Catlin made tangible some of the customs and places that, for them, defined Native life in the early nineteenth-century West.[25] In the 1850s, as the West assumed its continental dimensions, the Pacific Railroad Surveys documented the topography and resources along potential transcontinental railroad routes. Many of the volumes in these lavishly produced reports had depictions of landscapes, which art historian Robert Taft has called "the most important single contemporary source of knowledge on

FIG. I.20 Visualizing the West: Pacific Railroad Surveys, 1856. The widely reproduced landscape images in the government-sponsored Pacific Railroad Surveys completed during the 1850s offered the first comprehensive snapshot of the region. This view shows the landscape along western Montana's Clark Fork River south of Flathead Lake. *Source: United States,* Pacific Railroad Surveys, *vol. 12, book 1 (Washington, D.C.: Thomas Ford, 1860), plate 54, image courtesy of Montana Historical Society, Helena, MT.*

Western geography" of the period (fig. I.20).[26] Illustrators such as John Mix Stanley saw their work reproduced and copied in books and magazines promoting the region's development. Their images captured a landscape on the eve of an incredible environmental and cultural transformation.

In the second half of the nineteenth century, another generation of landscape painters, many schooled in traditions of European Romanticism, reimagined western landscapes to emphasize their sublimity and otherworldly beauty. Canvases by artists such as Albert Bierstadt and Thomas Moran captured scenes from Yosemite Valley and the Yellowstone region in paintings that emphasized the extraordinary over the everyday.[27] These artists contributed in enduring ways to notions of nature worship and wilderness aesthetics at a time when the nation was becoming more urbanized. By the end of the nineteenth century, artists such as Frederic Remington and Charles M. Russell celebrated the character and the passing of Native peoples and traditional cowboy life. Popular magazines such as *Harper's, The Youth's Companion,* and *The Atlantic Monthly* reproduced their sketches, which were important in defining national stereotypes about western landscapes and the characters who inhabited them.[28]

Photographers swarmed across the region in the late nineteenth century, producing panoramas, daguerreotypes, and galleries of images. Historian Martha Sandweiss concludes that the West's rapid late-nineteenth-century economic development was inextricably intertwined with the invention of

FIG. I.21 Promoting the West: Santa Fe Railroad, 1926. Railroad companies, hotels, and promotional organizations utilized appealing visual images to attract visitors to the Southwest. These images "off the beaten path" from the Santa Fe line show Native Americans, pueblos, and automobile tourists. *Source: author's collection.*

photography and that this new technology made more tangible the landscapes and settlements of the region. She demonstrates how photographs of the West were used to promote and define particular narratives of western expansion, imperialism, and nostalgia. Promotional images of western railroads, towns, and mines celebrated the nation's inevitable progress. At the same time, photographers such as Edward S. Curtis portrayed Native people as an exotic and vanishing race.[29]

Our notions about particular regional landscapes of the West have also been shaped by what we have inherited. Consider the deserts that stretch across interior Southern California, Arizona, and New Mexico. Long seen during most of the nineteenth century as an unlivable wasteland, this region came into its own after 1900. Writers such as John C. Van Dyke, George Wharton James, Mary Hunter Austin, and Charles F. Lummis published books and magazine articles that crafted new regional myths and aesthetics for the southwestern desert, rejoicing in its wide vistas, bare-bones topography, and outlandish cacti as well as in a romanticized portrayal of its Native inhabitants.[30] Zane Grey (whose *Riders of the Purple Sage* [1912] sold two million copies during his lifetime) and other novelists set their stories among the region's dramatic cliffs and canyons.[31] Regional artists such as Ernest Blumenschein, John Sloan, Peter Hurd, Maynard Dixon, and Georgia O'Keeffe added to the identity of the Southwest by serving up a visual feast of landscape images. Railroads and promoters reinforced these messages as they worked to attract Americans to localities such as Santa Fe, the Painted Desert, and the Grand Canyon (fig. I.21).[32] The arid-lands aesthetic is also celebrated in Edward Abbey's acerbic prose and Tony Hillerman's novels, set in Navajo country.[33]

Visual media used in advertising also have exerted a powerful influence on how Americans understand and experience the western landscape. Our twenty-first-century reinventions of the region are just as much molded by the Marlboro Man and Ralph Lauren as they are by Thomas Jefferson and Albert Bierstadt.[34] Advertising images glorify the drama of the western past and make use of iconic visual symbols. For decades, scenes of western mountains, horses, wildlife, cowboys, Indians, frontier life, and fly-fishing have been used to sell us everything from Ford trucks to high-end jewelry.

Performance art also helped invent the West.[35] Wild West shows (1880s–1920s) dramatized a glorified regional past of cowboys, Indian battles, stagecoach attacks, outlaws, and gunslingers. Buffalo Bill codified what west-

FIG. I.22 Inventing the West, Conover's Trading Post, Wisdom, Montana. This general store in the southwestern Montana community of Wisdom is a rich repository of masculine western stereotypes. The storefront features images of fishing, hunting, wildlife, and a female Native American.

erners wanted to remember as the frontier economy was replaced by an urbanized and industrialized West. More recently, filmmakers and television writers have created ways for Americans to experience the western landscape vicariously, making regional landscapes the backdrops for storytelling and adventure. Beginning in the 1920s and 1930s, locations such as the Alabama Hills near Lone Pine, California, and Monument Valley along the Utah-Arizona state line appeared in dozens of western films. Cast as a meeting place of savagery and civilization or as a wide-open psychological refuge, free from the pressures and boredom of modern urban life, the western landscape emerged as a cinematic personality. The 1992 film adaptation of Norman Maclean's story "A River Runs through It" illustrates the point. And what would John Wayne be without Monument Valley or *Thelma and Louise* (1991) without the welcoming abyss of the Grand Canyon?

From explorer accounts to Saturday afternoon matinees, these media created a picture of the western American landscape that has both reflected and reimagined the past. Combined with our own experiences, either as western residents or as visitors, these myths, images, and sensibilities, both accurate and fanciful, influence how we see the West (fig. I.22). That accumulated cultural memory also shapes the structure and content of this field guide and reminds us that landscape features and meanings are essential building blocks in how we construct our ideas about western places.

WHY WESTERN LANDSCAPES MATTER

Landscapes are important because they tell us who westerners are and where they came from. People arrived in the West from every corner of the world, and their vernacular landscapes reveal those stories: country schools in northern Montana reflect Midwestern roots (fig. I.23), and identities brought from Latin America are reflected in landscapes in northern New Mexico (fig. I.24). In other settings, landscapes reveal cultural innovation: California's patio lifestyle, which flowered after World War II, produced a new assemblage of

FIG. I.23 Cultural diffusion in the West, west of Havre, Montana. This abandoned country school reflects a building style used across much of the Midwest, then transplanted to Montana in the early twentieth century.

FIG. I.24 Vernacular landscape, northern New Mexico. Adorned with *ristras* (bundles of dried chiles), the Vigil Store offers Pepsis, Popsicles, and *santos*. A folk art preserved in the largely rural Catholic area, a *santo* is a carved statue depicting a saint, angel, or other religious figure.

landscape features that became popular nationwide. Cultural geographer Peirce Lewis put it best when he wrote that "all human landscape has cultural meaning" and that "our human landscape is our unwitting autobiography, reflecting our tastes, our values, our aspirations, and even our fears, in tangible, visible form."[36]

Landscapes matter also because they shape and structure our world.[37] They impose an everyday reality that most of us take for granted. When a busy freeway bisects an urban area, for example, its very presence controls the flow of traffic and movement through that part of the city (fig. I.25). The sprawling nature of most western American suburbs impels us to navigate them by automobile rather than on foot. "The shape of the land has the power to shape social life," cultural geographer Don Mitchell tells us.[38] By carefully reading a

FIG. I.25 Landscapes structure the world: Denver, Colorado. Denver's busy freeway system south of downtown shapes the flow of morning traffic, but the concrete structures and walls of the interstate also divide urban space and constrain movement.

landscape, you can learn about the social, political, and economic relationships that contributed to its construction as well as how these accumulated artifacts continue to define, structure, and constrain our world.

Landscapes are an integral part of our lives. Cultural geographer Donald Meinig argues that we can enrich and deepen our appreciation for landscape as a "humane art" just as we can cultivate our appreciation for music.[39] Learning about music can nurture our sensibilities and deepen our enjoyment of a symphony. In much the same way, learning to look more keenly at landscapes can enhance our connections with buildings, streets, communities, and countryside. Environmental historian William Cronon suggests that cultivating a sensibility for landscapes in our own backyards is a crucial step in improving our stewardship of the larger world.[40] Similarly, western writer Charles F. Wilkinson suggests that we need to develop a heightened "ethic of place" that combines a respect for the land, the cumulative record of human history in a locality, and the desire for creating a landscape that can be passed to the next generation.[41]

In that spirit, this field guide charts a course through western terrain with fresh ideas that render landscapes more tangible and accessible, that show us a view both larger and more intimate, and that foster a way of seeing and thinking that allows us to be better stewards of the West.

HOW TO USE THIS BOOK

You can use this field guide much as you would the guides by Roger Tory Peterson, who wrote more than a dozen books on birds, mammals, minerals, trees, and many other elements of North America's natural environment. The entries in his first *Field Guide to the Birds*, published in 1934, focused on four key aspects of birds, and these characteristics remain essential parts of any good field guide today.[42] First, Peterson clearly described the defining features of the bird and used illustrations to identify wing tips, beaks, body shapes, and colors that distinguish one species from another. Second, recognizing that variations within species would be important, Peterson noted the more subtle contrasts that might allow us to tell the difference between an ash-throated flycatcher, say, and a western kingbird. Third, he noted the broad geographical ranges of the species he described, often providing maps to indicate distribution. Finally, he recognized the importance of habitat, seeing a bird in its broader ecological setting. *How to Read the American West* takes a similar approach.

Peterson also encouraged his readers to wander widely and not limit themselves to a particular habitat. "One looks for meadowlarks in meadows, thrushes in woods," he wrote. "Therefore, those of the field-glass fraternity out to run up a list do not work only one environment."[43] It is good advice. Users of this guide are encouraged to explore and not to restrict themselves to the suburbs, the countryside, or alpine summits. The West has astounding variety, and every corner is worth exploring.

Peterson offered his readers other useful advice. Initially, spend some time thumbing through the guide and glancing at illustrations before going into the field. He also encouraged readers to make the guide a personal thing. "It is gratifying," he wrote, "to see a *Field Guide* marked on every page, for I know it has been well used."[44]

Finally, Peterson suggested taking in the larger ambient environment when in the field. In the case of birds, he stressed listening to their songs. Similarly, you can expand your visual observations of the western landscape by listening to the steady roar of an urban freeway or the silence of a mountain meadow, feeling the texture of pine bark in a Montana forest or smelling the food cooking in a Vietnamese restaurant in West Los Angeles. This broader, more encompassing sensory approach to landscape often shapes our experiences more memorably than simply looking at our surroundings.

In this guide, you will encounter one hundred numbered landscape elements, organized topically and geographically. There is nothing sacred about either the number or the subject of the features, and you can no doubt think of other possibilities to add to the list. The ones included here are meant to take you in fruitful directions and invite you to look carefully at what you encounter, wherever you are.

You can begin by looking at physical landscape features, examining their character and distribution as well as their role as symbols of regional cultural

FIG. I.26 Abandoned farmhouse, Roy, New Mexico. Many rural areas of the West have seen large population losses since 1920. Thousands of abandoned homes across the eastern plains of Montana, Wyoming, Colorado, and New Mexico reflect earlier unsustainable settlement in agriculturally marginal settings.

identity. Next, you will find those vast areas of the West that are dominated by farming and ranching. You can peer into irrigation ditches, grain elevators, farm towns, and more as you make your way across a landscape rich in history but increasingly impoverished in remaining residents and surviving settlements (fig. I.26). Extractive landscapes follow, reflecting the traditional importance of activities such as mining and lumbering in the West. Open-pit mines, ghost towns, and logging camps are reminders that hundreds of localities across the West have been shaped by such industries.

While every landscape has a human imprint, some carry persistent and enduring signatures of cultural diversity—of African Americans, Indians, Latinos, and Asians, as well as earlier waves of European settlers. Mormon villages, utopian colonies, and New Age haunts have all left their mark on the West. Movement and connections also remain defining elements of the West, and this guide explores the legacies of historic western trails and railroad lines, along with the contemporary imprints made by interstate highways and the electrical grid. Much of the West also remains influenced by federal largesse, an enduring imprint that has left its own peculiar stamp on the region.

You will see the landscapes where most westerners live, the cities and suburbs of major metropolitan areas, with their ranch houses, strip malls, and office complexes on the periphery. And finally, you can consider western playgrounds—hot springs, rodeo grounds, ski towns, and retirement communities—places where people have searched for entertainment, relaxation, and the good life (fig. I.27).

One final piece of advice: On your travels, take along a camera, a sketchpad, and a notebook. While field guides are great invitations to investigate the regional scene, the best way to learn about the landscape is to render it—sketch it, photograph it, or write about it from different angles. Capturing a landscape in those ways forces you to look at it differently, to come to terms with it, to explore its parts and larger character.

As you travel through the West, either on the road or from your armchair, seek out surprise. As Grady Clay once said, "To turn is to learn."[45] And relish variety. Explore suburbs as well as wildlife refuges, and visit strip malls and strip mines as well as golf courses and mountain streams. This field guide suggests a few things to look for along the way, but the real discovery begins in the world beyond these pages. ❋

FIG. I.27 Amenity landscape, central Colorado. Colorado's Copper Mountain resort is less than a hundred miles west of Denver and offers a four-season recreational escape for Front Range residents in search of fairways, condos, and ski slopes.

NOTES

1. See Carl Sauer, "Foreword to Historical Geography," in *Land and Life: A Selection from the Writings of Carl Ortwin Sauer,* ed. John Leighly (Berkeley: University of California Press, 1965), 365–66; Dan Flores, *Horizontal Yellow: Nature and History in the Near Southwest* (Albuquerque: University of New Mexico Press, 1999); and William Wyckoff, *On the Road Again: Montana's Changing Landscape* (Seattle: University of Washington Press, 2006), 149–53.
2. See Carl Sauer, "The Morphology of Landscape," in *Land and Life,* 315–50; J. B. Jackson, *A Sense of Place, A Sense of Time* (New Haven, CT: Yale University Press, 1994), 16–17; Peirce Lewis, "Axioms for Reading the Landscape: Some Guides to the American Scene," in *The Interpretation of Ordinary Landscapes,* ed. D. W. Meinig (New York: Oxford University Press, 1979), 24–26; Stephan Harrison, Steve Pile, and Nigel Thrift, eds., *Patterned Ground: Entanglements of Nature and Culture* (London: Reaktion Books, 2004); and Don Mitchell, "Battle/fields: Braceros, Agribusiness, and the Violent Reproduction of the California Agricultural Landscape During World War II," *Journal of Historical Geography* 36 (2010): 147–49. See also William Cronon, "The Trouble with Wilderness; or, Getting Back to the Wrong Nature," in *Uncommon Ground: Rethinking the Human Place in Nature,* ed. William Cronon (New York: W. W. Norton, 1996), 69–90.
3. Donald Worster, *Rivers of Empire: Water, Aridity, and the Growth of the American West* (New York: Oxford University Press, 1985); Marc Reisner, *Cadillac Desert: The American West and Its Disappearing Water* (New York: Penguin, 1986); Ed Marston, ed., *Western Water Made Simple* (Washington, D.C.: Island Press, 1987); Donald J. Pisani, *To Reclaim a Divided West: Water, Law, and Public Policy 1848–1902* (Albuquerque: University of New Mexico Press, 1992); and Char Miller, ed., *River Basins of the American West* (Corvallis: Oregon State University Press, 2009).
4. The scales of landscape analysis are explored in Kenneth I. Helphand, *Colorado: Visions of an American Landscape* (Niwot, CO: Roberts Rinehart, 1991), and Harrison, Pile, and Thrift, *Patterned Ground.*
5. See Don Mitchell, "New Axioms for Reading the Landscape: Paying Attention to Political Economy and Social Justice," in *Political Economies of Landscape Change: Places of Integrative Power,* ed. J. Wescoat and D. Johnston (Dordrecht, Netherlands: Springer, 2007), 29–50.
6. See Grady Clay, *Close-Up: How to Read the American City* (Chicago, IL: University of Chicago Press, 1973), and his *Real Places: An Unconventional Guide to America's Generic Landscape* (Chicago: University of Chicago Press, 1994). The dynamic urban-wildland interface is explored in Dolores Hayden, *A Field Guide to Sprawl* (New York: W. W. Norton, 2004), and Travis Paveglio et al., "Understanding Social Complexity within the Wildland-Urban Interface: A New Species of Human Habitation?" *Environmental Management* 43 (2009): 1085–95.
7. The notion of landscape density is illustrated in Hayden, *Field Guide to Sprawl;* Michael P. Conzen, ed., *The Making of the American Landscape* (New York: Routledge, 2010); Larry R. Ford, *Cities and Buildings: Skyscrapers, Skid Rows, and Suburbs* (Baltimore, MD: Johns Hopkins University Press, 1994); and Jack Williams, *E40°: An Interpretive Atlas* (Charlottesville: University of Virginia Press, 2006).
8. Power and control in the evolution of the West are examined in Patricia Limerick, *The Legacy of Conquest: The Unbroken Past of the American West* (New York: W. W. Norton, 1987); Richard White, *"It's Your Misfortune and None of My Own": A New History of the American West* (Norman: University of Oklahoma Press, 1991); and William G. Robbins, *Colony and Empire: The Capitalist Transformation of the American West* (Lawrence: University Press of Kansas, 1994). The shaping power of both the Old West and the New is assessed in Paul Robbins et al., "Writing the New West: A Critical Review," *Rural Sociology* 74 (2009): 356–82. More generally, see Mitchell, "New Axioms for Reading the Landscape," and William K. Wyckoff, "Imposing Landscapes of Private Power and Wealth," in *Making of the American Landscape,* 381–402.
9. See John B. Wright, *Rocky Mountain Divide: Selling and Saving the West* (Austin: University of Texas Press, 1993); Philip L. Jackson and Robert Kuhlken, *A Rediscovered Frontier: Land Use and Resource Issues in the New West* (Lanham, MD: Rowman and Littlefield, 2006); and William R. Travis, *New Geographies of the American West: Land Use and the Changing Patterns of Place* (Washington, D.C.: Island Press, 2007).

10 See Richard H. Jackson, "Federal Lands in the Mountainous West," in *The Mountainous West: Explorations in Historical Geography*, ed. William Wyckoff and Lary M. Dilsaver (Lincoln: University of Nebraska Press, 1995), 253–77.

11 Linda Flint McClelland, *Building the National Parks: Historic Landscape Design and Construction* (Baltimore, MD: Johns Hopkins University Press, 1998); Ethan Carr, *Mission 66: Modernism and the National Park Dilemma* (Amherst: University of Massachusetts Press, 2007).

12 The connection between American places and cultures is explored in Wilbur Zelinsky, *The Cultural Geography of the United States* (Englewood Cliffs, NJ: Prentice-Hall, 1992); Zelinsky, *Not Yet a Placeless Land: Tracking an Evolving American Geography* (Amherst: University of Massachusetts Press, 2011); Meinig, ed., *Interpretation of Ordinary Landscapes;* and Conzen, ed., *Making of the American Landscape*. Useful case studies are offered in Richard L. Nostrand and Lawrence E. Estaville, eds., *Homelands: A Geography of Culture and Place across America* (Baltimore, MD: Johns Hopkins University Press, 2001), and Chris Wilson and Paul Groth, eds., *Everyday America: Cultural Landscape Studies after J. B. Jackson* (Berkeley: University of California Press, 2003).

13 See Kent C. Ryden, *Mapping the Invisible Landscape: Folklore, Writing, and the Sense of Place* (Iowa City: University of Iowa Press, 1993); John Fraser Hart, *The Rural Landscape* (Baltimore, MD: Johns Hopkins University Press, 1998); Christopher L. Salter, "No Bad Landscape," *Geographical Review* 91 (2001): 105–12; and Richard L. Nostrand, *El Cerrito, New Mexico: Eight Generations in a Spanish Village* (Norman: University of Oklahoma Press, 2003).

14 See Daniel D. Arreola, ed., *Hispanic Spaces, Latino Places: Community and Cultural Diversity in Contemporary America* (Austin: University of Texas Press, 2004); Eliot Tiegel, *Latinization of America: How Hispanics Are Changing the Nation's Sights and Sounds* (Los Angeles: Phoenix Books, 2007); and Cristina Benitez, *Latinization: How Latino Culture Is Transforming the U.S.* (Ithaca, NY: Paramount Market Publishing, 2007).

15 Geographer Yi-fu Tuan has thought extensively about the meaning of *home*. See his *Topophilia: A Study of Environmental Perception, Attitudes, and Values* (Englewood Cliffs, NJ: Prentice-Hall, 1974), and *Space and Place: The Perspective of Experience* (Minneapolis: University of Minnesota Press, 1977). Also see Jackson, *A Sense of Place, a Sense of Time,* and Ryden, *Mapping the Invisible Landscape*. For a tour of the interior spaces of home, see Akiko Busch, *Geography of Home: Writings on Where We Live* (Princeton, NJ: Princeton Architectural Press, 2003). Landscapes of labor are assessed by Andrew Herod, *Labor Geographies: Workers and the Landscapes of Capitalism* (New York: Guilford Press, 2001), and Blake A. Harrison, *The View from Vermont: Tourism and the Making of an American Rural Landscape* (Burlington: University of Vermont Press, 2006).

16 See collections of Jackson's essays in *A Sense of Place, a Sense of Time,* and John B. Jackson, *Landscape in Sight: Looking at America* (New Haven, CT: Yale University Press, 1997).

17 See Meinig, ed., *Interpretation of Ordinary Landscapes;* Denis Cosgrove and S. Daniels, eds., *The Iconography of Landscape: Essays on the Symbolic Representation, Design, and Use of Past Environments* (Cambridge: Cambridge University Press, 1988); William Wyckoff, "Postindustrial Butte," *Geographical Review* 85 (1995): 478–96; Arthur Krim, *Route 66: Iconography of the American Highway* (Santa Fe, NM: Center for American Places, 2005); Richard H. Schein, ed., *Landscape and Race in the United States* (New York: Routledge, 2006); David Robertson, *Hard as the Rock Itself: Place and Identity in the American Mining Town* (Boulder: University Press of Colorado, 2006); Kevin Blake, "Imagining Heaven and Earth at Mount of the Holy Cross, Colorado," *Journal of Cultural Geography* 25, no. 1 (February 2008): 1–30; Mitchell, "New Axioms for Reading the Landscape"; and Conzen, ed., *Making of the American Landscape*.

18 See D. W. Meinig, "The Beholding Eye: Ten Versions of the Same Scene," in Meinig, ed., *Interpretation of Ordinary Landscapes*, 33–48.

19 A fine analysis of western American landscape meanings is in Thomas R. Vale and Geraldine R. Vale, *Western Images, Western Landscapes: Travels along U.S. 89* (Tucson: University of Arizona Press, 1989), 1–14, 153–69.

20 See Howard Lamar, "An Overview of Westward Expansion," in *The West as America: Reinterpreting Images of the Frontier, 1820–1920*, ed. William H. Truettner (Washington, DC: Smithsonian Institution, 1991), 1–26.

21 For links between images of the West and ideas of economic opportunity, see Henry Nash Smith, *Virgin Land: The American West as Symbol and Myth* (Cambridge, MA: Harvard University Press, 1950), and William Cronon, "Telling Tales on Canvas: Landscapes of Frontier Change," in *Discovered Lands, Invented Pasts: Transforming Visions of the American West,* ed. Jules David Prown et al. (New Haven, CT: Yale University Press, 1992), 37–87. For themes of imperialism, racism, and economic inequality, see Limerick, *Legacy of Conquest,* and White, *"It's Your Misfortune and None of My Own."*

22 See Truettner, *West as America*, esp. 27–53, 96–147.

23 Anne Farrar Hyde, *An American Vision: Far Western Landscape and National Culture, 1820–1920* (New York: New York University Press, 1990).

24 See John Logan Allen, *Passage through the Garden: Lewis and Clark and the Image of the American Northwest* (Urbana: University of Illinois Press, 1975), and James P. Ronda, *Finding the West: Explorations with Lewis and Clark* (Albuquerque: University of New Mexico Press, 2001).

25 For examples, see Patricia Trenton and Peter Hassrick, *The Rocky Mountains: A Vision for Artists in the Nineteenth Century* (Norman: University of Oklahoma Press, 1983), and Larry Curry, *The American West: Painters from Catlin to Russell* (Los Angeles and New York: Los Angeles County Museum of Art and Viking, 1972).

26 Robert Taft, *Artists and Illustrators of the Old West, 1850–1900* (New York: Scribner's, 1953), 5.

27 See Truettner, *West as America*; Trenton and Hassrick, *The Rocky Mountains*; Joni Kinsey, *Thomas Moran's West: Chromolithography, High Art, and Popular Taste* (Lawrence: University Press of Kansas, 2006); and Gordon Hendricks, *Albert Bierstadt: Painter of the American West* (New York: Harry N. Abrams, 1974).

28 Richard Etulain, *Re-Imagining the Modern American West: A Century of Fiction, History, and Art* (Tucson: University of Arizona Press, 1996); Peter Hassrick, *Remington, Russell, and the Language of Western Art* (Washington, D.C.: Trust for Museum Exhibitions, 2000).

29 See Martha A. Sandweiss, *Print the Legend: Photography and the American West* (New Haven, CT: Yale University Press, 2002). See also Timothy Egan, *Short Nights of the Shadow Catcher: The Epic Life and Immortal Photographs of Edward Curtis* (Boston: Houghton Mifflin Harcourt, 2012).

30 See Hyde, *An American Vision*, 191–243; Richard Francaviglia, "Elusive Land: Changing Geographic Images of the Southwest," in *Essays on the Changing Images of the Southwest,* ed. Richard Francaviglia and David Narrett (College Station: Texas A and M University Press, 1995), 8–39; Etulain, *Re-Imagining the Modern American West*; and Emily Neff, *The Modern West: American Landscapes, 1890–1950* (New Haven, CT: Yale University Press, 2006).

31 See Kevin Blake, "Zane Grey and Images of the American West," *Geographical Review* 85 (1995): 202–16.

32 The role of railroad promotion in the Southwest is explored in Hal K. Rothman, *Devil's Bargains: Tourism in the Twentieth-Century American West* (Lawrence: University Press of Kansas, 1998), 29–112; Marta Weigle and Barbara A. Babcock, eds., *The Great Southwest of the Fred Harvey Company and the Santa Fe Railway* (Phoenix, AZ: Heard Museum, 1996); Carlos Schwantes, *Railroad Signatures across the Pacific Northwest* (Seattle: University of Washington Press, 1996); and Carlos Schwantes and James P. Ronda, *The West the Railroads Made* (Seattle: University of Washington Press, 2008).

33 See Edward Abbey, *Desert Solitaire: A Season in the Wilderness* (New York: McGraw-Hill, 1968), and *The Monkey Wrench Gang* (New York: Perennial Classics, 2000). Hillerman's novels include *Coyote Waits* (New York: Harper and Row, 1990), and *The Shape Shifter* (New York: HarperCollins, 2006).

34 See Elliott West, "Selling the Myth: Western Images of Advertising," *Montana, the Magazine of Western History* 46, no. 2 (Summer 1996): 36–49. Carlos Schwantes discusses contemporary western stereotypes in *So Incredibly Idaho! Seven Landscapes That Define the Gem State* (Moscow: University of Idaho Press, 1996), 21–23.

35 See Louis S. Warren, *Buffalo Bill's America: William Cody and the Wild West Show* (New York: Knopf, 2005). For the links between western myths and films, see Robert G. Athearn, *The Mythic West in Twentieth-Century America* (Lawrence: University Press of Kansas, 1986), and White, *"It's Your Misfortune and None of My Own,"* esp. 613–32. The geographies of

movie-making are summarized in Gary J. Hausladen, "Where the Cowboy Rides Away: Mythic Places for Western Film," in *Western Places, American Myths: How We Think about the West,* ed. Gary J. Hausladen (Reno: University of Nevada Press, 2003), 296–318. For the Rockies, see Rick Newby, ed., *The Rocky Mountain Region: The Greenwood Encyclopedia of American Regional Cultures* (Westport, CT: Greenwood Press, 2004), 161–92.

36 See Lewis, "Axioms for Reading the Landscape," 12.

37 Richard H. Schein developed this argument in "The Place of Landscape: A Conceptual Framework for Interpreting an American Scene," *Annals of the Association of American Geographers* 87 (1997): 660–80, and in "Normative Dimensions of Landscape," in *Everyday America*, 199–218.

38 Mitchell, "New Axioms for Reading the Landscape," 43.

39 Meinig presented this view in "Environmental Appreciation: Localities as a Humane Art," *Western Humanities Review* 25 (1971): 1–11. The avenues of place appreciation are further explored in his "Beholding Eye: Ten Versions of the Same Scene."

40 Cronon, "Trouble with Wilderness," 87–90.

41 Charles F. Wilkinson, "Toward an Ethic of Place," in *Beyond the Mythic West,* ed. Stewart L. Udall et al. (Salt Lake City, UT: Peregrine Smith, 1990), 71–104.

42 Roger Tory Peterson, *A Field Guide to the Birds* (Boston: Houghton Mifflin, 1934).

43 Roger Tory Peterson, *A Field Guide to Western Birds* (Boston: Houghton Mifflin, 1961), xviii.

44 Peterson, *Field Guide to Western Birds*, xvi.

45 Grady Clay, "Crossing the American Grain with Vesalius, Geddes, and Jackson," in *Everyday America*, 120.

1

NATURE'S
FUNDAMENT

FIG. 1.1 Eastern Sierra north of Bishop, California. Mountain-building processes have shaped the western landscape for millions of years. The steep slopes of California's eastern Sierra Nevada reveal granite (igneous) intrusions, patterns of rapid uplift along faults, and subsequent erosion.

JOHN CHARLES FRÉMONT, THE GREAT PATHFINDER, CONQUERED PLENTY of mountain ranges in his nineteenth-century explorations of the West. Perhaps none was more thrilling than his August 1842 ascent of Wyoming's Wind River Range. Champion of Romantic prose and his own heroic persona, Frémont articulated the sheer drama of nature's fundament in the West and realized the role it played in defining the region's character in the national imagination:

> We continued climbing, and in a short time reached the crest. I sprang upon the summit, and another step would have precipitated me into an immense snow-field five hundred feet below. . . . It is presumed that this is the highest peak of the Rocky Mountains. The day was sunny and bright, but a slight shining mist hung over the lower plains. . . . On one side we overlooked innumerable lakes and streams, the spring of the Colorado . . . and on the other was the Wind river valley, where were the heads of the Yellowstone branch of the Missouri; far to the north, we just could discover the snowy heads of the Trois Tetons, where were the source of the Missouri and Columbia rivers. . . . Around us, the whole scene had one main striking feature, which was that of terrible convulsion. (Frémont, *Report of the Exploring Expedition to the Rocky Mountains in the Year 1842*, 69–70)

Much of the West's appeal remains connected to its physical musculature, its sheer material, visceral presence. The size and heterogeneity of the West can be overwhelming, exhausting, and exhilarating. Some of the West's most dramatic, rugged landscapes can be enjoyed (with perhaps less drama than Frémont describes) on a trip along U.S. 395 between Reno, Nevada, and Los Angeles, California. The road passes just east of the great, granite-faced Sierra Nevada, probing the mountains at times near Mono and June lakes, and then making the long descent toward Bishop into the arid, mountain-rimmed Owens Valley. Looking south from near Sherwin Summit, you can still be reminded of one of the West's great defining elements, the omnipresence and scale of its landforms (fig. 1.1). Frémont had it right: mountains, open space, and the overwhelming visible signatures of geological forces are the quintessential elements of the region's physical geography and enduring parts of why many people are drawn to the West.

Nevertheless, nature's fundament in the West has everywhere been modified by human action and intent, often over long periods of time. Even what seems at first glance to be a natural landscape has been shaped by culture and technology: California's rolling hills are populated by exotic European grasses, and high-country conifers in the Rockies may have been modified by human-caused fires and nineteenth-century logging. At the same time, the West's cultural landscapes are embedded within a larger environmental context. A California farm field still needs to be anchored in the requisites of soil and water, and every western suburb creates its own ecology of plants and animals. This section of the field guide focuses on elements of the natural world that have contributed to the West's regional identity, shaped our daily lives, and become interwoven into the region's cultural fabric.

What are the defining elements of the West's natural environment? The region's geological setting and topography (terrain and landforms) are

FIG. 1.2 Salmon River, Idaho. Idaho's Salmon River (near White Bird) is a tributary of the larger Columbia system. Note how many north-facing slopes (facing right) are forested, while warmer, sunnier south-facing slopes (facing left) remain open and grassy.

extraordinarily complex, a function of the West's complicated and often-fractured position on the Earth because of the ways tectonic plates have collided (the region is a contact zone between the North American and Pacific plates) and fragmented to produce the diversity we see today.

Whether you are looking at coastal Washington's Olympic Mountains or southeastern New Mexico's Pecos Plains, it is helpful to keep some basic ideas in mind. What is visible today is only a snapshot in geological time, the most recent expression of processes of mountain building, erosion, and change that have been at work for millions of years. As the geologists remind us, if all geological time since the beginning of the Cambrian period (almost 600 million years ago) were embraced in a calendar year, the first mammals would appear in late August, the first primates in early December, and modern humans on New Year's Eve. When we look at that Victorian-era house in the "old" part of town, we should remember how recent such cultural signatures are in comparison to the bedrock they are sitting on.

While you can find exposed examples of rocks that are more than one billion years old (such as the 1.7-billion-year-old Vishnu schists at the bottom of the Grand Canyon), much of the ancient geological foundation in the West has been covered by more recent sedimentary and igneous deposits. Overall, much of the West's surface geology is relatively young. In terms of our geological year, for example, the Rocky Mountains don't show up until late November (about 60 million years ago) and the glaciers don't appear until about December 28 (within the past 2.5 million years). The distinctive granite core of California's rugged Sierra Nevada (fig. 1.1) did not experience its major thrust skyward until about 4 million years ago. Similarly, the lava beds that cover much of the Columbia Plateau and Snake River Plain in the interior Pacific Northwest are recent additions, dating from about 2 million to about 30 million years ago. Indeed, active Cascade volcanoes, such as Mount Saint Helens in Washington (which erupted in 1980), continue to make new landscapes today.

Water and major drainage basins also play key roles in organizing the region's physical and human geography (fig. 1.2). Much of all life in the West

FIG. 1.3 Annual precipitation in the West. Complex regional patterns of annual precipitation suggest the role of the Pacific Ocean and of interior mountain ranges in shaping the distribution of moisture across the West.

is concentrated within reach of the fragile ribbons of surface water that trace their paths across this largely arid land. In addition, the Continental Divide snakes across the West, from southern New Mexico through northern Montana, separating the drainage of the region into the Pacific and Atlantic basins. Being west or east of that line can have profound implications for weather patterns, vegetation, and agriculture.

The varied climates, unpredictable weather, and overall atmospheric conditions create additional landscape signatures in the region. Wide-open spaces are defined by the clarity of the air, the prominence of sunshine, and a high, well-defined sky. Clouds gather along west-facing mountain ranges as moisture is forced upward and condenses, while nearby valleys to the east may remain dry. A simple map of western precipitation is a reminder of how much moisture is squeezed out by the region's higher uplifts and how plants, animals, and people within the Columbia Plateau, Great Basin, and High Plains areas must inevitably adjust to their position on the drier, leeward side of the nearby mountains (figs. 1.1 and 1.3). The wet sides of the Cascades, Sierra Nevada, and Rocky Mountains accumulate between twenty-five and fifty inches (and sometimes more) of moisture annually, much of it in the form of snowpack, while portions of eastern Oregon, southeastern California, and Colorado must make do with fewer than ten inches a year.

Western vegetation is an elegant expression of the complex, nuanced relationships between climate and topography as well as a result of the local effects of soil type, slope, and drainage. Some plants have become iconic regional landscape symbols—consider cacti and Joshua trees, sagebrush, and conifers—and changes in vegetation, often influenced by people, are a part of the story in almost every western landscape. Wildfire, for example, is a significant natural and cultural element across the western landscape, and its frequency in grasslands or forests profoundly influences the species composition of these environments. Similarly, exotic and invasive plants introduced by accident or on purpose can rapidly rework the vegetative landscape.

More generally, what explains the difference between the oak-studded grasslands of central California's Coast Ranges (fig. 1.4) and the subalpine fir and spruce forests of Montana's Glacier National Park (fig. 1.5)? On a regional scale, both latitude and proximity to the Pacific Ocean are pivotal considerations. Different seasonal patterns of temperature and precipitation shape opportunities for vegetation—and agriculture—at 33 degrees north latitude (southern New Mexico) and at 48 degrees north latitude (north-central Montana). Similarly, maritime settings such as western Oregon or the northern Salinas Valley in California offer ecological niches that reflect the modifying impact of the sea—frequent fogs, cooler summers, warmer winters—while continental settings such as eastern Arizona or central Wyoming offer niches that reflect greater daily and annual variability, especially hotter summers and colder winters.

Elevation is an essential ingredient in understanding western landscapes, particularly vegetation. Western topography profoundly influences average temperatures, which generally fall about three or four degrees Fahrenheit with every thousand-foot rise in elevation, and precipitation, which is generally higher with elevation and more abundant on the windward side, less on the leeward side, an effect known as the rain shadow. Differences in slope and aspect (the direction a slope faces) can profoundly alter vegetation patterns. In western Idaho, for example, the cooler, north-facing slopes are typically forested, while warmer, drier, south-facing slopes are open grasslands. The result is an intricate landscape pattern (see fig. 1.2).

FIG. 1.4 Oak woodlands near Williams, California. These hardy grasses (mostly European exotics) and native oaks are common to the California Coast Ranges and are well adapted to long, hot summers and cool, wet winters.

FIG. 1.5 Subalpine fir and spruce forest, Glacier National Park. These hardy trees surround a meadow near Two Medicine Lake in Montana's Glacier National Park. Short, cool summers alternate with long, cold winters; nearby peaks (above the tree line) are often dusted with snow in July.

Physical geographers point to general relationships between elevation and natural vegetation, and the terminology they use provides a useful lexicon of life zones (fig. 1.6). While the species may vary with the setting, you might encounter with increasing elevation a complex transition between Lower Sonoran mesquite or cacti, Upper Sonoran sagebrush, Transition pinyon and juniper woodlands, Montane forests of ponderosa pine (fig. 1.2), higher Canadian and Hudsonian zones of Douglas fir, white fir, Engelmann spruce, and aspen (seen in fig. 1.5), and a true Alpine zone of tundra, open meadows, and exposed rock (also visible in fig. 1.5).

The distribution of wild animals is related to these environmental variables, as well as to a myriad of human influences. An animal's place in the western ecosystem depends on a web of interdependent relationships with plants and other animals, including people. Wild animals form an especially significant component of the regional landscape because they have taken on cultural and political value. For many people, the survival of animals says a great deal about their quality of life and the region's environmental health. For others, some species may be considered a threat to sheep and cattle or an impediment to economic development. Either way, wildlife remains an essential part of the region's physical geography, and the cultural and political significance of particular species exemplifies the intermingling that inevitably welds the natural world to its human occupants. ❋

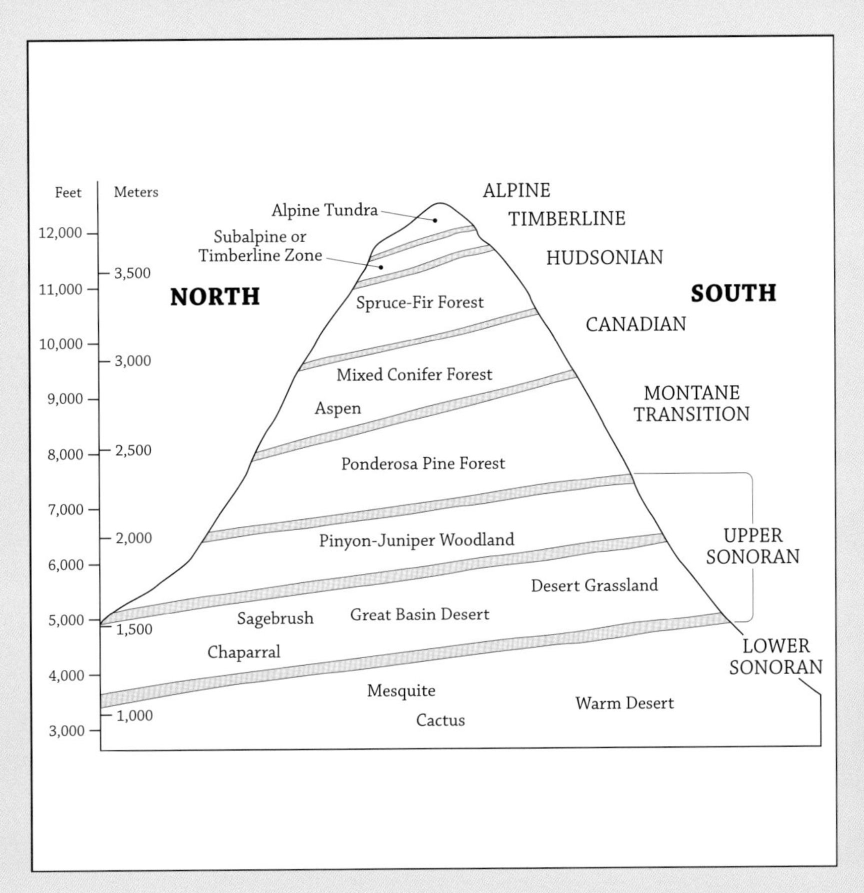

FIG. 1.6 Vegetation life zones in the West. Links between elevation and plant life are suggested in this example from the Southwest. Vegetation shifts from desert species to pine and mixed conifer forests. At timberline, hardy subalpine trees give way to tundra and snowfields. Note differences between south- and north-facing slopes.

1 WIDE-OPEN SPACES

FIG. 1.7 *Open Country*, Thomas Hart Benton (1952). Benton's canvas captures the classic characteristics of the wide-open western landscape including sparse population, a broad, sweeping horizon, and associations with freedom and independence. *Art © T. H. Benton and R. P. Benton Testamentary Trusts/ UMB Bank Trustee/Licensed by VAGA, New York, NY.*

AMERICAN ARTIST THOMAS HART BENTON'S *OPEN COUNTRY* (1952) captures the feeling (fig. 1.7). While most westerners live in cities, the region's wide-open spaces remain one of its most recognized and celebrated qualities. What are its essential characteristics?

First, these are places of low population density (fig. 1.8). Benton's archetypal scene might be in many western settings, but a key attribute is that cows far outnumber people.

Second, wide-open spaces possess physical qualities that include a largely nonforested setting, a sweeping horizon, and a sky defined by the dryness and clarity of the air (**11**). Benton's imagery is simple and compelling: sky, land, and weather form the view. Similarly, Monument Valley (in northern Arizona and southern Utah) defines a dramatic foreground but derives its power from the boundless distances beyond the buttes (fig. 1.9).

Third, Americans invest wide-open spaces with cultural meaning. Benton's painting is grounded in well-established artistic and cultural traditions: the figure of the lone cowboy and his horse celebrate freedom and independence, acknowledge a lifestyle strongly wedded to place, and hint at human insignificance framed by a landscape overwhelming in its scale. These qualities began appealing to urbanizing Americans late in the nineteenth century and were

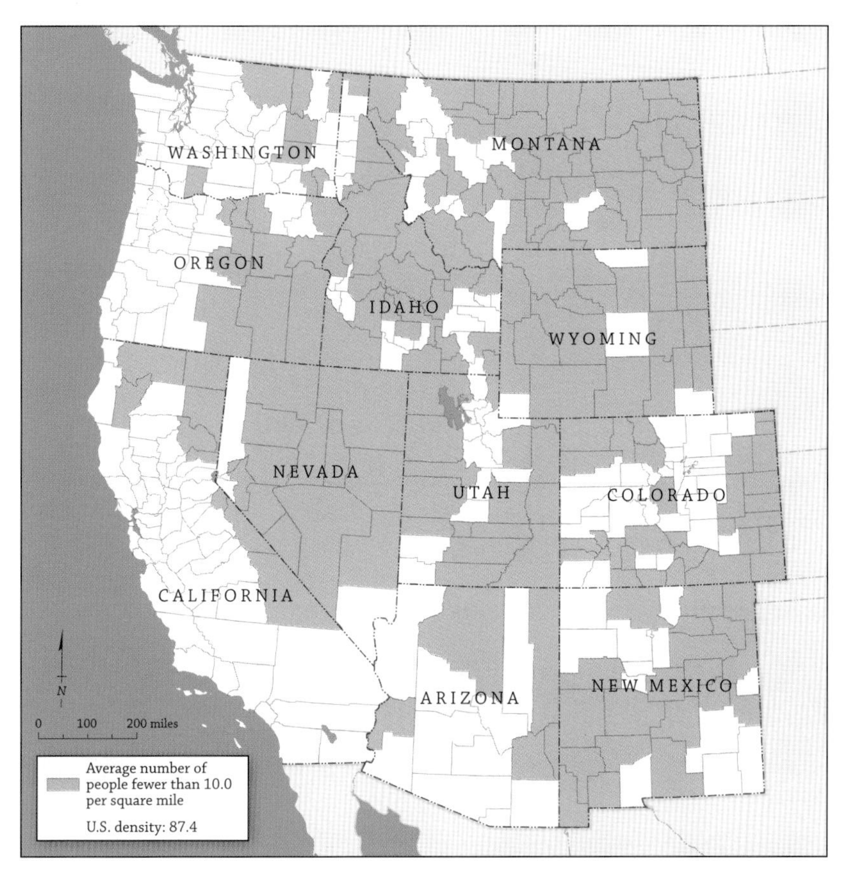

FIG. 1.8 Counties with fewer than ten persons per square mile, 2010. The average population density for counties across the nation is eighty-seven persons per square mile. But large areas of the Great Basin, Rockies, High Plains, and interior Southwest retain their wide-open spaces. *Source: U.S. Census Bureau, 2010 Census.*

incorporated into cultural values associated with western tourism, the appreciation of scenery, and preservation of open space. The celebration continues today, codified in the value of view lots, viewsheds, and planning mandates that seek to maintain wide-open spaces where they are most threatened, typically on the edge of growing western cities (**83**).

Wide-open spaces include the eastern plains of Montana, where the town of Jordan is true Big Sky Country; much of Wyoming; and Colorado, east from Fort Collins, Denver, or Trinidad. You might also visit the Four Corners country, described in Edward Abbey's *Desert Solitaire;* probe the Great Basin by driving U.S. 50 across Nevada; or explore Steens Mountain in eastern Oregon. ❋

FIG. 1.9 Monument Valley. The simple visual elements of distance, space, and sky define the character of Monument Valley, located within northern Arizona and southern Utah in the Navajo Nation. The richly layered sedimentary rocks, deposited and eroded over geological time, add to the timeless quality of the view.

2

MOUNTAIN AND VALLEY TOPOGRAPHY

FIG. 1.10 Mountain and valley topography, San Luis Valley, Colorado. This view, south of Poncha Pass, shows the steep, west-facing slopes of the Sangre de Cristo Mountains. A ranch (foreground) occupies the valley floor.

MOUNTAINS AND VALLEYS DOMINATE MUCH OF THE WEST, SHAPING landscapes in predictable ways. Long, narrow valleys (often trending north to south) are bordered by mountain ranges in places like southern Colorado's San Luis Valley and Sangre de Cristo Mountains (fig. 1.10). The complex geology behind this pattern relates to the region's location in a dynamic, tectonically active portion of the Earth's crust. Stresses produce different responses: in some places, the crust folds upward (anticlines) or downward (synclines), creating structures that can produce uplifts and basins. Elsewhere, forces cause more brittle responses, creating faults or breaks in the crust (**3**).

Mountain and valley topography appears near the Pacific coast, where residents of the Puget Sound (Washington), the Willamette Valley (Oregon), and the Central Valley (California) regions live between the Coast Ranges and the Cascades or Sierra Nevada. Parallel ranges (horsts) and basins (grabens) characterize Nevada and western Utah, where early cartographers were struck by the predictable alignment of alternating mountains and valleys (fig. 1.11). A

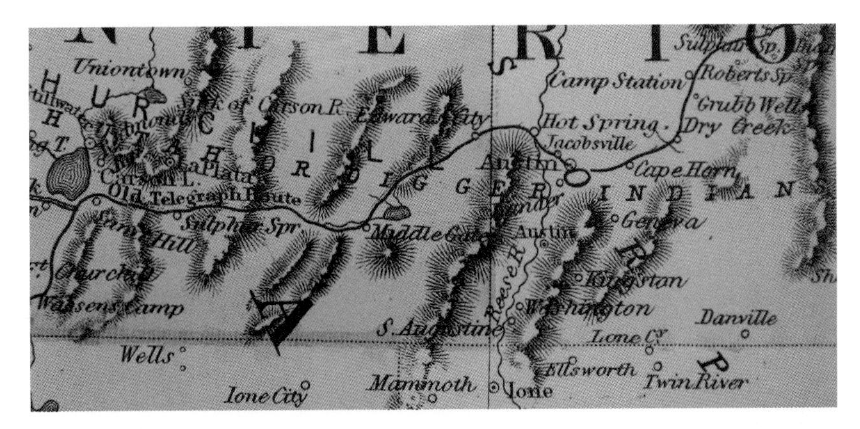

FIG. 1.11 Basin and range topography, Nevada. The regular pattern of north-south-trending mountain ranges can be seen in this detail of an 1872 map of central Nevada. These horsts (mountain blocks) and grabens (structural valleys) are evident to modern-day travelers, especially along U.S. 6 and U.S. 50. *Source: Warner and Beers, Map of Nevada and California, 1872.*

FIG. 1.12 Alluvial fan, southwestern Montana. The large Cedar Creek alluvial fan sprawls beyond the base of the Madison Range. Small streams are lined with riparian vegetation. Terraces in the foreground suggest that tectonic forces continue to adjust the levels of the valley floor in relation to the nearby mountains.

parallel assemblage of valley bottoms and mountain summits also occurs in Colorado and western Montana. In the Southwest, the pattern across southeastern Arizona and southern New Mexico reflects similar geological forces at work: isolated ranges such as the Chiricahuas and the Sacramento Mountains are bordered by large desert lowlands.

This topographic pattern can be divided into three elements. Begin by assessing the valley floor. Is it marked by a dry lake bed (**9**), a thin line of riparian vegetation (**7**) amid scattered farms (fig. 1.10), or a denser pattern of settlement? Valley bottoms in the West are often home to water, animals, and people, mainly on private land. Next, examine the transition zone between valley and mountain. In some cases, foothills extend for miles, as on the west-facing slopes of California's Sierra Nevada, producing subtle changes from valley to mountain. Elsewhere, steeply sloped sidehills mark abrupt changes in terrain and often indicate seismically active range fronts (fig. 1.10). Large alluvial fans, created by streams unloading sediment as they enter the valleys, create a sloping, sometimes terraced zone between valley bottom and mountain front (fig. 1.12). Finally, the mountains themselves vary tremendously, with changing slopes and elevation producing different opportunities for vegetation, wildlife, and land use on mainly public acreage (**67** and **95**). ❋

5

FAULT SCARPS AND QUAKE ZONES

FIG. 1.13 Fault scarp near San Bernardino, California. The alluvial fans (the sloping lands of material deposited by water flow) west of the San Bernardino Mountains are sharply broken by a fault scarp (foreground) that parallels the mountain front. These more steeply sloped hills are evidence of recent movement along the fault.

MANY PARTS OF THE WEST ARE IN EARTHQUAKE COUNTRY. EARTHquakes reshape the landscape, producing subtle shifts in stream courses as well as dramatic breaks along the surface that can rupture the ground, collapse buildings, and disrupt freeway systems.

Fault scarps (surface breaks) are common, but not always easy to recognize because vegetation or rock materials may obscure them. Sometimes, however, they appear as more or less linear features along the base of mountains (fig. 1.13). Over time, faults such as those along the eastern edge of the Sierra Nevada (fig. 1.1) can produce thousands of feet of vertical displacement over a short distance. Normal faults, such as northern Utah's Wasatch fault, produce steep vertical movement along a scarp as they extend the crust. Reverse faults, such as western Wyoming's Wind River fault, produce similar movement as they compress the crust.

In other settings, strike-slip faults cause lateral movement as rocks slide past each other, such as along California's San Andreas Fault. These forces produce landscape features that include highly deformed rock layers (look

for them in road cuts), abrupt contact zones where two dissimilar rock types are adjacent (also visible in road cuts), small sag ponds and narrow valleys in zones of recent movement, and streams that have been offset by fault movement (fig. 1.14).

The distribution of earthquake hazards varies considerably across the West. Seismic risk is high near the Pacific coast, along the eastern slope of the Sierra Nevada, in northern Utah and eastern Idaho, and across the Yellowstone Plateau and nearby areas of southwestern Montana. Thus Salt Lake City, Los Angeles, and San Francisco are at much higher risk for a major quake than are Denver and Boise.

Mandates for new construction standards and required seismic retrofitting of older structures have profoundly reshaped the region's urban landscapes, especially in high-hazard earthquake areas (fig. 1.15). While retrofitting is not very visible to casual observers, the use of reinforced steel, vibration control techniques, and base isolation technologies has reworked the interiors (and increased the construction costs) of thousands of buildings, highways, and bridges across the region, making them a bit less vulnerable when the next big quake hits. ❋

FIG. 1.14 Wallace Creek, California, looking northwest. This unusual bend along Wallace Creek (west of McKittrick) reveals lateral movement along the San Andreas Fault. The North American (or Sierran) Plate (right) is moving southward relative to the northward-moving Pacific Plate (left). Over time, the channel of Wallace Creek has been shifted in the seismic action.

FIG. 1.15 Retrofitted landscape near Los Angeles. Thousands of western buildings and bridges have been constructed or retrofitted to meet earthquake safety standards. These soaring freeway ramps along Interstate 5 near San Fernando, northwest of Los Angeles, must meet rigorous guidelines to withstand temblors.

LAYERED ROCKS

FIG. 1.16 View from Grand Canyon's North Rim, looking south. While exposed rocks at the very bottom (not visible) of Arizona's Grand Canyon are more than 1.5 billion years old, the bands of sedimentary rock above are younger, including the lighter-toned deposits of Kaibab Limestone (about 260 million years old), visible just below the rim in the distance.

WHEN ROADRUNNER OUTFOXES WILE E. COYOTE IN THE CLASSIC Warner Brothers cartoons, the pair leap off spectacular plateaus or speed by natural stone bridges and arches—a reminder that most of the Earth's surface is composed of sedimentary rocks. These are the many-layered, multihued deposits you can see in many places, from coastal bluffs near Los Angeles to Rocky Mountain summits in Glacier National Park.

Layered rocks tell many stories, revealing clues about the age and the environment in which they were created (fig. 1.17). Different layers conform to different time periods and episodes of deposition. They also reveal whether they were created on a sea floor (some limestones), a sandy shoreline (sandstone), a lake (mud shale), or a streambed (conglomerate).

If you step back and assess the larger picture (figs. 1.9 and 1.16), you can often see how sediments were pressured or cemented into rock. Were layers folded or thrust upward into mountains or high plateaus? How did they erode? Did minerals such as iron oxides chemically weather, creating colors along exposed surfaces? In Arizona's Grand Canyon National Park (**66**), the region was uplifted, and the Colorado River gradually exposed older deposits at the canyon's bottom.

The Colorado Plateau is home to what geologists call the Grand Staircase, a monumental layering of sedimentary rock sequences that spans nearly two billion years. A journey up the staircase—and forward in time—begins amid the ancient, 1.7-billion-year-old Vishnu schists along the Colorado River at the bottom of the Grand Canyon (fig. 1.18). As you climb up through the canyon, you hike toward the geological present (fig. 1.16), passing layers of rock that tell tales of tropical oceans, limy ooze, river alluvium, and volcanic ash. By the time you reach the lighter-hued bands near the South Rim (figs. 1.16 and 1.18), you find Kaibab Limestone, deposited only 260 million years ago. To follow the staircase to its more recent past, you can venture to Zion National Park in Utah (fig. 1.17), where Mesozoic-era (about 65–250 million years ago) sandstones and shale remain visible today. The red cliffs and pinnacles on the high plateaus of nearby Bryce Canyon National Park expose even younger Cenozoic-era rocks, all deposited and eroded within the past 65 million years. ❋

FIG. 1.17 Sedimentary rocks, Zion National Park, Utah. In many portions of the West, a layered landscape of sedimentary rock is readily visible, as in this cut made along Zion's Watchman Trail. These Mesozoic-era sandstones were laid down successively between 65 million and 250 million years ago.

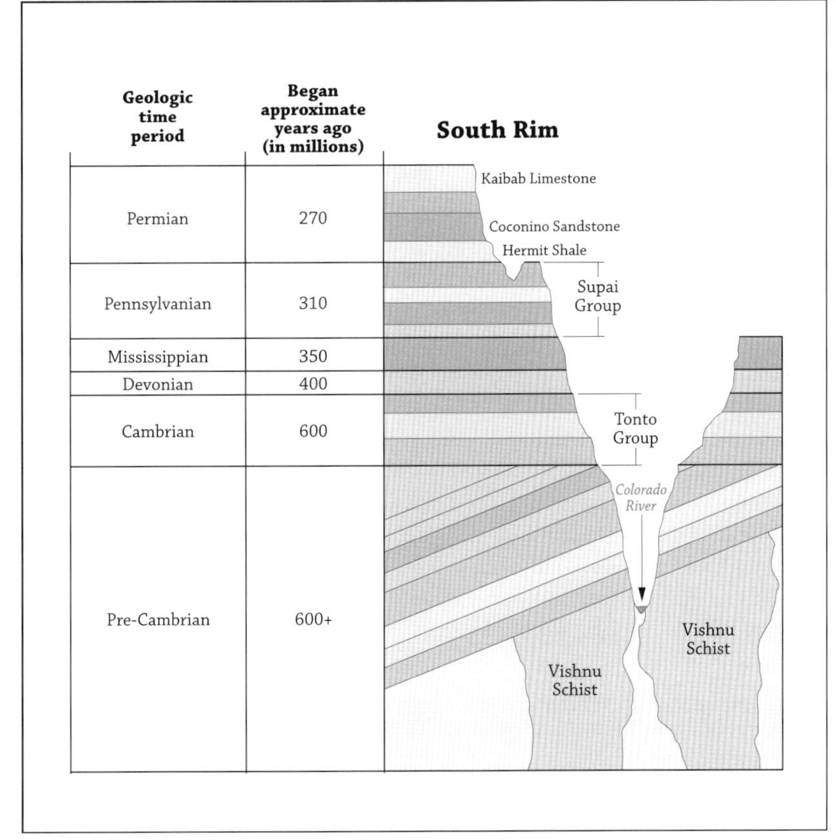

FIG. 1.18 Layered rocks through time, Grand Canyon National Park. Arizona's Grand Canyon remains one of the best localities in the West to ponder the relation between geological time and the sedimentary rock record. A hike down the canyon is a hike back in time.

5 IGNEOUS LANDSCAPES

FIG. 1.19 Volcanic landscape, Shiprock, New Mexico. Igneous landscapes often contain erosion-resistant rocks that degrade slowly. Watch for examples of old volcanic necks (right) and long, narrow dikes (left) that are now exposed on the earth's surface.

NO ONE LIVING IN THE PACIFIC NORTHWEST AT THE TIME WILL EVER forget the morning of May 18, 1980. In moments, a volcanic blast removed 23 square miles from southern Washington's Mount Saint Helens, triggering landslides, floods, and mudflows that killed 57 people, 12 million salmon, and 7,000 big game animals. An 80,000-foot-high ash cloud deposited thick layers of material downwind from the volcano, changing day to night in places such as Yakima, Washington.

Mount Saint Helens is an example of how igneous landscapes have shaped the American West. Igneous rocks form when magma or lava cool and solidify, a process that can occur beneath the surface (intrusive igneous rocks) or at the surface (extrusive igneous rocks). The West has experienced repeated episodes of magma intrusion and volcanism, with most igneous landscapes forming within the past eighty million years and many just within the past few million years.

We can see intrusive igneous rocks in places where they have been uplifted and erosion has exposed their surfaces. The light-colored granites of California's Sierra Nevada, central Idaho, and southwestern Montana are examples of batholiths that first crystallized beneath the surface. Smaller necks—as seen at Shiprock, New Mexico, or Devils Tower, Wyoming—are

exposed interiors of igneous bodies, and dikes are similar linear intrusions of magma revealed by erosion (fig. 1.19).

Extrusive igneous rocks appear in every western state. Southwestern Colorado (San Juan Mountains), western New Mexico (Mount Taylor), and Arizona (San Francisco Peaks) offer prominent examples. Volcanoes in the Cascades dominate the landscape from Mount Baker to Lassen Peak (fig. 1.20). In the interior Pacific Northwest, large flows of dark basalt are visible on the Earth's surface in the Columbia Basin. Formed between twelve million and seventeen million years ago, the sheetlike lavas escaped from fissures, cooled, and eroded. These basalt landscapes vary tremendously, depending on chemical composition, cooling rates, and weathering. Certain patterns of fracturing can cause columnar formations in some landscapes (fig. 1.21), while in others jagged chunks of volcanic material can make hiking hazardous.

Igneous and sedimentary landscapes (**4**) penetrate each other in several ways: younger magmas can be injected into older sedimentary deposits; lava can cover sedimentary surfaces; ash falls can appear as layers in sedimentary rock; and igneous material erodes and becomes part of a sedimentary landscape. ❋

FIG. 1.20 Mount Shasta from McCloud, California. Volcanic peaks that include (south to north) Lassen, Shasta, Jefferson, Hood, Saint Helens, Adams, Rainier, and Baker offer the most dramatic signatures of igneous landscapes in the West. All are considered active and capable of erupting.

FIG. 1.21 Columnar basalt near Antelope, Oregon. Note how the chemical structure of columnar basalt typically produces an orderly, predictable pattern of fracturing and erosion. The cliff shown here exemplifies this dark gray and black rock, seen in many portions of the igneous West.

6 HIGH COUNTRY

FIG. 1.22 Donner Lake (center) from Donner Pass, looking east. This view, taken west of Reno along old U.S. 40, captures the rugged, granite-defined high country typical of the Sierra Nevada and is the location where members of the ill-fated Donner Party became stranded as they attempted to cross the mountains in the snowy autumn of 1846.

QUINTESSENTIAL PLACE-NAMES OF THE HIGH COUNTRY ARE REVEALing. Both the Sierra Nevada ("snowy mountains" in Spanish) and the Rocky Mountains, for example, identified exotic terrain to the Euro-Americans who named them. These landforms bore little resemblance to the landscapes that nineteenth-century explorers knew in New Spain or eastern North America. The West's high country usually meant danger and delay to those who were attempting to cross it.

For those who headed west to California, few more terrifying landscapes existed than the steep, stony slopes of the Sierra crest on California's Donner Pass (fig. 1.22). In the winter of 1846–1847, more than thirty pioneers perished there in the ill-fated Donner Party expedition. Their notorious story of high-country cannibalism no doubt cooled the heels of thousands of potential migrants.

Today, the high country is a magnet for tourists and outdoor enthusiasts, and Donner Pass is home to ski resorts, campgrounds, trails, and nearby wilderness. It is part of a much larger alignment of Pacific ranges, from California's Sierra Nevada to the high Cascade peaks and Washington's Olympic Mountains. A more diverse array of high-country landscapes marks the Rockies in Montana, Idaho, Wyoming, Colorado, Utah, and northern New Mexico.

You can find other isolated fragments across central Nevada, in the Southwest, and east of Los Angeles.

Topographically, these boundary zones define and divide the region's major watersheds. Each mountain pass takes you from one river system to another, and you can see some of the West's most spectacular topography in landscapes of rushing streams, steep slopes, hairpin turns, and high peaks (**53**). There is nothing quite like switchbacking up an alpine trail and taking in the sweep of the landscape, feeling a mountain breeze, and seeing further mountain ranges miles away. The popularity of climbing Colorado's "fourteeners"—peaks higher than fourteen thousand feet—and alpine hiking across the West in general suggest the appeal of the experience (**89**).

Much of the high country has been sculpted by alpine glaciers, and their advances and retreats between 2.5 million and 12,000 years ago (the Pleistocene) radically reshaped the landscape. Watch for cirques, hollowed-out, concave-shaped erosional features near the source or headwalls of glaciers; U-shaped valleys, eroded and broadened by glacial movement; moraines, deposits of material along the sides and at the terminus of former alpine glaciers; tarns, small ponds found near cirques; and larger glacial lakes, near moraines where drainage is blocked or disrupted (fig. 1.23).

The high country also offers an amazing gradient of plant life (fig. 1.24). If you climb any Colorado fourteener, you will experience the biological transition zones that demonstrate how difficult it is for plants to survive on top of a mountain. The timberline, the upward edge of continuous forest cover, varies considerably in the West, tending to be higher in the south (about 11,000 feet in Colorado) and lower in the north (about 5,500 feet in northern Montana and northwestern Washington). Smaller,

FIG. 1.23 Two Medicine Lake, Glacier National Park, Montana. A landscape carved by glaciers, this view includes eroded cirques (rocky alpine headwalls of former glaciers in the distance), U-shaped valleys (including the depression occupied by the lake), and moraines, which are deposited materials (the long, low forested hills beyond the boat docks) from earlier advances of the ice.

FIG. 1.24 Tree line near Loveland Pass, west of Denver, Colorado. Note the ragged, uneven upper edge of forest growth across the alpine landscape near the crest of Loveland Pass (11,990 feet).

often wind-contorted trees called krummholz dot slopes until the tree line is reached, the ragged, uneven upper boundary of the forest cover. Above that elevation, only alpine meadows, rocky slopes, and tundra vegetation separate you from the summit.

The West is filled with what geographer Kevin Blake calls "peaks of identity," summits and other elevated western terrain that hold special meaning for residents and visitors. Many western peaks are sacred to Native Americans and play enduring roles in Native beliefs, cultural practices, and traditions (**38**). Navajos, for example, believe that the Creator placed their people on land among Tsisnaasjini' (southern Colorado's Mount Blanca), Tsoodzil (New Mexico's Mount Taylor; fig. 1.25), Doko'oosliid (Arizona's San Francisco Peaks), and Dibé Nitsaa (southern Colorado's Mount Hesperus).

Many contemporary western places are defined by high country. What would Denver be without the Front Range or Colorado Springs without Pikes Peak? Other western high points play similar roles, including Taos Mountain (Taos, New Mexico), the Sangre de Cristo and Sandia mountains (Albuquerque, New Mexico), the Santa Catalina Mountains (Tucson, Arizona), the San Francisco Peaks (Flagstaff, Arizona), Mount Timpanogos (Provo, Utah), Mount Rainier (Seattle), and Mount Hood (Portland).

Similarly, the much-celebrated reputation of "the mountain," wherever it may be, typically imparts its identity and character to a nearby ski town (**92**). These topographic signatures become icons of identity that are

reimagined by urban and resort-area boosters, real estate promoters, and local advertisers. The result is a western landscape in which mountains play a visible and symbolic role, even though relatively few residents call the high country home. ❋

FIG. 1.25 Tsoodzil. New Mexico's Mount Taylor (right) remains a sacred mountain to the Navajo people. The volcanic peak (11,305 feet) is northwest of Laguna in the western portion of the state.

7 RIVERS AND RIPARIAN CORRIDORS

FIG. 1.26 Little Blackfoot River, near Garrison, Montana. The quiet waters of the Little Blackfoot River are lined with cottonwoods, willows, and grass, creating a fine riparian habitat for fish, birds, insects, and mammals.

THERE IS ONE SUREFIRE WAY TO LEARN ABOUT WESTERN RIVERS AND the riparian corridors that border them. Find a quiet stretch of water, close your eyes, and listen. The stream offers an organizing melody of moving water over rocks, sand, and gravel bars. The wind rustles the trees and the nearby sedges and grass. And there are birds, a quick high-pitched song from a yellowthroat or the shrill cry of a red-winged blackbird. There is the whine of insects, trout rippling the surface of the water, field mice rustling in the bushes, and sometimes the deep, slow thrashing downstream of a browsing deer or moose. You might also hear a distant jet, a passing truck, voices from a campground, or the curse of a frustrated fisherman (fig. 1.26).

For centuries, rivers have organized and shaped human encounters with the West. For both Native Americans and Euro-Americans, western history has revolved around its river geography: migrations parallel their courses, settlements line their banks, politics define their significance, and cultural meanings swirl within their waters. Six great watersheds divide much of the West into the basins we see on the map (fig. 1.27). Flowing predominantly

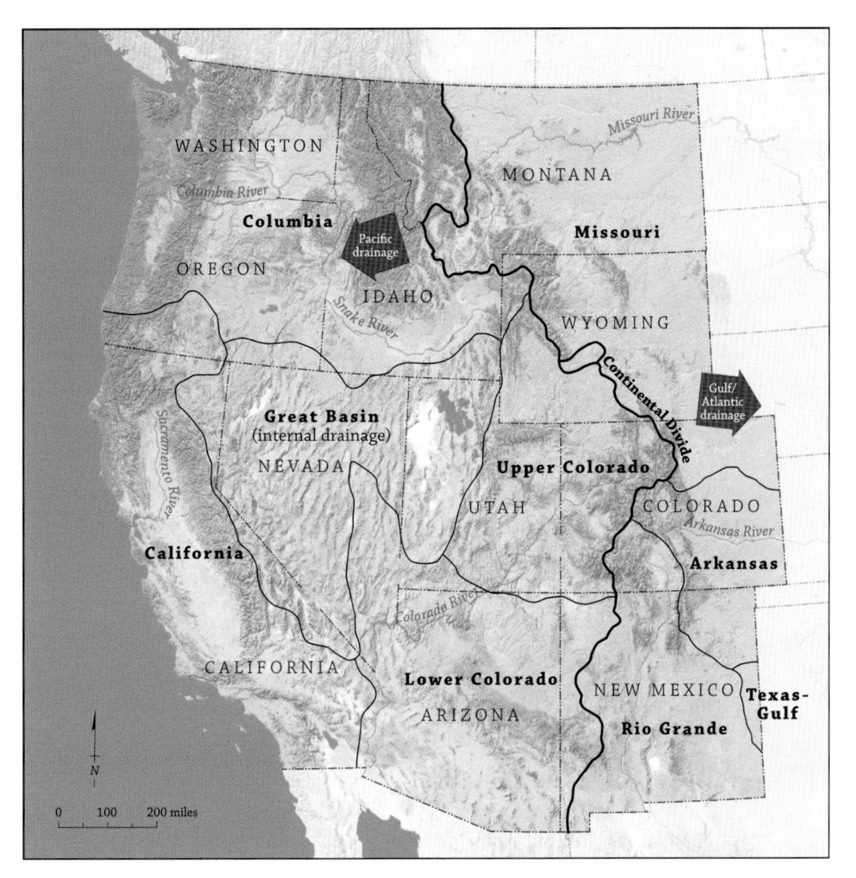

FIG. 1.27 Drainage basins of the West. Six great watersheds divide the West, with the Rio Grande, Arkansas, and Missouri rivers draining toward the Atlantic and the Colorado, Columbia, and Sacramento systems emptying to the Pacific.

west, the Colorado, Sacramento, and Columbia drainages quench the thirsts of some of the largest populations in the region, while the Missouri, Arkansas, and Rio Grande flow east and south toward the Gulf of Mexico and the Atlantic Basin. Complicating the story are sizable portions of the West that have interior drainage—that is, their rivers and streams have no outlet to the sea—a fact of western geography that confounded generations of early Euro-American explorers and mapmakers.

Every western river tells four stories. First, rivers are fluvial geomorphic agents: they organize the movement of water and in the process shape the land and the people. You can watch the changing character and gradient (rate of elevation loss) of rivers from high mountain streams in the Rockies to ocean-bound trunk lines nearing the Pacific (fig. 1.28). Different rivers also have different drainage patterns. Watch for a dendritic (treelike) pattern in many river systems as small channels converge into larger ones. Sometimes, local geology or terrain produces rectangular patterns (along faults), radial patterns (around volcanic peaks), or other geometries that defy easy description.

The work a river does is a function of many properties: the width and depth of its channel, the volume and velocity of its discharge, the amount of sediment suspended in the water, and the rocks and environment through which it flows. All of these characteristics are interrelated and suggest how

FIG. 1.28 Columbia River Gorge, east of Portland, Oregon. The Columbia River drains much of the Pacific Northwest and nearby western Canada. By volume, it is the West's largest river. Here, it has eroded through the volcanic highlands of the Cascade Range, creating a spectacular gorge on its path toward the Pacific.

and why a river does what it does. When you stop alongside a rushing western stream, look at the intimate microgeography of channel and bank features. Rapids, pools, sand bars, gravel bars, muddy banks, and more are present in a single stretch of water.

On a larger scale, many of the West's transportation routes are determined by the drainage patterns of rivers. Many Native American trade routes paralleled major rivers (**38**). Pioneer trails (**49**), transcontinental rail lines (**50**), and interstate highways (**55**) are laid out in relation to the geographies that are largely aligned with great river systems.

Second, every river creates habitat, an intricate riparian world of plants and animals, often in an otherwise arid land. The visual impress of the riparian zone in drier parts of the West is striking. Standing in sharp contrast to the tans and browns of the surrounding countryside, a dark corridor of sedges, rushes, grasses, willows, and trees defines the riparian zone, along with the aquatic environment of the river itself (figs. 1.12, 1.26, and 1.29). Riparian zones are great sponges that absorb floodwaters, hold nutrients, and act as nurseries for animal populations from mayflies to moose. The more diverse the habitat, the more species diversity is likely to be present. Watch for how aquatic insects, fish, mammals, birds, reptiles, and amphibians vary along a stream, depending on habitat, food sources, and competition. Similarly, plant communities are shaped by soil and moisture conditions as well as the width of the valley bottom.

A third characteristic of western rivers has redefined their roles as geomorphic agents and habitats: rivers have become commodities. Capitalism has refashioned and revalued water in the West, forming what historian Donald Worster calls a "hydraulic civilization." Water is thus redefined as a material asset, every drop a credit on the balance sheet. The results are often sobering. No wonder 90 percent of the West's riparian environment (much of it on public lands) has been identified by the federal government as badly degraded.

As one of the West's most precious resources, water has been manipulated and valued for its role in economic activities as diverse as beaver trapping, logging, metals mining, manufacturing, farming, urban living, and recreation. As a result, human activities and settlements are concentrated along its waterways. Every river is a hybrid expression of natural processes and human intent (fig. 1.30). Government-sponsored dam-building imperatives and the concrete-lined channels of California's Los Angeles River are the ultimate expressions of this impulse toward artifice and control (**23** and **64**). You can see the ways, both subtle and profound, that people have reworked western rivers and how they have responded by adjusting their courses, gradients, and sediment loads, sometimes in unpredictable ways.

Fourth, rivers are corridors of political power and cultural meaning, precisely because they play such central roles in defining both the narrowly material and the broader human value of western lands. Politically, water often provokes passionate territoriality and has led to legal battles. Beginning with nineteenth-century Euro-American settlement, water was frequently at the center of political conflict, primarily over the doctrine of prior appropriation (the idea that earlier users of water have senior rights to that water), and it

FIG. 1.29 Riparian flats along the Gallatin River near Three Forks, Montana. The rocky, tan-toned hills of the lower Gallatin Valley stand in sharp visual contrast to the lush tall grasses, shrubs, and trees that border the river.

FIG. 1.30 Snake River, near Twin Falls, Idaho. Large segments of Idaho's Snake River have been dammed, creating a carefully managed riparian landscape of irrigated farms (distance, right), golf courses (center, left), and nearby suburban neighborhoods (on rim, left). Note the river's erosion of the Snake River Plain's igneous rocks.

is a common nexus for disputes among local, state, and federal governments; Native American interests; and foreign governments (Mexico and Canada). Water issues often organize political constituencies and define political battle lines: consider the people of Colorado's Western Slope versus those of Denver; Northern versus Southern Californians; Las Vegas developers versus central Nevada ranchers. Every mile of every river in the West is a politicized space over which competing interests have fought and upon which is etched a complex legal geography of rights, constraints, and responsibilities. You can see these power struggles in the signs, technology, and fences that suggest who controls the water and for what purpose.

More broadly, people develop attachments to river landscapes. As early as the 1860s, John Wesley Powell recognized that river basins in the West should play a central role in how lands were settled and how the West devel-

oped. When it comes to questions of water use and water conservation today, residents strongly identify with their local interests, and almost every western river has associated organizations to advocate for its use. The Platte River Parkway Trust (in Casper, Wyoming) and the Blackfoot Challenge (in Ovando, Montana), for example, bring together landowners, environmentalists, and governmental organizations dedicated to promoting sustainable-use practices along their respective waterways.

Or take a more informal approach: listen to conversations at any small-town café in the West and it probably won't be long before you hear the name of a stream or river. For many residents "their" river is a part of their notion of home, neighborhood, and community, and you can see that powerful emotional tie in local politics, recreational activities, and the affections they share for that nearby stretch of water. ❋

8

DRY WASHES AND GULLIES

FIG. 1.31 Flash flood area south of Needles, California. Sudden desert cloudbursts can turn dry washes in this normally arid landscape into raging torrents of water and mud. This sign along California's highway 62 south of Needles warns unsuspecting drivers that in bad weather the road ahead may be covered by water.

TRAVELERS CROSSING THE ARID SOUTHWEST HAVE ALL SEEN ROAD signs warning motorists about potential flash floods (fig. 1.31). Usually, the cloudless sky and dry streambeds make such threats seem ludicrous, but those dry washes and gullies have been known to carry away cars and cows. They are important parts of a regional drainage pattern dominated by seasonal or intermittent streams. Surface moisture may also be discontinuous: small amounts of water may flow within a channel and perhaps gather in a pool or two, separated by long stretches of desert sand. Water might be flowing beneath the surface, making only short, ephemeral appearances where conditions permit. But after a short, heavy summer downpour, these dry channels can foam with dirty, sediment-filled water, quickly eroding the floor of the channel and carrying away just about anything in its path. In these circumstances, dramatic landscape changes can occur in a few minutes.

Heavy livestock grazing (**20**), brush and timber harvesting, and exotic plant species (**17**) can contribute to rapid erosion and the deepening of dry washes (arroyos) and gullies, especially in higher, more steeply sloped arid and semiarid mountainous localities in Arizona, New Mexico, Nevada, and Utah (fig. 1.32). Historically, in various places such as California and Colorado,

FIG. 1.32 Gully near Chimayó, New Mexico. This steep, narrow gully carries a good deal of water when mountain thunderstorms visit or spring snowpacks rapidly melt. Extensive livestock grazing in the area also contributes to more rapid runoff. Most of the year, however, the channel is dry.

FIG. 1.33 Tanque Verde Wash, Tucson. This broad, sandy wash has been left unsettled on the east side of Tucson. Planners and developers know the wash is a major runoff channel when heavy precipitation falls in the nearby Santa Catalina Mountains.

hydraulic mining activities (**28**)—where water is applied to gravelly slopes at high pressure to wash out valuable metals—added to gullying and produced spectacular eroded landscapes that remain visible today.

People have adjusted to these unpredictable, potentially dangerous features on the landscape. In the Southwest, in cities like Tucson, Phoenix, and Albuquerque, broad, largely linear swaths of open, undeveloped land along dry washes can be seen on the edge of developed areas (**83**), sometimes near housing developments and shopping centers (fig. 1.33). This is a sure sign that these corridors are subject to flash flooding. Municipalities and residents often add a rock and rubble riprap along the edge of dry washes to control flows when cloudbursts arrive. Across the Southwest, the flash floods occur most often when summer thunderstorms rumble between early July and late August. ❋

9

DRY LAKES

FIG. 1.34 Owens Lake, California. Eastern California's Owens Lake (near Lone Pine) features a white, alkali- and salt-rich lake bed that remains following extensive evaporation and drainage of the former lake. Mining operations in these settings can often harvest salt, brine, borax, and soda ash.

THE DISTRIBUTION OF LAKES IN THE PLEISTOCENE ERA (2.5 MILLION to 12,000 years ago) across Utah, Nevada, southeastern Oregon, and eastern California constitutes one of the most remarkable maps in the natural history of the West (figs. 1.34 and 1.35). Dozens of sizable lakes covered more than fifty thousand square miles of the Great Basin, marking the cooler and wetter climates of earlier glacial periods. Largest by far of these landlocked, rain-fed (pluvial) lakes were prehistoric Lake Bonneville (a thousand feet deep) in western Utah and Lake Lahontan (nine hundred feet deep) in western Nevada. Slowly, these great relics dwindled in size and depth, leaving a scattering of smaller lakes (Great Salt Lake and Pyramid Lake are remnants of Bonneville and Lahontan), ancient shorelines (visible in the Salt Lake City and Provo areas), and salt-encrusted dry lakes.

Dry lakes (known as playas) are an important contemporary landscape signature of Pleistocene climates in the Great Basin. Today, you can still see many small playas across arid stretches of eastern California, Nevada, western Utah, and southeastern Oregon. You can find smaller-scale examples in arid portions of the Southwest, such as Willcox Playa, Arizona, and Tularosa Basin, New Mexico. The characteristic white earth that defines these features, containing varying amounts of alkali salts and clays, represents the surface

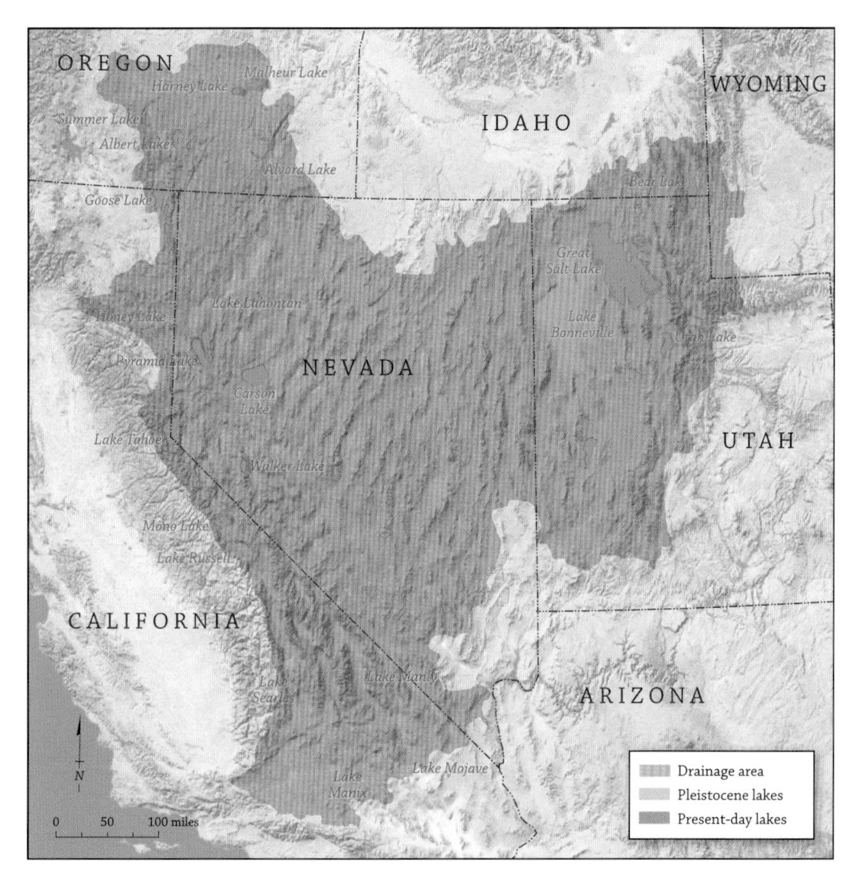

FIG. 1.35 Pluvial lakes, Great Basin. This map shows the approximate location of some of the West's former Pleistocene-era (2.5 million to 12,000 years ago) lakes. Prehistoric Lake Bonneville (Utah) and Lake Lahontan (Nevada) were the largest. Today, playas (dry lakes) are all that remain of most of this ancient pluvial landscape.

residue that followed the evaporation of moisture (fig. 1.34). When it is wet, you can spot salty brine-filled remnant ponds in the playas, but they soon disappear when drier conditions return.

Vegetation is almost entirely lacking on dry lake beds. Surrounding crusts are typically sprinkled with salt-resistant species such as saltbush or salt grass, which absorb and then excrete excess salt. Halophilic (salt-loving) bacteria add toning and color to the playa's palette. Pinnacle-like tufa columns (carbonate-rich deposits) also form in these settings—California's Mono Lake and Trona Lake offer world-famous examples. On windy days, alkali dust storms (**12**) are another common feature of these unusual western environments.

There are signs of human activity on these dry lakes. At Owens Lake in California, for example, industrial operations have mined the salt and brine to extract chemicals. Nearby Searles Lake is home to brine wells, borax mining—first hauled out in the late nineteenth century with twenty-mule-team wagons—and other salt- and soda-ash operations. Some dry lakes became magnets for the military (**62**) because of their flat surfaces and relative isolation. Much of Edwards Air Force Base in the Mojave Desert, for example, makes use of Muroc Dry Lake. Other dry-lake environments, such as Nevada's Black Rock Desert and the Bonneville Salt Flats, have been used for rocket experimentation and for setting land speed records. Increasingly, playas are being identified as areas for protection (**68** and **69**), as these fragile settings are threatened by increases in population and growing demands for desert land. ❋

10
COAST

FIG. 1.36 Coastal bluffs and beach, Cambria, California. Typical of rugged and emergent coastlines (coastal lands exposed by a relative change in sea level), these eroded headlands north of San Luis Obispo offer visitors an attractive sample of ocean scenery that includes view lots, sandy beaches, and rocky tide pools (in the rocks at the distant center right).

MORE THAN THIRTEEN HUNDRED MILES OF PACIFIC COAST SEPARATE Imperial Beach south of San Diego from Cape Flattery on the Olympic Peninsula. These settings are integral, dynamic parts of the regional landscape and westerners have transformed these special places in myriad ways (**57** and **87**). But begin by focusing on the raw physical setting. The transect between land and sea, often referred to as the littoral zone, includes nearby coastal cliffs, beaches, the intertidal areas that lie between low and high tide marks, and the near-shore environment just beyond the breakers. The beach absorbs incredible amounts of energy as waves crash onshore, and it accommodates large, moving accumulations of sediment, typically sand (fig. 1.36).

As you wander from land to sea, enjoy the seabirds, smell the sea air, and listen to the surf zone between wave sets and between low- and high-tide cycles. You can sense the transitional character of the coastal zone. You first encounter the splash zone, which may receive occasional doses of seawater. Then you might encounter sea life attached to rocky outcrops and in tide pools as you pass from the high-tide zone (underwater at high tide) to the low-tide zone (above water only at low tide). At low tide, you are more likely to see barnacles, mussels, seaweed, starfish, sea urchins, and kelp.

FIG. 1.37 Venice Beach, California. Sandy beaches attract sunbathers, surfers, and development. Venice Beach, near Los Angeles, draws a crowd on a sunny weekend afternoon. Note how nearby housing crowds the beach. The priciest units face the sunset.

The Pacific coastline is an especially dynamic geological setting, with classic elements of an emergent erosional coastline where the coast marks the leading edge of an advancing continental plate (fig. 1.36). Typically, these coasts are marked by unstable sea cliffs, rugged headlands, rockbound coves, eroded terraces (marking earlier wave-cut platforms) of sedimentary rock (**4**), and offshore sea stacks.

Coasts inevitably are varied landscapes. Some areas of persistent sediment accumulation have become important recreational zones, such as the sandy beaches of Southern California (fig. 1.37) or the dunes on the central Oregon coast. Seasonal cycles can be dramatic, too: a sandy summer beach can become an eroded, rocky cove during stormy winter months. Coastal estuaries, transitional settings where rivers meet the sea, are zones where fresh water mingles with salt water (the combination is known as brackish water), often in complex wetland, mud flat, and salt marsh environments. You can see these zones in San Francisco Bay, at Humboldt Bay, on the lower Columbia River, and at Puget Sound. Elsewhere, portions of the Oregon coast feature dark, jagged basalts, formed by lava that escaped from underwater vents or flowed overland near the shoreline (**5**).

Most important, sample the coast at sunset. Let your eyes discriminate shades of pink, lavender, violet, blue, and gray. Parse sky and clouds, find silhouettes of fishing boats. You might even catch what is known as the green flash, a rare optical treat as the sun's last sliver slips below the sea. ❋

11 CLOUDSCAPES

FIG. 1.38 Anvils and virga, San Luis Valley, Colorado. These summer cumulonimbus clouds in southern Colorado feature cirruslike anvil tops that generally point in the storm's direction of movement. Rainfall evaporating in the valley's dry air produces feathery virga beneath.

IN THE WEST, CLOUDS HAVE HELPED DEFINE THE COSMOS FOR NATIVE Americans, raised the hopes of western farmers, and sparked the imagination of modern artists (figs. 1.38 and 1.39). Even as they drift a mile or two above the earth, clouds are an integral element of the larger landscape and the meanings we derive from it. Combined with the region's dry atmosphere and its terrain (**1**), cloudscapes offer a dynamic geometry of shapes and a shifting array of shadow and light.

Clouds are nourished and fed by surface moisture, largely from the Pacific Ocean. Rising moisture cools and condenses, forming a variety of stratus (uniform, featureless), cumulus (varied, billowy), and cirrus (high, wispy, feathery) cloud types. Under the right conditions, the hydrologic cycle is completed when precipitation returns to the surface, although in the West that falling moisture sometimes fails to reach the ground, producing virga, which taunts farmers and ranchers (fig. 1.38). Other striking regional variants include billowing pyrocumulus clouds, which sometimes reach twenty thousand feet high and mark the spread of summer and fall wildfires (**16**; fig. 1.40).

Three fundamental factors shape western weather and its clouds; the first is latitude. The West lies mostly in the mid-latitude zone of the westerlies, meaning that air masses from the Pacific are the dominant weather makers.

The most common and effective delivery systems of clouds, rain, and snow within the westerlies are cold fronts, which pass across the region from west to east, especially between October and April. These advancing wedges of cool, unstable air, often birthed in the Gulf of Alaska, are key producers of precipitation that can make or break mountain snowpacks, affecting water supplies in cities and on farms. But conditions vary from north to south: Mediterranean climates along the Pacific coast south of 40 degrees north (central California) have wetter winters and very dry summers, while the marine West Coast climates farther north see more year-round moisture.

Occasionally, other air masses play their part. On the eastern plains, from Montana to New Mexico, the winter weather can originate in the dry, cold Canadian interior. In summer, the Gulf of Mexico is an important source for seasonal monsoons, which bring clouds and showers to the Southwest (fig. 1.39).

The second major factor controlling western weather is the land-sea effect. Areas west of the Sierra Nevada and the Cascades are strongly shaped by this effect, wherein the proximity of the Pacific Ocean (**10**) provides rela-

FIG. 1.39 Maynard Dixon's *Open Range* (1942). This dramatic painting captures the role clouds can play in shaping the western landscape. Here, a towering cumulus cloud above the buttes suggests the possibility of afternoon showers. *Courtesy of Joslyn Art Museum, Omaha, Nebraska, Maynard Dixon (American, 1875–1946), Open Range, 1942, oil on canvas, collection of William C. Foxley, L-2005.19.*

FIG. 1.40 Pyrocumulus clouds north of San Luis Obispo, California. Many western skies are filled with smoke in summer and fall. Active blazes, like this summer grass and woodland fire in central California's Coast Range, create their own signature cloudscapes.

FIG. 1.41 Coastal stratus near Carmel, California. These low, filmy clouds drift for hours along the Pacific Coast, often hugging the immediate shoreline.

FIG. 1.42 Lee wave clouds near Reno, Nevada. These stacked, lens-shaped clouds can stay in place for hours and reveal the interaction between high winds aloft and a major mountain range, in this case the Sierra Nevada.

tively warmer weather in winter and cooler weather in summer. The coastal cloudscape found near Pacific shores on a summer afternoon exemplifies the pattern: a stiff breeze, chilly air mass, and fast-moving stratus skims the headlands and makes you reach for a jacket in July (fig. 1.41). Relatively mild but cloudy winters in western Oregon and Washington also reflect the moderating presence of the Pacific.

Finally, the western sky is wedded to the land beneath it by the interplay of atmosphere and terrain. You can see evidence of orographic lifting in the western sky, whereby mountains mechanically force moisture to rise, cool, and condense (**2**). Summer thunderstorms (with anvil-shaped tops indicating storm movement) gather strength above mountain ranges (fig. 1.38), and smaller cap clouds wrap their mist around higher mountain peaks. When high winds aloft combine with higher terrain, stacked layers of lens-shaped clouds form, particularly in winter, near and just east of major mountain ranges such as the Sierra Nevada, Cascades, and Rockies (fig. 1.42). These standing lenticular clouds can remain a fixture in the sky for hours and indicate increasing upper-level winds and potential for a storm. ❋

12 DUST STORMS AND DUST DEVILS

FIG. 1.43 Dust storm, southeastern Colorado. This historical image, taken in the Dirty Thirties, captures some of the ferocity of the storms that engulfed communities for hours in thick, choking dust. Modern-day dust storms are most common in the desert Southwest. *Courtesy of Denver Public Library, Western History Collection, image no. X-17626.*

DUST IS AN INTEGRAL AND UNWELCOME PART OF LIFE IN THE WEST, AN element of its natural landscape that residents have seen, smelled, and tasted for centuries. People have contributed to the legacy of this dust, particularly with the widespread plowing-up of the Great Plains in the early twentieth century. In the Dirty Thirties, huge dust storms blew across the plains (fig. 1.43), and ranchers found their cattle choked to death, their stomachs half-filled with sand. Thousands of residents suffered from respiratory ailments, and many died from something called "dust pneumonia." For most westerners today, the Dust Bowl experience of the 1930s and again in the 1950s is no more than a chapter in the history books, but continuing dry years across the Plains and predictions of longer-term droughts in the Southwest portend more dust ahead.

Dust storms—sometimes called haboobs (from the Arabic word *habb*, meaning "wind") in southern Arizona—remain a regional hazard. They occur in sparsely vegetated areas when the winds are stronger than twenty-five miles per hour, and can develop along frontal systems—dry cold fronts are good candidates—or near the gusty outflow boundaries of thunderstorms, such as in the Southwest when summer monsoon conditions occur (**11**). These huge, rolling clouds of dust-filled air, like their Depression-era ancestors, can sweep across a highway or into a community and plunge visibility to zero within minutes. Dust storms have caused highway accidents that involve dozens of vehicles.

Southern Arizona is particularly prone to dust storms. Phoenix experiences about three haboobs a year, most occurring between June and September. They also happen in the deserts of Southern California and Nevada, in California's Central Valley, on the Columbia Plateau, and across the eastern plains. Near the Colorado Plateau, persistent dust storms have been blamed for early runoff from the Rockies, where darker layers of accumulated dust mix with snow and lead to speedier spring melts.

Longer-term hazards can lurk within the dust. An airborne fungal infection called "valley fever" (or coccidioidomycosis) plagues many places such as California's San Joaquin Valley, southern Nevada, Arizona, and southern New Mexico. The fungus thrives in dry, alkaline desert soils and is spread when disturbed. Rapid urban growth and suburban sprawl (**83**) have contributed to the diffusion of mold spores that carry the disease (construction workers are most often exposed to it).

Dust devils are also common in the West (fig. 1.44) but are less dangerous than dust storms. Driven by strong, uneven surface heating, dust devils are small convective whirlwinds of rapidly rising air that are made visible by the dust, dirt, and debris they pick up. Unlike dust storms, which can rage for hours, dust devils rarely last long and seldom inflict property damage. ❋

FIG. 1.44 Dust devils. Uneven surface heating produces swirling convective currents of rapidly rising air. The resulting dust devils carry sand, dirt, and trash hundreds of feet aloft. Hunt for dust devils in thinly vegetated areas on a sunny, hot day between 11:00 A.M. and 3:00 P.M. These dust devils are picking up bare soil near Ritzville, Washington.

15

CACTI AND JOSHUA TREES

FIG. 1.45 Cacti in the wild, southern Arizona. This tall saguaro cactus south of Ajo sits amid smaller cholla cacti (light-toned, prickly-headed plants beneath the saguaro) and a ground covering of buffelgrass (a recently arrived invasive plant). To see saguaros, visit Saguaro National Park or (for cacti generally) the outstanding Boyce Thompson Arboretum east of Phoenix.

THE GIANT SAGUARO (SUH-WAH-RO) CACTUS, *CARNEGIEA GIGANTEA,* HAS been celebrated as an exotic, humorous, multiarmed cartoon character (figs. 1.45 and 1.46), but it is also a symbol of how plants survive in a marginal landscape. Attaining heights of more than fifty feet and weighing up to ten tons, most of it in stored water, the saguaro is one of the most widely recognized plants that defines the boundaries of the Sonoran Desert (fig. 1.45). With their distribution limited by freezes in winter (the Southwest lies on the far northern range of many cacti), they are most commonly found in southern Arizona between sea level and 4,000 feet elevation (fig. 1.47).

Mature saguaros can be 150 to 175 years old. Spiders, lizards, birds, and rodents live among their folds and nesting holes. For centuries, Native people used the fruit of the saguaro for food and its woody ribs for building materials.

Like many cacti (part of the Cactaceae family), saguaros store water in their tough skin. They lie dormant for many months each year and have extensive root systems that efficiently harvest what little rain falls. Surviving saguaros typically space themselves out on the landscape, never crowding thirsty

FIG. 1.46 Reg Manning's *What Kinda Cactus Izzat?* (1941). Manning's humorous books on Arizona and the Southwest often explored the lighter side of cacti. Actually, this "field guide" contained a great deal of useful material about these bizarre desert plants.

neighbors. Other cacti have similar adaptive strategies for storing and conserving water (fig. 1.48). Common regional examples include organ pipe and senita cactus, which you can see in Organ Pipe Cactus National Monument; prickly pear cactus, which has classic yellow blooms in the spring; various cholla cacti; and barrel cactus.

The Joshua tree (*Yucca brevifolia*), a yucca and member of the agave (Agavaceae) family, is another quintessential desert plant. Nineteenth-century Mormon pioneers named the tree, observing that its arms raised toward heaven reminded them of the biblical leader of the Israelites.

Ecologists identify the Joshua tree as a defining element in many high desert environments of eastern California and portions of Nevada, Utah, and Arizona (fig. 1.49). The range of the Joshua tree is often used to define the borders of the Mojave Desert, and you can see the trees in the Mojave National Preserve, in Joshua Tree National Park, and along the dry north-facing slopes of the San Gabriel and San Bernardino mountains of California. They also grow along western Arizona's Joshua Forest Parkway between Kingman and Wickenburg and outside Searchlight, Nevada, along lonely Nipton Road (fig. 1.49).

A warming climate is altering the distribution of saguaros and Joshua trees in the region. Saguaros are steadily climbing southern Arizona's foothills as the climate becomes warmer and drier. Under similar circumstances, Joshua tree enthusiasts have less to cheer about. Many Joshua trees may die as regional conditions become too extreme for the flower-bearing yucca. Fur-

FIG. 1.47 Distribution of saguaros and Joshua trees. Saguaros prefer low-desert settings while Joshua trees thrive in higher-desert localities.

thermore, Joshua tree seeds don't migrate easily, so the species may not be able to make it to cooler, wetter environments.

Nurseries in the Southwest are good places to explore the ecology and cultural meaning of desert vegetation (fig. 1.48). In these way stations between wild nature and manicured human landscapes, you can discover an amalgam of ecological imperatives and accumulated cultural meanings. Once desert vegetation leaves the nursery, it heads for the front yard (**77**), where millions of western homeowners proudly acknowledge their arid-lands identity (fig. 1.50). ❋

FIG. 1.48 Cacti in a nursery, Tucson, Arizona. These varied cacti are being prepared for their journey to front yards around the Tucson metropolitan area.

FIG. 1.49 Joshua trees west of Searchlight, Nevada. These plants are typically seen in many higher-altitude settings across the Mojave Desert.

FIG. 1.50 Cacti in a front yard, Lakewood, California. This homeowner obviously has an eye for cacti, though allegiances are split. Half the front yard remains in lawn.

14 SAGEBRUSH

FIG. 1.51 Sagebrush west of American Falls, southern Idaho. Tiny yellow flowers mark sagebrush in bloom. Note the scattered juniper trees, riparian vegetation (along the Snake River), and basalt cliffs (in the distance).

PUNGENT, AROMATIC, TOUGH, ADAPTABLE, SAGEBRUSH CREATES A landscape whose beauty is subtle, featuring generously spaced plants that leave plenty of room to tap available moisture (fig. 1.51). Nevada is rightly named the Sagebrush State, but the distribution of *Artemisia tridentata* (so named because its tiny leaves feature three teeth on their upper edges) and its regional variants make sagebrush one of the West's most common arid-land shrubs. The "sagebrush sea," a hundred-million-acre expanse of arid and semiarid lands, is one of North America's largest, most recognizable landscapes (fig. 1.52).

North of Las Vegas, the pale green colors of sagebrush gradually replace the darker green hues of the creosote bushes that grow in the Southwest's lower deserts. Sagebrush dominates the Great Basin across Nevada, western Utah, and southeastern Oregon. In Idaho's Snake River Valley and in the volcanic Columbia Plateau, hardy sagebrush thrives among lava fields. In irrigated areas near the Snake and Columbia rivers, the verdant green of well-watered pastures and croplands (**23**) contrasts with surrounding sagebrush, creating a mosaic of color. Farther east, in central Wyoming, you can see *Artemisia tridentata* var. *wyomingensis* and other varieties that thrive on the high, dry plateaus (fig. 1.53). Spilling out onto the plains, other species

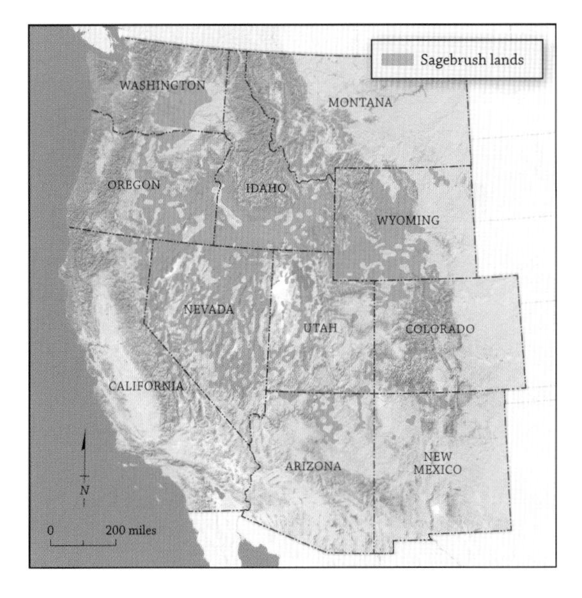

FIG. 1.52 Distribution of Great Basin sagebrush (*Artemisia tridentata*) across the interior West.

(silver sagebrush is common) grow across untilled portions of eastern Montana and Wyoming.

People and sagebrush coexist, but sometimes uneasily. For centuries, Native people valued sagebrush's curative powers, and the plant became an integral part of herbal medicines for ailments from digestive problems to infection (**38**). Early Euro-Americans typically avoided what they called the "sage barrens," often delaying agricultural development because they believed the lands were infertile. In later years, however, sagebrush land was often plowed under and successfully irrigated.

In the twenty-first century, the plant has become a battleground among environmentalists (who want to protect sagebrush habitat for pronghorn antelope and sage grouse), off-road vehicle enthusiasts, ranchers, the energy industry, and public-lands officials (**68**) who manage more than half of the West's remaining sagebrush acreage. In coming years, portions of the "sagebrush sea" may survive, but likely in smaller patches than before. ❋

FIG. 1.53 Sagebrush (*Artemisia tridentata* var. *Wyomingensis*) southeast of Riverton, Wyoming.

15 CONIFERS

FIG. 1.54 Ponderosa pine near Seeley Lake, Montana.

CONIFERS ARE EVERGREEN, CONE- AND NEEDLE-BEARING TREES THAT cover many of the West's higher and better-watered landscapes (fig. 1.54). The region's coniferous landscape remains the most complex on Earth. The distribution of boreal (northern coniferous) and Mexican species reflects broad patterns of precipitation (fig. 1.3), varying elevation (fig. 1.6), and local differences in slopes, soils, and drainage patterns (fig. 1.2).

As you travel among the West's conifers, you can consider what limiting factors explain why Rocky Mountain junipers *(Juniperus scopulorum)* grow in one setting, ponderosa pines *(Pinus ponderosa)* in another, and subalpine fir *(Abies lasiocarpa)* and Engelmann spruce *(Picea engelmannii)* in yet another (figs. 1.5, 1.54, 1.55). Add to these variables the complex effects of wildfires (**16**), insect and disease infestations, and regional warming, and it is no wonder that botanists write entire volumes on a single tree species.

People have profoundly shaped the coniferous landscape. Obvious examples include decoratively planted conifers in parks, cemeteries, and front yards and holiday conifer cropping at Christmas tree farms in western Oregon and Washington. Large expanses of second-growth timber, which are evident on hillsides with many trees of similar age, are more subtle signs of management that has taken place after logging or fire (**16** and **33**). The aesthetic and symbolic value of old-growth forest (fig. 1.56) and its importance to native plant and animal species have also made management of species such as the coast redwood *(Sequoia sempervirens)* and Sitka spruce *(Picea sitchensis)* increasingly contentious, particularly in light of the high commercial value of these trees, the homes they provide for rare animal species (**18**), and their limited ranges along the northern Pacific coast.

Every forest landscape in the West is a snapshot in time, destined to change. Conifers are often on the move, both as a natural process of vegetation succession and in response to human action. You can see examples in places where young trees are filling in the edge of a meadow (fig. 1.5) or an aspen forest is being replaced by fir and spruce, particularly following a wildfire. Across the West, tree invasion is also occurring where conifers are populating new hill slopes in response to fire suppression, livestock grazing (where hooves help open the soil for cone reproduction), and particularly heavy spring precipitation (fig. 1.57).

Two examples of western conifers hint at the diversity across the region. In warmer, drier landscapes are pinyon-juniper woodlands (fig. 1.55), a mix of hardy evergreens that grow widely across the semiarid interior. The dark green, often bushlike trees average only twenty to forty feet in height, and many distinctive subspecies characterize these open woodlands. Among the junipers (also called cedars), you can see scaly foliage, berries (blue-, brown- or copper-toned fruits), and varieties that include the Utah juniper *(Juniperus osteosperma)* in Utah, Nevada, and northern Arizona; Rocky Mountain juniper on the lower slopes of the Rockies and across hillsides in the Southwest; and western juniper *(Juniperus occidentalis)* in eastern Oregon and California.

Pinyon (or piñon) pines, *Pinus edulis,* which often intermix with junipers,

FIG. 1.55 Rocky Mountain juniper near Las Vegas, New Mexico.

FIG. 1.56 Old-growth forest, central Oregon coast. In its unlogged state, this thick forest of Sitka spruce, Douglas fir, and western hemlock allows the sun to penetrate only in narrow shafts of midday light.

have slender gray-green needles, singly or in bunches of two or three. Common across eastern California, Nevada, the Southwest, and portions of Colorado, pinyon forests are valued for their tasty pine nuts and the fragrance of the trees' tough, brittle wood.

Ponderosa pine (also called yellow pine) grows in all eleven western states. Hardy and adaptable, ponderosas thrive in the warmer montane environments of eastern Washington (near Spokane, for example), across the middle slopes of California's western Sierra Nevada, and along Arizona's Mogollon Rim (fig. 1.58). Ponderosa pines vary tremendously in height but can grow to be 150 to 180 feet tall. They feature long needles (in clumps of two or three), robust cones, and a notable vanilla-scented bark that forms a varied multihued pattern, particularly on the lower trunk (fig. 1.54).

Many ponderosa and other western varieties of pine struggle with periodic outbreaks of mountain pine beetle. Beetle infestations have resulted in the loss of millions of trees and can quickly change the look of the forest landscape. Similar challenges with spruce budworm have increased with regional warming, affecting thousands of acres of pine, spruce, and Douglas fir *(Pseudotsuga menziesii)*. You can see these infestations on hillsides of standing dead and diseased trees (central Colorado's high country offers widespread examples), where their browned, yellowed, or reddish foliage stands in sharp contrast to the deep green of their healthier coniferous neighbors. ❋

FIG. 1.57 Tree invasion near Alturas, California. Changing climate conditions, land use histories, and wildfire frequency can lead to tree invasion at lower timberline (into nearby dry meadows and hillsides). Watch for advancing small young trees downslope from the mature forest. This scene is in the foothills of northeastern California's Warner Mountains.

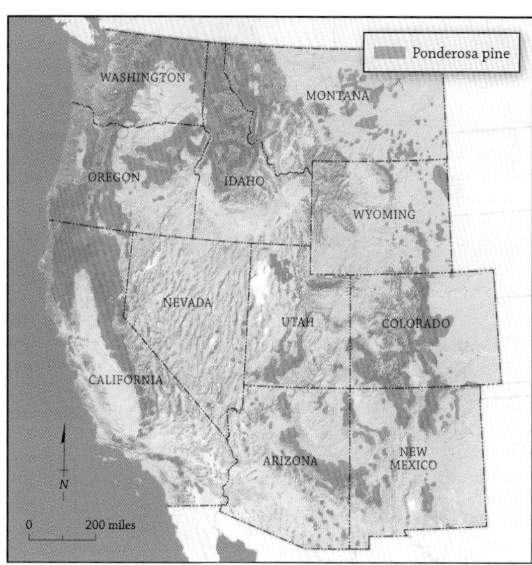

FIG. 1.58 Distribution of ponderosa pine.

16 WILDFIRE

FIG. 1.59 Brush clearing, Orange County, California. In the fire-prone Santa Ana Mountains above Irvine, new homes are buffered from potential blazes by brush-cleared fire breaks. Residents are encouraged to avoid planting large trees near structures, but some clearly choose to ignore the advice.

FIRE LOOKOUT TOWERS, RED FLAG WARNINGS (ALERTS FOR DRY, WINDY weather), signs warning "Extreme Fire Danger," and posters of Smokey Bear are all reminders that westerners have been trying to prevent wildfires for more than a century. Preventive measures are especially visible along the urban-wildland interface (**83**), where suburban landscapes display manicured fire breaks designed to slow approaching flames (fig. 1.59). Nonflammable roofing materials are mandated and few trees are allowed next to houses. In remote areas, seasonal restrictions also limit activities—such as the prohibition against open campfires—and fire management agencies (**67**) use intentional fires, called prescribed burning, to limit large, devastating blazes, known as megafires.

Lightning-caused fires are common across the West, but people are overwhelmingly responsible for wildfires. For centuries, for example, Native Americans used fire to drive game and expand desirable habitats for hunting and gathering (**38**). Other fires are not purposefully set, being started by sparks from rail cars, downed power lines, and burning cigarettes. Blazes spread unpredictably, creating their own wildfire front of advancing flame and smoke shaped by shifting winds, topography, humidity, and fuel sources (**11**). In tall forests, spectacular crown fires leap through treetops at breakneck speed.

FIG. 1.60 Aftermath of a wildfire, northern Montana. This standing dead timber in the Lewis and Clark National Forest may be harvested for sale. An abundant undergrowth of magenta-toned fireweed covers the slope, an early colonizer marking a new cycle of vegetation growth.

FIG. 1.61 Wildfire landscape, Mount Lemmon, Arizona. A devastating fire destroyed timber and homes near the community of Mount Lemmon in southern Arizona's Santa Catalina Mountains. Charred conifers dominate the ridgeline, but a new home and young aspen trees (midslope, right) have recolonized the foreground.

You can see how wildfires shape landscapes where there are fire-scarred trees (fig. 1.60), deep runoff channels and gullies (debris flows and flash floods are common after burns), streams with increased sediment (as well as altered nutrients that affect fish populations), and a new cycle of vegetation. In the high country, fireweed *(Epilobium angustifolium)*, with its brilliant magenta-toned petals, is a fast colonizer after a burn, and aspen *(Populus tremuloides)* is an ecological pioneer that provides shade for the conifers (**15**) that will ultimately crowd them out (fig. 1.61). Fire can be a welcome ecological event, promoting biological diversity and forest health, but it can also open the door for exotic vegetation (**17**) and more frequent blazes. For example, cheatgrass *(Bromus tectorum)*, native to Europe, has occupied many western rangelands, and its fast-maturing and -drying stalks often create the conditions for more fast-moving wildfires. ❋

17

EXOTIC AND INVASIVE PLANTS

FIG. 1.62 Russian olive northeast of Denver, Colorado. This exotic species from Eurasia has been widely used around the West as an ornamental and windbreak, but today it also grows wild in a variety of settings and is considered by many to be a pest, often competing with more desirable species for space and moisture.

IN THE EVOCATIVE LANGUAGE OF ECOLOGY, THE WEST'S NONNATIVE plants are referred to as "exotics" or "invasives." Exotics are plants that evolved elsewhere but have established themselves in a new area (fig. 1.62). In the coastal Oregon forest, for example, you might encounter an exotic from Eurasia, the fingerlike foxglove, which is used to make digitalis for heart patients (fig. 1.63). Sometimes migrations happen intentionally, such as when European grapevines are imported to California, and sometimes they happen by accident. The spotted knapweed, whose delicate purple-to-white flowers dot more than four million acres of the West, probably arrived from Eurasia in soil used for ship ballast.

Invasive plants are exotics that become an unwanted, costly part of an ecosystem. You have encountered an invasive if you ever sprayed weed killer on a patch of dandelions (many varieties from Europe) or had your socks covered with the sticky seed spikes of cheatgrass, a native of southern Europe and North Africa that now covers more than seventeen million acres of public lands in the West. Invasives disrupt, outcompete, and overwhelm native species, imposing economic and ecological costs. They can reduce crop yields and wildlife habitat (**18**), encourage more frequent wildfires (**16**), and change patterns of erosion and drainage (**8**).

In the arid Southwest, buffelgrass crowds out native grasses and threatens giant saguaro and other cacti communities (**13**). The plant competes for water and is much more prone than cacti to ignite and lead to range fires. It also reestablishes quickly after a fire, increasing its dominance. It is easy to spot the dry mantle of buffelgrass that has crept in among many of southern Arizona's cholla and saguaro forests (fig. 1.45).

Invasive plants create environments that can be both harmful and beneficial. An invading shrub may limit the production of a valuable grass or hoard large amounts of water, but it may also offer habitat for rodent and bird populations or help control streambank erosion. Russian olive, a thorny shrub or small tree with dull, light greenish-silver leaves, was imported from Eurasia as a windbreak and ornamental plant. While some now consider it a pest that has crowded out native cottonwoods and other plants, it offers a home to songbirds, quail, and pheasants (fig. 1.62).

FIG. 1.63 Foxglove, Cape Perpetua Scenic Area, Oregon. This attractive exotic from Eurasia can now be found deep within the old-growth forests of the Pacific Northwest, a quiet reminder that even our most "natural" landscapes are filled with nonnative species.

Most obviously, you see exotic plants in a Napa Valley winery, an eastern Montana wheatfield, and a San Diego yard planted in palms and poplars. But there are many more examples. Along roadsides and in other disturbed settings such as construction projects, heavily grazed areas, or recently burned zones, you can see invasives such as Canada thistles, tall, prickly plants with lavender to rose-purple flowers; tumbleweeds from Siberia, which neatly autodetach and disperse their seeds as they roll along; and Jim Hill mustard, which some believe came west along the Great Northern rail line.

Riparian areas are particularly susceptible to invasives. Most streambeds and dry washes in the Southwest are home to tamarisk (also called salt cedar), a tenacious water-loving shrub that has disrupted the stream ecology of thousands of watercourses in such places as northeastern Arizona's Navajo country (fig. 1.64). Intriguingly, one way to control tamarisk has been to introduce an exotic animal, a beetle native to Crete and Kazakhstan. Ecologists know that the *Diorhabda elongata* fancies the tiny, scaly leaves of the tamarisk and so have introduced them along the Virgin and Colorado rivers. Skeptics predict that other exotics such as Russian thistle will simply move in to take the place of the dying tamarisks and that native riparian species may continue to wane.

New combinations of native and nonnative species have produced even broader changes in topography, drainage, and habitat (fig. 1.65). The grass-covered, oak-dotted hills southeast of Salinas, California, are a land-

FIG. 1.64 Invasive tamarisk east of Kayenta, Arizona. Watch for this sturdy, moisture-loving invasive across much of the Southwest. It crowds out native species and decreases the flow of surface water but is very difficult to remove effectively once it is established.

scape that is replicated across thousands of square miles of the Golden State, but that landscape also represents an intricate amalgam of ecological and historical forces. The grasslands are 70 percent to 90 percent exotic species (including wild oats, cheatgrass, foxtail barley, and ripgut brome) that often arrived in connection with cattle grazing (**20**) during the Spanish and Mexican periods (1769–1848) and then were reinforced by Eurasian arrivals during the American era (especially after 1860). Native oaks on the flats and slopes and riparian species (willow, coffeeberry, and coyote bush) along the creek are also part of the mix. Trampling by exotic grazing animals—cattle, sheep, and horses—combined with the hybrid grassland cover to create exposed ground, more rapid runoff, and erosion. Thus, the pattern of hillside gullies and more deeply incised streams reflects the influence of exotic species on the landscape.

The lessons from California and elsewhere are clear: as nonnative plants have found new homes in the West, they have changed the landscape and ecology in ways no one could have predicted. Ultimately, both types of vegetation are part of the natural world in the West; some simply arrived on the scene more recently than others. ❋

FIG. 1.65 California coast range southeast of Salinas. This scene reveals a hybrid landscape of native and nonnative species. Even basic patterns of drainage and erosion have been shaped by the presence of exotic plants and animals in this setting.

18 WILD ANIMALS

FIG. 1.66 Steven Ball's mural is titled *Rocky Hill Guardian*, and it celebrates the beauty and majesty of the western mountain lion. The painting appears on a building in Exeter, California. Note the red-toned Native American pictographs that appear on nearby rocks.

WOLVES, EAGLES, BEARS, AND MOUNTAIN LIONS ARE ESSENTIAL parts of the West's identity, and our fascination with wild animals has shaped how we think about the region (figs. 1.66–1.73). We name our towns after them—Beaver, Utah; Deer Lodge, Montana—our sports teams—Colorado Buffaloes, Washington State Cougars, Montana Grizzlies, Nevada's Wolf Pack—and use their images on state quarters—Montana's bison skull, Idaho's Peregrine falcon, Nevada's wild horses.

The habitats of charismatic wildlife are incredibly diverse. Some creatures require undisturbed ecological conditions to thrive (northern spotted owls, desert tortoises, and grizzly bears), whereas others, such as coyotes, are happy in both wilderness areas and suburban backyards (fig. 1.67). Some animals, such as the grizzly bear, are keystone species, which means they play a pivotal role in their ecosystem.

You can see evidence of wild animals in their tracks (fig. 1.71), scat (inspect owl pellets and bear droppings for clues on diet), and nests and burrows. They turn over rocks on a hillside (bears looking for food), leave claw marks on trees (mountain lions marking territory), and create wallows or bowl-like depressions in the soil or mud (bison do this to keep cool and control insects). A browse line on vegetation indicates where animals have nibbled away at a

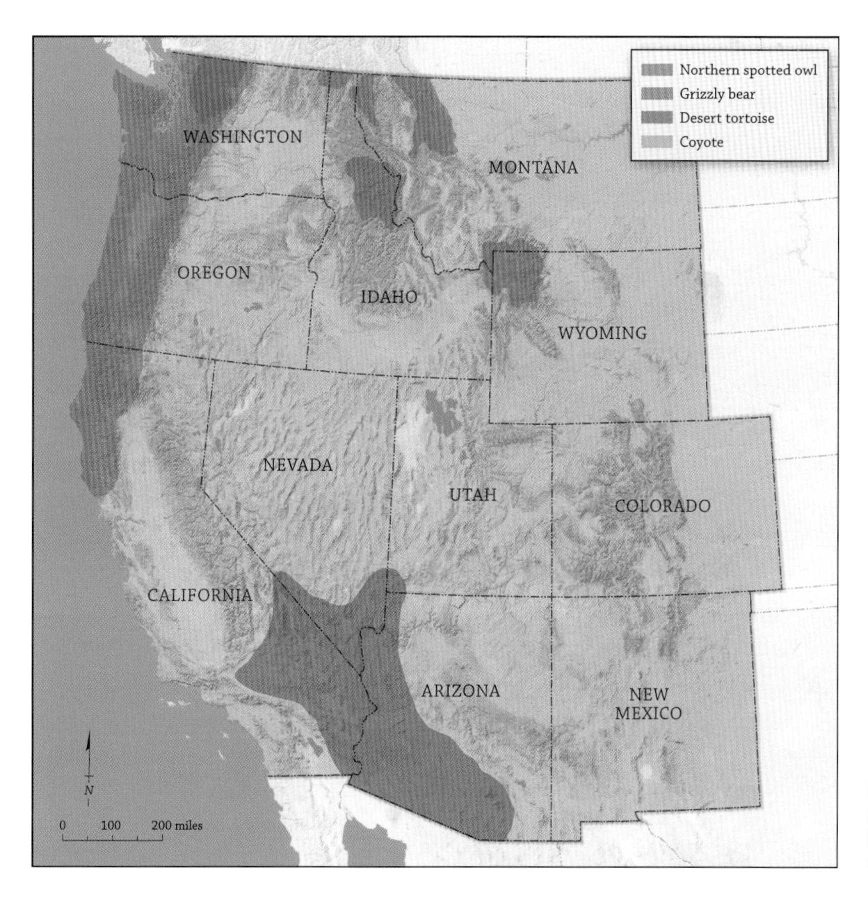

FIG. 1.67 Distribution of the northern spotted owl, grizzly bear, desert tortoise, and coyote.

plant (often a sign of animal overpopulation). Have beavers been at work on trees, or do an abundance of broken branches and matted-down grasses suggest a recent bedding ground?

Political battles involving western wildlife have been spirited, particularly since the Endangered Species Act was passed in 1973. There is also a larger economic and political context to consider, because species survival depends on creating suitable habitat and on managing human activities on both public and private lands. In addition, the limits of suitable habitat are evolving, especially in rapidly changing exurbs near national parks and forests (**95**) and along the urban-wildland interface (**83**). In Orange County, California, for example, mountain lions find their territories carved up by suburban developments and laced with mountain-biking trails (fig. 1.72).

FIG. 1.68 The desert tortoise is native to southeastern California, western Arizona, and southern Nevada. Many tortoise habitats, however, are increasingly threatened by suburban development, recreational activity, expanded use of military lands, and new wind and solar farms being constructed across the region.

Some of the highest-profile battles over pivotal wild animals in the West include protecting old-growth forests for northern spotted owls in the Pacific Northwest, finding undisturbed habitat for desert tortoises in the Mojave Desert, reintroducing and managing wolves across the northern Rockies, maintaining the viability of wild salmon in the Columbia River basin, protecting grizzly bears in the Greater Yellowstone Ecosystem, and ensuring

FIG. 1.69 Black bear, Yellowstone National Park. This young black bear, browsing by the road near the Grand Canyon of the Yellowstone, is representative of some of the charismatic megafauna that draw several million visitors to the park annually.

FIG. 1.70 At home in the high country, mountain goats are synonymous with the alpine wilderness. Look for them along rocky cliffs in the North Cascades and across the summits of the northern and central Rocky Mountains.

the survival of California condors in California and the Southwest. Even very noncharismatic species, such as California's Delta smelt (a small fish native to the San Joaquin–Sacramento delta), have been at the center of enormous controversy, especially when legal rulings have protected such species by curtailing agricultural and residential uses of that state's limited water supplies.

The role of wild animals as cultural symbols is often displayed on the landscape. Public sculptures celebrate eagles, bears, and wolves, and regional art galleries are filled with images of wild animals (**97**). More vernacular variants might end up as lawn ornaments (fig. 1.73). Deer, elk, moose, and fish trophies pepper the walls of many homes, bars, and local Elks lodges (**90**). All of this is a testament to the powerful, if varied, connections people feel toward the wild-animal populations in the West and how these creatures have successfully negotiated common ground with their human neighbors. ❋

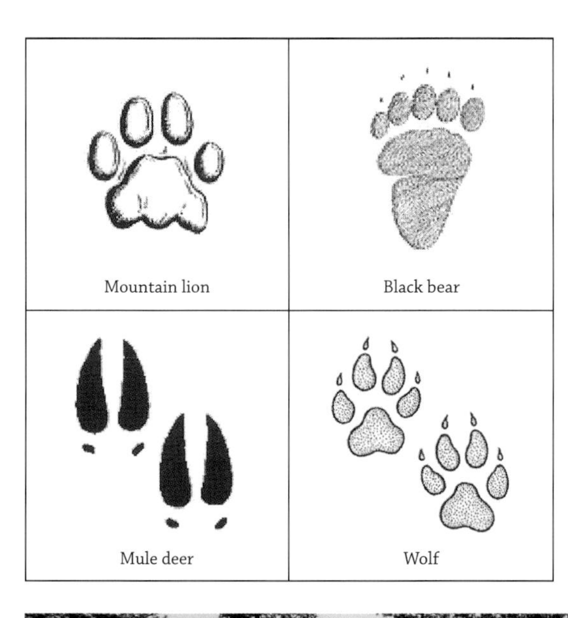

FIG. 1.71 Wildlife tracks. Watch for these distinctive tracks (not to scale) left by mountain lions, black bears, mule deer, and wolves.

FIG. 1.72 Mountain lion warning sign, Orange County, California. Rural parks in the Santa Ana foothills are also prime mountain lion habitat. After a number of attacks, local officials have clearly indicated potential hazards at many area trailheads.

FIG. 1.73 Bighorn sheep as lawn ornament, Darby, Montana. This friendly fellow is available for use in a front yard near you. Many of the West's wild animals have become powerful cultural symbols and are strongly associated with regional place identity.

2

FARMS AND RANCHES

FIG. 2.1 Sunflowers and sky, eastern Colorado. This scene near Akron, about 120 miles east of Denver, is a reminder that agricultural land uses still dominate many portions of the western American landscape. Here, the farms are large and populations are small.

THOMAS JEFFERSON, CHAMPION OF RURAL LIFE, BELIEVED IN THE VIRtues of an agricultural West, a land of small farmers where democracy would flower. He shared his vision with John Jay in a letter dated August 23, 1785: "Cultivators of the earth are the most valuable citizens. . . . They are the most vigorous, the most independent, the most virtuous, and they are tied to their country . . . by the most lasting bonds." More than two centuries later, the reality of farming in the twenty-first century West hardly conforms to that ideal, and yet the region's agricultural roots directly extend from that agrarian vision. Large portions of the West are rural. While the region's agricultural population continues to ebb, the land devoted to farming and ranching far exceeds the acreage taken up by cities and suburbs (fig. 2.1). Today you can explore the contemporary echo of Jefferson's vision, identifying the region's characteristic agricultural landscapes and how they connect people to place.

Farmland takes up more than 275 million acres of land in the eleven western states (about 35 percent of the total acreage), more than 20 percent of it in Montana. Using the U.S. Census Bureau's liberal definition of a *farm*—any establishment that sells a thousand dollars or more of agricultural products annually—the West had more than three hundred thousand farms in 2012. That number includes everything from a small family-run ranch to a corporate agribusiness operation.

Subsistence agriculture, in which farmers eat what they grow, was first practiced in the West by Native Americans. Early irrigation was especially notable in the Southwest and along the major eastward-flowing rivers of the Great Plains. Once Euro-Americans arrived, the geography of *improved land*—a term with Jeffersonian overtones that refers to land that is used, occupied, and "made better"—developed by 1860 in localities such as the Rio Grande Valley, Utah's Wasatch Front, western Oregon, and California's Central Valley. With some local exceptions, western farming quickly became commercially based, a system in which farmers sold their crops and animals to other consumers either in nearby urban centers or, increasingly after 1880, to buyers nationwide. In many parts of the interior West, the best years of the agricultural frontier in terms of cultivated acreage were between 1910 and 1920, when good crop years combined with high prices to expand dry farming (rain-fed agriculture) into marginal places such as southeastern Oregon, northeastern Nevada, and eastern Colorado.

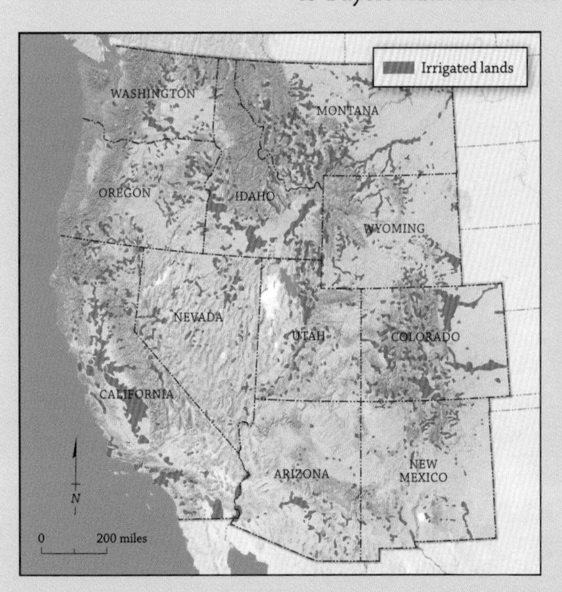

FIG. 2.2 Irrigated lands in the West. This map shows the distribution of irrigated agriculture in the arid West, clustered particularly in the interior of California; in the Columbia, Snake, and Missouri basins; and across northeastern Colorado.

Many of the marginal zones that saw brief booms around World War I later reverted to pasture and rangeland. Elsewhere, intensive irrigated agriculture expanded in localities such as California, the Snake and Columbia river valleys, and eastern Colorado (fig. 2.2). At the same time, urban growth transformed orchards and fields into suburbs and strip malls. More subtly, many traditional cattle holdings were subdivided into the ranchette landscape of the "gentrified range," where livestock and horses are present largely for personal enjoyment.

FIG. 2.3 Sheep drive along U.S. highway 89 north of Gunnison, Utah. The seasonal movements of sheep and cattle occasionally slow traffic in the interior West. Motorists must wait their turn as livestock, cowboys, and herd dogs mosey down the road.

FIG. 2.4 Wheat harvest, Gilliam County, eastern Oregon. Once covered with sagebrush, these flats north of Condon (east of the Cascades) proved fertile ground for dryland wheat cultivation. Successful operators invest large amounts of capital in labor-saving machinery. The average farm size in the area is more than four thousand acres.

The West has an amazing variety of agricultural landscapes—emu farms, cotton fields, sheep drives (fig. 2.3), fields of small grains (especially wheat), and a dizzying variety of fruit and vegetable crops. In particular areas, however, you will also encounter the reality of monocropping, a kind of agricultural single-mindedness in which only one crop is worked intensively, often with large capital investments and labor demands. You can see this at work in the wine country of Napa and Sonoma counties, in eastern Washington's wheatfields, and in central California's vast citrus groves. In the less-populated interior, you will also see acres of pasture crops—hay and alfalfa fields—and rangeland devoted to the raising and fattening of livestock.

The region's agricultural diversity and zones of specialization are a function of the West's varied ecological niches as well as how people have learned to use these landscapes to their own ends. Growing lettuce in California's Imperial Valley, one of North America's hottest and driest spots, is a case in point, part of an ongoing process of experimentation and the manipulation of land, labor, and technology. Growing wheat in the semiarid portions of the West worked only because late-nineteenth-century farmers were first willing to give it a try, often under harsh conditions. Today, wheat country is a land of large farms, major capital investments in equipment, and very few people (fig. 2.4).

FIG. 2.5 Dairy operation east of Fort Collins, Colorado. Look for large telltale Harveststores (dark blue steel and glass storage tanks), concrete bunkers, or huge plastic bags for silage (feed). Cows are usually packed into high-density sheds (center and right) for milking.

How can you make sense of this complex visible signature? Each agricultural adaptation—from cattle ranching to strawberry growing—has a set of signatures on the landscape that are related to the production of a particular crop or animal. So as you pass through the rural West, watch for five interrelated landscape elements: the crops (plants or animals); farm density and organization; the farms' structures, equipment, and labor; the support infrastructure in the area; and related symbols of place identity that connect an agricultural practice with a particular locality and its residents.

Plants and animals. Ask what is being produced and how it appears on the landscape. Signs, company names, and a friendly chat with a local by the side of the road can tell you a good deal. Think about where the crop might be in its annual cycle of production. Are fields lying bare? Is there stubble, suggesting a recent harvest? Are animals thinly spread across the range, or are they clumped together in close quarters, maybe on a dairy farm or in a fattening operation before being sent to the slaughterhouse (fig. 2.5)? Also think about how the fundamental physical platform of soils and water has been modified to maximize yields and returns.

Farm density and organization. On a larger scale, the density and layout of an agricultural landscape can be seen in the arrangement of repeating elements. Much of the West is divided by the rectangular regularity of the federal Public Land Survey System (or township-and-range survey system), which created a cardinally oriented grid of ownership parcels, field boundaries, and roads. In some landscapes, you will see farms that are similar in size and regularly spaced. Typically, dryland farming and ranching feature larger operations, while irrigated agriculture produces a denser, more intensively utilized rural landscape of smaller, carefully managed fields. There are real differences between a landscape in which only one or two farmhouses are on the horizon and one with smaller, more closely spaced operations. How large are indi-

FIG. 2.6 Cattle ranch northwest of Helena, Montana. Western farms and ranches produce an accumulation of material artifacts, from the old wagon wheels on display at the entrance to the ranch to the assorted structures, both old and new, visible in the distance.

vidual fields? Are they square (preferable for managing a grazing operation), long and narrow (often to minimize machinery use in cropped areas), or irregularly shaped to conform to terrain?

Farm structures, equipment, and labor. Farms and ranches represent points of capital investment by individual families and corporate entities. The agricultural landscape is a physical storehouse of those investments, and deciphering its elements helps tell the story of how they have accumulated in a place over time (fig. 2.6). Older buildings on a farm or ranch were often hand-built, creating a "folk landscape" of log cabins, outbuildings, and fences. Geographers, historians, and folklorists have studied these features for clues to their cultural origins. Often isolated rural settings in central Idaho, western Montana, northern Nevada, eastern Oregon, central Utah, and northern New Mexico are prime hunting grounds for surviving remnants of this agricultural landscape.

Elsewhere, newer farmhouses, migrant worker facilities, prefabricated barns, and metal storage units represent the changes brought about by mass production and the imposition of large-scale capitalism on the rural economy. Agricultural equipment can vary from hand-hewn tools to large mechanized irrigation systems.

You can also see the evidence of labor on the landscape. Is there a combine operator in a distant wheat field or a large assembly of fruit and vegetable pickers harvesting a crop? Each form of western agriculture has a different set of labor requirements that vary depending on the annual cycles of crop and livestock production (fig. 2.7).

Support infrastructure. When we move beyond the boundaries of an individual farm or ranch, we can see an elaborate support network that enables them to survive, connecting them with services, storage and distribution facilities, and community infrastructure. Trucks and railroads move products from fields to markets, and livestock, grain, and other crops need to be stored, sold, or processed (fig. 2.8). Farm towns supply bank loans, crop insurance, and farm equipment and provide the schools, stores, and community services necessary to support farm families and workers. Farmworkers often have separate settlements, both temporary and fixed.

FIG. 2.7 Farmworkers, Salinas Valley. Strawberry picking in central California's Salinas Valley remains tough physical labor. With Latin music blaring in the background, these workers put in long days to pick maturing fruit at the right point in the growing cycle. Note the portable bathrooms, mandated by law, in the background (right).

FIG. 2.8 Trackside grain elevators, Ritzville, Washington. Nearby wheat farmers bring their grain to these facilities, and then rail cars begin the long voyage to national and global markets.

FIG. 2.9 Hereford on wheels, Clearwater Junction, Montana. This prominent landscape symbol along the highway in west-central Montana leaves no doubt what's for dinner, and it demonstrates how a side of beef can be an effective billboard.

FIG. 2.10 César Chávez mural, San Francisco, California. Celebrating California's farmworkers and the role played by César Chávez in promoting farmworker rights, this mural along a quiet San Francisco alleyway is a reminder of that city's close cultural and economic ties to the land.

Symbols of place identity. As Jefferson might have hoped, Americans are still tied to the land. The visual expressions of cultural traditions of rural life in the West are an important part of place identity, reflected in a complex repository of symbols, myths, and values. Some of the larger examples of these symbols embrace the independence and freedom of cowboy culture, the conservative values of the family farm and its social institutions (such as the 4-H youth development organization), and the edenic landscapes of abundance promoted by California's citrus growers. As you explore the rural landscape, you will find place-names, historical sites, statues, local sports teams, and advertising that display symbols of identity, from a steer on wheels (fig. 2.9) to a mural celebrating César Chávez and farmworkers' rights (fig. 2.10).

The region's agricultural landscape is a dynamic one, so don't expect stability. The steady hemorrhaging of people and services from small farm and ranch towns continues to redefine hundreds of rural settlements across the West. Closed stores and empty streets suggest the larger challenges of population loss in the agricultural West, as machinery has replaced labor and as farm families lose their children to opportunities elsewhere. At the same time, mushrooming suburbs lead to a retreating rural landscape on the urban fringe, a process known as "greenfield development" that is steadily removing farm and ranch land from production (fig. 2.11).

FIG. 2.11 Urban-farmland interface, Nampa, Idaho. New high-end suburban homes are crowding the cornfields on the suburban fringe. Soon these fields will be a distant memory amid a landscape of sprinkler-fed lawns, backyard barbecues, and trampolines.

Shifting cultural preferences and food distribution systems are also shaping agricultural landscapes. The love affair with locally grown food—known as the locavore movement—is having a substantial impact on the popularity of organic foods, community gardens, farmers' markets (fig. 2.12), and community-supported agriculture (in which consumers purchase shares of locally produced livestock and farm products). Large-scale global forces will also continue to shape the rural landscape, just as they have since 1850. How will climate shifts, changing immigration trends, or new international markets for western farm products rework tomorrow's rural scene? Surely in ways that Thomas Jefferson never would have imagined. ❋

FIG. 2.12 Farmers' market, Salem, Oregon. The popularity of farmers' markets across the West highlights the value placed on organically raised, locally produced food. These trends reshape agricultural geographies, particularly near larger urban areas, where these products are increasingly in demand.

19
ISOLATED FARMSTEADS

FIG. 2.13 Rural landscape in the Yakima Valley near Ellensburg, Washington. This largely irrigated landscape in central Washington supports a dense collection of relatively smaller farms, many oriented toward the raising of forage crops for livestock. Several local vineyards also grow grapes for commercial production.

DISPERSED FARMSTEADS HAVE BEEN A PART OF AMERICAN LIFE SINCE colonial times. Perhaps it was the abundant, inexpensive land or the independence of America-bound migrants. Perhaps the particulars of the land survey and disposal systems encouraged a dispersed pattern of settlement (**58**). No single factor explains why Europeans repositioned themselves on isolated farmsteads all across the North American landscape. But they did. The agricultural West, with few notable exceptions, remains a region of remarkably dispersed settlement.

The precise geometry of the pattern varies, as do its local characteristics. Farmstead density reflects the underlying agricultural economy and the size of operating units that developed there (compare figs. 2.1 and 2.13). In the drier intermountain West and across the Great Plains, farmhouses are often set amid mature, planted trees that serve as shade and windbreak. Trees, planted linearly as a shelterbelt, can offer protection from wind on the windward side of the house, often to the west (fig. 2.14), or they may line an approach road or field boundaries.

Farmsteads are fascinating expressions of history, changing styles, and relative economic success (see fig. 2.14). Often you can see both earlier and more recent versions of the family home and outbuildings. A hand-hewn

FIG. 2.14 Farmstead, eastern Colorado. Located south of Kit Carson, near the Kansas border, this isolated farmstead includes (from left to right) a main farmhouse, several hand-built, folk-style outbuildings, a long-retired double-wide, and a pair of old metal storage bins. A patchy windbreak offers protection from prairie gales.

homestead-era original may sit alongside a modern upgrade. Sometimes older homes were purchased from mass-merchandising outfits such as Montgomery Ward and shipped west by train. House styles reflect the popular tastes of the time they were built: a Victorian-style home (1880–1900), a bungalow (1910–1930), or a ranch-style house (1950–1980 or later) each suggests a style fashionable in the nation during different eras (**75**, **76**).

Next to the house, you might see a garden plot and wash on the line, suggesting it is not a "suitcase farm," an operation managed from afar. You might also see satellite dishes, propane tanks, and vehicles, linking the family to the larger world. Farm equipment, barns and outbuildings—some old, some recycled, others new—are also close by.

There are implications to living in this kind of setting. In pioneer times, women were especially isolated, and even today, schoolchildren may ride the bus for three hours a day and a trip to the dentist or the nearest farm town (**26**) is at least as long. In many small rural high schools, it may be difficult to scare up a six-man football team. But there is also the support gained from a rural social fabric of mutual interdependence.

Economic downturns and persistent population losses shape the scene as well, and you can see those changes in abandoned homesteads (see fig. I.26), feral windbreaks, and weed-infested foundations. Overgrown cemeteries and empty schools (see fig. I.23) are part of an earlier era when rural life in the West was in many ways less isolated than it can be today. ❋

20

CATTLE RANCHING

FIG. 2.15 Cattle ranch west of Wisdom, Montana. Fenced irrigated pastures, Angus beef, and the main ranch house (surrounded by trees) are all typical signatures of cattle ranching in the Mountain West.

CATTLE RANCHING HAS LEFT A UBIQUITOUS BUT SUBTLE SIGNATURE on the rural West (fig. 2.15). The quintessential cowboy is often low-key and soft-spoken, and the same can be said of the ranching landscape, which is often quietly dominant across much of the rural West, both on private and on public land. Today, however, many family-run cattle operations struggle to make ends meet and the ranching lifestyle may gradually fade, replaced by a gentrified landscape subdivided into exurban ranchettes, where cattle and horses are kept for pleasure, not for work (**95**).

The roots of ranching are complex. As the geographer Terry G. Jordan-Bychkov demonstrates, three traditions shape patterns we see today. First, the Texas-based system (an amalgam of southern and Mexican influences) of all-year, open-range grazing, combined with seasonal roundups, expanded across the plains and the Mountain West between 1865 and 1885. The longhorn cattle drives to Kansas railheads or to the mines of southwestern Montana are part of this tradition. Cattle were minimally tended and grazed on the free grass of the public domain, but the system was ill-adapted to western weather. Overstocked ranges, falling beef prices, and vicious winters combined to wipe out this way of ranching between 1885 and 1890.

Second was a regional extension of the pastoral California system. Rooted

FIG. 2.16 Haying near Salida, Colorado. Central Colorado's upper Arkansas Valley produces excellent summer hay. These circular bales are stored and used as winter feed. Trees (distant left) provide a windbreak near the main ranch headquarters (center).

in Mexican and Anglo ranching practices developed in the grass-covered valleys of California between 1821 and 1860, this tradition included the practice of transhumance, the seasonal movement of cattle to cool mountain pastures in the summer and to protected valley bottoms in the winter. The California system spilled into northern Nevada and eastern Oregon, eventually diffusing eastward into Idaho, Montana, Utah, and western Colorado. Elements of this system survive today.

The third and most influential adaptation was the midwestern stock-raising system. Rooted in Anglo-Celtic livestock traditions, this method of cattle ranching emphasized livestock breeding and stock growers' associations. It also featured a seasonal management of stock, fenced pasturelands, a storehouse of winter feed, and irrigated meadows to produce hay. By the late nineteenth century, this system had replaced the open-range approach and blended with elements of the California tradition.

On the modern cattle-ranching landscape, you still spot some longhorns, but you are more likely to find brown-and-white Herefords (fig. 2.9), red or black Angus (fig. 2.15), or red-and-cream-toned Simmentals. Whatever the breed, you often see a cow-calf operation, in which ranchers maintain a breeding herd and raise calves for sale. These businesses have a predictable annual round of activities that shapes the landscape.

Much of the ranching landscape is organized around the production of forage (fig. 2.16). Cows are not picky eaters, and they munch on a wide vari-

FIG. 2.17 Haying in the Big Hole Valley, southwestern Montana. The traditional large haystacks in the meadow are built with the help of Beaverslide hay stackers (left), a local innovation that later diffused to similar settings in the interior West.

ety of plant life, including summertime needlegrass and fescue on mountain slopes and white amaranth (or winterfat) in desert areas. You can also find them in irrigated pastures, feeding on stubble (where an earlier crop has been harvested), or in a field where they feed on bales of hay, probably produced nearby.

Hay lands can generate multiple cuttings for winter feed. On hayscapes, you can see traditional haystacks (or bread loaf stacks) built with the help of large old-fashioned stackers (fig. 2.17); but automatic balers have transformed most hayscapes, creating large rolled bales (fig. 2.16) or smaller squared-off bales of compressed hay.

Fences and gates (figs. 2.18 and 2.19) are a reminder that cattle ranching is all about the availability and control of land and water. These landscape elements are integral parts of an intricate system of range and pasture management, designed to feed cows and retain the value of the land. You can find older, hand-built, folk wood fences (fig. 2.18), newer versions of simple wood and wire fencing (fig. 2.20), and mobile, steel-pipe fences and corrals that are put in place wherever they are needed.

As western geographer Paul Starrs observed, ranch layouts can vary widely, but look for a home ranch of corrals, barns, repair shops, worker housing, and a headquarters, often surrounded by hay pastures (figs. 2.6 and 2.16). There is usually water nearby, a land acquisition strategy that protects legal claims to that precious resource.

Farmers frequently supplement their core landholding with additional acreage, either by acquiring parcels elsewhere or by leasing private land. Where they are available, grazing allotments managed by the Bureau of Land Management (**68**) and the U.S. Forest Service (**67**) offer essential access to public rangeland (fig. 2.18).

Satellite "line camps," often miles from the home ranch, offer temporary worker housing along with fenced pastures, water tanks, and holding corrals (fig. 2.20). Seasonal ranges vary widely in their carrying capacity—that is, their ability to support livestock. One cow grazing in central Nevada, for example, may require the equivalent of hundreds of acres of land a year to survive.

You may also spot cattle packed tightly on the landscape, with two hundred to four hundred animals per acre (fig. 2.21). These CAFO enterprises (concentrated animal feeding operations) are the industrialized version of "ranching" in which animals are fattened for sale and slaughter. The Simplot feedlot in Grand View, Idaho, for example, can support 150,000 head of cattle on its 750-acre facility. These farm factories take advantage of scale economies, technology, and global markets to turn a profit. Many feedlot operations don't own cattle but charge a fee to fatten the herds before slaughter. The cattle are released on irrigated pastures or fed a diet designed to quickly add "carcass yield" to the finished product. Central California, southern Idaho, and eastern Colorado support the largest number of cattle in the West and are good localities for seeing these fattening operations.

There are other regional variations. In the highlands of northern New Mexico, the four-hundred-year-old tradition of ranching in many Hispano communities is supplemented by crop production (**39**). Elsewhere, niche operations thrive. Dude ranching (**85**) can be an extra source of income for operators who capitalize on the appeal of life in the saddle. You may also see ranches that specialize in raising breeding stock—a seedstock operation—and other ranches produce only organically grown beef.

Ranching continues to shape popular culture, and cattle ranching and the

FIG. 2.18 Worm fence near May Creek, Montana. Summer grazing allotments provide forage on public lands for many western ranching operations. Fences, gates, and cattle guards help manage impacts of stock on upland meadows and pastures. Folk-style worm fences are gradually disappearing from the landscape.

FIG. 2.19 Ranch gate, San Luis Valley, Colorado. A few strands of wire, a simple gate, and a hand-painted sign prove essential elements in controlling stock on range-land near the Rio Grande.

FIG. 2.20 Watering hole north of Burlington, Colorado. This small windmill pumps well water to the surface (left), providing a cool drink for the herds. The wood and wire fencing helps control the movement of stock through the area.

mythic figure of the cowboy still symbolize western self-determination and independence (see figs. I.15 and 1.7). Many suburban westerners still enjoy the informal living of their ranch-style homes (**76**). The Marlboro Man, the global appeal of westerns, the enduring tradition of summertime rodeos (**91**), and the popularity of cowboy poetry and country music also suggest that the cultural cachet of cattle ranching is thriving even as many of its real-world practitioners struggle to make their lifestyle pay. ❋

FIG. 2.21 Cattle feeding operation, San Joaquin Valley, California. Concentrated animal feeding operations (CAFOs) play an important role in moving cattle from the range to their final markets. California's Central Valley is home to many of these facilities. Sensitive noses beware!

21
DRY FARMING

FIG. 2.22 Dryland wheat north of Cut Bank, Montana. Alternating strips of wheat and fallow, a pattern used to reduce soil erosion from the wind, give the landscape a characteristic banded appearance in many of the West's grain-producing districts.

DRY FARMING DEPENDS ON NATURAL RAINFALL AND HAS ALWAYS BEEN a risky proposition in the interior West. Dry farming began in the Southwest (and is still practiced by the Hopi today) once agriculture arrived from Mesoamerica centuries ago. In northern Arizona, you can still spot small plots of dryland corn, beans, squash, and melons in washes and fields below mesa tops (**38**).

The principles of modern, commercialized dry farming are simple enough: rather than bring in irrigation water (**23**), farmers store moisture in the soil itself and try to preserve it there until it is needed. In addition, farmers grow crops that are adapted to the semiarid conditions of the western interior.

Strip farming, the practice of alternating long, narrow strips of summer fallow (resting) land and grain, is commonly used in many dry-farming districts (fig. 2.22). This technique conserves moisture in the fallow land, particularly when old stubble and plant residue limit wind erosion. In hilly country, contour plowing on slopes also limits erosion and moisture loss (fig. 2.23). Many dry farmers have shifted to more continuous cropping practices that rotate pasture, grain, and legume crops in ways designed to conserve moisture and maintain soil quality.

Between 1860 and 1920, many semiarid settings hosted dry-farming booms. Early experiments by farmers in Utah and California suggested the

FIG. 2.23 Wheat near Waitsburg, Washington. Contour strip farming across the Palouse produces a unique visual signature in the interior Northwest.

possibilities of growing drought-resistant crops, especially small grains, without irrigation. With the encouragement of land and railroad companies, town promoters, and boosters like Hardy Campbell—who was instrumental in developing the Campbell systems of dry farming and in organizing the first Dry Farming Congress in Denver in 1907—dry farmers in later decades, especially between 1900 and 1920, tried their luck from the plains of eastern Colorado and central Montana to the deserts of western Utah and northeastern Nevada. A few good years were typically followed by drought, leaving a legacy of abandoned homesteads and ghost towns (**32**).

But commercial dry farming persisted in a variety of settings, the result of experimentation, scientific farming practices, and stubborn hard work. The inland Pacific Northwest still contains some of the West's most productive winter wheat land (planted in the fall), with much of the harvest shipped to markets in Asia (**22**; figs. 2.4 and 2.23). In eastern Washington's Palouse region, the wind-deposited soils are particularly fertile, and the wheat-covered hills create an archetypal agricultural landscape that exists nowhere else in the West. Farmers in the wheat-growing districts in Montana's Golden Triangle north of Great Falls and the Hi-Line along the Canadian border grow rapidly maturing spring wheat (planted in the spring), often visible in strips and dependent on late-spring rain in May and June (fig. 2.22). Another sizable zone of dry farming is in eastern Colorado, where farmers grow a mix of winter wheat, barley, sunflowers, and other forage crops (fig. 2.1). ❋

22 GRAIN ELEVATORS

FIG. 2.24 Country grain elevators, Straw, Montana. Sitting alongside an abandoned rail line, these aging, empty wooden grain elevators are quiet reminders of how shifts in the rural economy reveal themselves on the landscape.

GRAIN ELEVATORS PUNCTUATE THE SKYLINE IN THE WEST. THEY ARE often the first indication of an approaching town as you travel across areas such as north-central Montana, eastern Colorado, or eastern Washington. Designed to store grain safely in a cool, dry, pest-free environment, elevators multiplied along rail lines (**50**) as collecting points in the commercial grain economy (**21**). Farmers needed handy facilities to store and market their grain, and wholesale and commercial buyers required places where grain could be purchased and then shipped along key transportation corridors.

Country grain elevators were built throughout the West after 1870 as railroad connections made commercial grain farming possible. Typically built of wood and often covered by metal siding, these facilities store between ten thousand and fifty thousand bushels of grain. You still see examples of older gable-style elevators, often evenly spaced every five to ten miles along existing or abandoned rail lines (fig. 2.24). After 1900, concrete silo elevators proved to be a more efficient alternative. Today, commercial elevator operations are dominated by long parallel lines of larger-scale storage facilities that can hold more than a million bushels. Multiple styles and eras can be represented at the same trackside location (figs. 2.8 and 2.25). Also look for cylindrical grain

bins of corrugated metal, which vary in size from small, single-farm units to mammoth collecting facilities (fig. 2.26).

At most grain elevators, a sheltered offloading area allows farmers to deliver grain out of the weather, and workers perform weighing and accounting functions in nearby scale rooms and offices. A bucket elevator (or leg) lifts the grain into interior storage bins, and conveyors, chutes, and spouts deliver the grain to a waiting rail car or ship. Look for signs or lettering that identify former and current owners, perhaps a local farmers' cooperative or a milling company.

Grain elevators are indexes of a farming community's vitality and identity (**26**). Rural and small-town residents take great pride in their grain elevators and in the farm economy they represent. The closing of a small country elevator is like seeing the local high school shut its doors. The old buildings are either moved or left to weather slowly away. ❋

FIG. 2.25 Grain elevators, Holyoke, Colorado. Larger, more centralized elevator operations remain essential to the West's wheat economy. This complex displays a mix of concrete silo elevators (right), conical-topped metal grain bins (distant left), and older country grain elevators (visible in the distance, center).

FIG. 2.26 Grain for the globe: north of Walla Walla, Washington. This huge storage facility in the Palouse helps funnel grain down the Snake and Columbia rivers. From there, it flows to consumers in every corner of the world.

25 FIELD IRRIGATION SYSTEMS

FIG. 2.27 Center-pivot irrigation, southeastern Oregon. The radial geometry of center-pivot irrigation has created characteristic signatures across the rural West. This haying operation is south of Burns, Oregon. Steens Mountain is in the distance.

IRRIGATION HAS TRANSFORMED MORE THAN TWENTY MILLION ACRES of the West. From an airplane or the top of a hill, you can see how large center-pivot irrigation systems have inscribed elaborate arcs and circles across the landscape (fig. 2.27). On the ground, you can see signatures of irrigation in California's Central Valley, southern Idaho's Snake River country, and eastern Washington's Columbia Basin. Canals and pipelines transport water to irrigated fields, where smaller ditches and sprinkler systems deliver moisture to waiting crops. In eastern New Mexico and northeastern Colorado, deep-water wells tap directly into aquifers hundreds of feet deep.

Some of the most elaborate ancient irrigation systems in the Americas were built in Arizona's Gila and Salt river drainages by Hohokam people, beginning about two thousand years ago. By 1400, their irrigated fields covered more than 250 square miles near modern-day Phoenix. Spanish-style irrigation systems built between the seventeenth and nineteenth centuries shaped landscapes from northern New Mexico (**39**) to coastal California, and many elements of Spanish water law became part of later western adaptations. By the mid-nineteenth century, Mormon farmers in the Great Basin also developed their own system of communally based irrigation, which became a defining part of their distinctive cultural presence in the interior West (**42**).

After 1880, the scale of irrigated agriculture expanded as populations grew and as the federal government became directly involved with the financing and construction of large-scale irrigation projects (**64**). Legislation such as the Desert Land Act (1877), the Carey Act (1894), and the Newlands Reclamation Act (1902) offered financial support for irrigation. Investments in federal water storage projects such as the Hoover (Colorado River) and Grand Coulee (Columbia River) dams provided more downstream opportunities for irrigation.

FIG. 2.28 Irrigated corn, Idaho. Young corn plants depend on flood irrigation from a nearby canal in this field northeast of Idaho Falls.

Today irrigation imposes a visual order on the rural West. On the regional scale, you can see irrigation as a carefully managed, interconnected system of water storage, transport, and distribution. Irrigation is all about getting water to a particular place at a particular time to produce a valuable commodity such as grass, lettuce, or fruit. To make water delivery more predictable, westerners have developed storage reservoirs, dams, transbasin water transfers, huge canals, and elaborate delivery systems. The West is crisscrossed with this megascale infrastructure, much of it financed through federal and state initiatives (fig. 2.2).

On a local scale, you can follow canal-and-ditch systems as they deliver water to fields. In fields of irrigated corn (fig. 2.28), for example, water can be distributed via a concrete-lined ditch that borders the field. Water flows directly by gravity to the young plants. Fields are leveled and graded with laser-driven accuracy, often with above- and below-ground pipes, drip lines, and drainage systems. The precise timing and rates of water application in the fields, measured in acre-feet, are reminders that such operations are carefully managed high-tech factories.

Two basic varieties of water application are most common in the West. Surface irrigation is still used in many settings: water introduced at one end of a field flows across the surface by gravity. Watch for main feeder canals (usually lined with concrete), border ditches and laterals, headgates, and siphon pipes to control the release of water; drop structures to control flow at different elevations; furrows to irrigate row crops; and smaller corrugations to water alfalfa and grains (fig. 2.29). For high-value fruit and vegetable crops, water may be delivered directly to plants through plastic pipes, both on and beneath the surface. Known as drip irrigation, this variation is more costly to set up but is efficient in getting water directly to plants and uses less water overall than other systems.

Sprinkler irrigation produces other landscape signatures. Since 1950, inexpensive, lightweight aluminum piping, efficient pumps and sprinklers, and innovative labor-saving application methods have made sprinkler systems more economical. Sprinkler irrigation is the predominant method of irrigation in the region, particularly where there are porous soils, erosion-prone terrain, and water limitations. Some systems have sprinklers at fixed points, but you are more likely to see a mobile system that moves pipes and sprinkler heads across a field at regular intervals. There are many variations of this technology, including movable straight-line lateral sprinklers (fig. 2.30), half-mile-long, rotating center-pivot systems (fig. 2.27) with pipes mounted on side-roll wheels, and portable single-boom outfits for smaller parcels.

Irrigation has changed western environments: at the local level, the new ecological settings created along irrigation canals and ditches support plants and animals that do well with a moist environment. Soil chemistry, topography, and pollen counts shift as well. Environmental historian Mark Fiege refers to these as hybrid landscapes, neither quite natural nor completely fashioned by people. Irrigation has also fundamentally altered the distribu-

FIG. 2.29 Managing water, southern New Mexico. Allocating irrigation water in the Rio Grande Valley involves timely releases from the mainline canal (foreground) to flood nearby fields, all controlled by an elaborate system of smaller laterals and headgates (on canal, left).

FIG. 2.30 Water on wheels, southern Idaho. These sugar beets east of Twin Falls are being showered by a mobile irrigation system that can be readily moved and offers an even delivery of water across a field.

tion of water in the West: river systems such as the Colorado and San Joaquin have seen their in-flow volumes sharply reduced by demands from irrigators, and groundwater levels in many western settings (particularly the Ogallala Aquifer in the plains) have been severely affected by the persistent drawdown of that precious resource for the benefit of current agricultural interests.

The western landscape also reveals the larger transformative power of irrigation in radically new agricultural geographies, including the ability to grow tree crops near arid Fresno (**24**) or high-value vegetables in the Imperial Valley. Irrigation, intensive agriculture, and the need for large numbers of laborers have also shaped the social geography of the region (**27**). In addition, food-processing industries and agriculturally based settlements (**26**) are located near major zones of irrigated farming. Historian Donald Worster has described the West as a "hydraulic society" in which the technology, capital, and power relationships necessitated by large-scale irrigation have profoundly shaped the region's politics, culture, and economy. Water and money in the West have long been connected, and the region's vast irrigation infrastructure suggests that the relationship will endure. ❋

24 ORCHARDS

FIG. 2.31 Cherry trees east of Flathead Lake, Montana. Local microclimates offer a niche for these trees in northwestern Montana. Cool lake temperatures help delay premature blossoms that might be killed by the region's late spring frosts.

PULLING OFF THE HIGHWAY NEXT TO AN ORCHARD TO BUY A PEACH OR a box of freshly picked cherries is certainly one of life's simple pleasures. But consider the complex conditions that make such transactions possible. Begin with nature. Fruit and nut trees typically require a delicate balance of heat and coolness, dependable moisture, and the right soil chemistry. Montana's Flathead Valley cherries, for example, thrive on only eight hundred acres of land on gently sloping benches east of Flathead Lake (fig. 2.31), a microclimate defined by summer warmth, bearable winter chill, and cool springs that prevent premature budding.

Orchards are long-term investments, and capital, technology, and labor are essential to sustain them (**27**). In the cherry orchards, pruning crews prepare trees for their next growth cycle and mow between the rows to keep down weeds. Fertilizers and pesticides are applied by hand around each tree. More workers are hired during harvesttime, in the hectic month between mid-July and mid-August, accounting for about one-third of the cost of growing the crop. Finally, the cherries leave the orchard, shipped to wholesalers or perhaps a local fruit stand. This process of planting, growing, maintaining, harvesting, and marketing tree crops is repeated in hundreds of western landscapes where orchards predominate.

Spanish settlers initiated tree cultivation in New Mexico and California, and Anglos experimented with orchards in Utah, western Colorado, and central Washington. With transcontinental rail connections, refrigeration, and national markets, commercial orchards expanded between 1880 and 1940 and selectively penetrated new niches, encouraged by irrigation projects (**23** and **64**), experimentation, and low-paid migrant (mostly Mexican) labor.

Today the Central Valley, western coastal valleys, and the Imperial Valley in California together account for over half of the nation's orchard acreage. Washington has 6 percent, and significant orchard landscapes are also found in western Oregon, southwestern Idaho, western Colorado (near Grand Junction), and southern New Mexico and Arizona.

Orchard geographies depend on environment, patterns of investment, access to labor, and consumer demand. Many of Southern California's citrus trees, for example, were paved over after 1950, as the land became more valuable for nonagricultural uses. Today most of the West's industrial-scale citrus groves—a modern-day Orange Empire—are concentrated in California's San Joaquin Valley, where large, mostly Latino populations (**41**) make such operations possible (fig. 2.32). Avocados thrive in selected southwestern California valleys, particularly northeast of San Diego, where the distinctive dark-leaved trees clothe slopes once covered in oaks and chaparral. Apples dominate the Yakima Valley, taking up more than 175,000 acres of Washington east of the Cascades. Nut trees cluster in their own distinctive settings, with pecans being grown on irrigated lands in southern Arizona and New Mexico and almonds and pistachios being cultivated almost exclusively in the Central Valley. ❋

FIG. 2.32 Orange groves north of Orange Cove, California. These large-scale fruit-making factories thrive in the warm summers of California's eastern San Joaquin Valley. The area serves as one of the country's leading production zones for oranges. Rows offer access for people and machinery and allow for the proper spacing of trees.

25

VINEYARDS AND WINERIES

FIG. 2.33 Central coast vineyard, south of Salinas, California. These favored well-drained gravelly soils are typical of the Arroyo Seco AVA in Monterey County. Summer heat is moderated by an afternoon sea breeze. The area is well known for its Zinfandel, Bordeaux, and Rhône varietal wines.

WITH AMERICANS' THIRST FOR WINE INCREASING, VINEYARDS (where grapes are grown) and wineries (where wine is manufactured, stored, and often sold) are an increasingly common sight across the West. Both a Spanish-era fondness for the vine and an accommodating Mediterranean climate encouraged early grape production in California, and that state still accounts for about 90 percent of U.S. wine. Still, a vigorous regional expansion of vine-producing districts has been under way since the 1970s.

The regional wine movement in the West has propelled vineyards far beyond California's Napa and Sonoma counties to Monterey and Paso Robles south of San Francisco (fig. 2.33), the North Coast districts in Mendocino County, and large portions of the Central Valley (where much of the vineyard acreage is devoted to jug-wine, table-grape, and raisin production). These California settings are home to both small estate vineyards and large commercial operations (such as the Gallo Winery, the world's largest family-owned wine producer). Central Washington's Columbia and Yakima valleys are the West's second largest grape-growing region, followed by Oregon's Willamette Valley, where Pinot Noir wines have won recognition since the 1980s. Trendy frontier zones are multiplying, and you can now try a glass of southern Arizona Petit Verdot, western Colorado white Merlot, or Idaho dry Riesling.

FIG. 2.34 Selected grape production districts. California remains dominant as a western wine producer, but the growing importance of Oregon and Washington is notable, along with selective contributions from other specialized zones in localities such as western Colorado. California's Central Valley remains the West's most important producer of table and raisin grapes.

Examine the label of any bottle of wine made in the West. If the wine is described as being from a particular geographical location, it is probably from an American Viticultural Area (AVA). Since 1980, the federal Alcohol and Tobacco Tax and Trade Bureau and the U.S. Treasury have mandated that wine marketed as originating from a particular place must contain at least 85 percent grapes grown in that AVA. Some AVAs are tiny (one in California's Mendocino County is less than one hundred acres), while others can be thousands of square miles. Petitioners (single or multiple growers) are required to describe and justify why each AVA possesses a unique growing environment.

As with all things agricultural, consider the precise physical setting for a vineyard (fig. 2.33). The French term *terroir* captures the hard-to-quantify combination of soils, climate, and topography that produces a particular wine. The best vineyards are found in localities that have summer heat but cool nights, some winter chill (when vines are dormant), good drainage (both air and water), and enough soil moisture to let vines thrive.

Vineyards are not casual enterprises. They combine the science of industrial agriculture with the art of creating a wine that embodies and successfully markets a unique sense of place. The spacing of vines, the technology of trellising (the way grapes are trained to grow), and the degree of pruning are all important (fig. 2.35). Then consider the different varieties of grapes that

FIG. 2.35 Vineyard under construction near Famoso, California. A reminder of all the careful engineering that supports grape production, this sturdy open-gable Y-trellis system, ready for the vine, is often used to grow table grapes.

might be selected for production. Successful vineyards often are the product of decades of testing and tinkering. Managing the vineyard's seasonal need for labor is also complex, dependent on shifting flows of migrant and temporary workers (**27**).

Wineries can be surrounded by vineyards, but they don't need to be. Boyd Teegarden, for example, owner and lead vintner of northern Oregon's tiny Natalie's Estate Winery, produces a small amount of local Chehalem Mountains Pinot Noir (from an AVA near Dundee, approved in 2006), but most of the grapes for his Merlot, Cabernet Sauvignon, and Chardonnay are grown elsewhere, primarily in Washington. The location of wine manufacturing, warehousing, and direct sales operations is often strongly tied to consumer markets rather than production areas. Woodinville, Washington, thirty minutes north of Seattle, is not in the state's grape-growing heart-

FIG. 2.36 Outdoor tasting room of Stoller Vineyards near Dundee, Oregon. Sipping estate-bottled Chardonnay or Pinot Noir rarely gets better than this. Visitors can relax on the lawn as they enjoy the carefully landscaped rural scenery beyond.

FIG. 2.37 Tasting room, Wilbur, Washington. This retooled gas station attracts travelers along U.S. 2 in eastern Washington. The Whitestone Winery produces a fine Merlot and Cabernet Sauvignon from its nearby Lake Roosevelt Shores Vineyard.

land on the east side of the Cascade Range, but it is close to more than a million potential consumers and now is the base of operations for more than 130 wineries.

Tasting rooms, often located at the end of a winding lane in a pleasant rural setting, can make up a significant percentage of wine sales and can include everything from sumptuous outdoor sipping opportunities (fig. 2.36) to more ordinary settings (fig. 2.37). ❋

26 FARM TOWNS

FIG. 2.38 Welcome to Chester, Montana. This northern Montana farm town, seat of Liberty County, is home to about a thousand residents. Motorists along U.S. 2 (on the Hi-Line) can spot the grain elevators as they approach. Time to slow down.

WHEN YOU DRIVE THROUGH CALIFORNIA'S CENTRAL VALLEY OR along Montana's Hi-Line, you will experience a predictable landscape punctuated by farm towns, often five to ten miles apart. As the road approaches town (fig. 2.38), speed limits fall and the small main street flashes by with its parked pickups, storefronts, and café. Then it's back to more open road.

These farm towns—many modeled on towns in the Midwest—multiplied between 1880 and 1920, their size and spacing shaped by agricultural economics, the railroad, and boosters. Some towns such as Walla Walla, Washington, or Bozeman, Montana, predated the railroads and were linked to nineteenth-century forts and trails, but railroads dramatically affected the region's commercial agricultural potential and financed many town-promotion efforts (**50**). Railroads sold town lots, and longer-term profits were made by controlling business in nearby agricultural areas. Bad roads and the local scale of rural life encouraged many small farm towns to blossom along rail lines. Towns were spaced close together in densely settled zones and farther apart in sparsely peopled hinterlands.

Today farm towns remain most numerous where agriculture predominates. To see these landscapes, consider six transects: any of a half-dozen

FIG. 2.39 Manville, Wyoming. This eastern Wyoming town has about a hundred residents. Its rectangular plat is laid out just south of the railroad tracks (diagonal, upper left) and just north of U.S. 20 (lower right corner). Most of the town's commercial district has been abandoned. *Photo courtesy of Wyoming State Archives, Department of State Parks and Cultural Resources, image no. 35/522.*

east-west highways across eastern Colorado; a course through southern Idaho's Snake River Valley; the north-south-trending valleys of California and western Oregon; eastern Washington's dryland wheat, orchard, and vineyard country from Yakima to Reardan; central Utah's Mormon country (U.S. 89, for example; **42**); or Montana's Big Sky country (U.S. 2, U.S. 12, or Montana 200).

Imagine an aerial approach to a farm town (fig. 2.39). Typically, the street layout (called a plat) is rectilinear, often oriented to the cardinal directions and organized by the township-and-range survey system (**58**). Complicating variables include terrain, cultural traditions (**39**), and railroad lines (**50**), which often run at angles across an otherwise gridded landscape. One common pattern is a *T*-town, in which the tracks form the top of the *T* and the main commercial thoroughfare defines its shaft (fig. 2.40).

Now come down to earth (fig. 2.38). The vertical harbingers on the skyline include water towers and grain elevators (**22**). You can see connections between rail alignments and street patterns. You can also examine street names and find farm-related businesses (fig. 2.8) and agricultural equipment retailers on the edge of town. At the commercial core, you will see long, narrow lots designed for maximizing commercial access by street traffic. Stone and brick structures, often called blocks, suggest that their owners imagined permanence, even metropolitan possibilities (fig. 2.41). Count banks, and notice how busy the streets are. What sorts of businesses are there and

FIG. 2.40 Top of the *T*, Haxtun, Colorado. This view toward downtown Haxtun (population about 1,000) looks north across the railroad tracks (crossing left to right). In a *T*-town, the commercial district (straight ahead) runs perpendicular to the rail line. Note the modern water tower in the distance.

FIG. 2.41 Wasatch Block and bank building, downtown Mount Pleasant, Utah. Brick and stone structures traditionally symbolized stability and permanence in a farm town.

how many buildings are empty (fig. 2.42)? You might see old, even elegant hotels or movie theaters, but are they occupied? The timing of boom eras is often evident in the construction dates chiseled along the street front or above doorways (dates between 1885 and 1920 are likely). In the residential districts, you will find parks, schools, and churches—the centers of community and social interaction.

If the town is a county seat, near an interstate highway (**55**), or within driving distance of a major city, then there is a good chance that it is healthy and stable. In some towns, cultural shifts have reworked the community; for example, the demand for inexpensive Latino labor has redefined the cultural geographies of farm towns in southern Colorado and the Central Valley of California (**27** and **41**).

The prognosis for many farm towns is dim, as declining populations mean less demand for services and as highways and large retail chains make it easier to shop elsewhere. Surviving landscapes are rich in memory and community values, even as they offer few job prospects for the people who remain. Many farm towns seem destined to dwindle in size, victims of an economy that has left them behind. ❋

FIG. 2.42 Downtown Ritzville, Washington. This eastern Washington farm town (population about 1,700) is the Adams County seat. Downtown businesses struggle. Note the ornate Gritman Building (left distance), with its corner turret and conical roof. Also ponder the creative mix of retail services still available to residents (right).

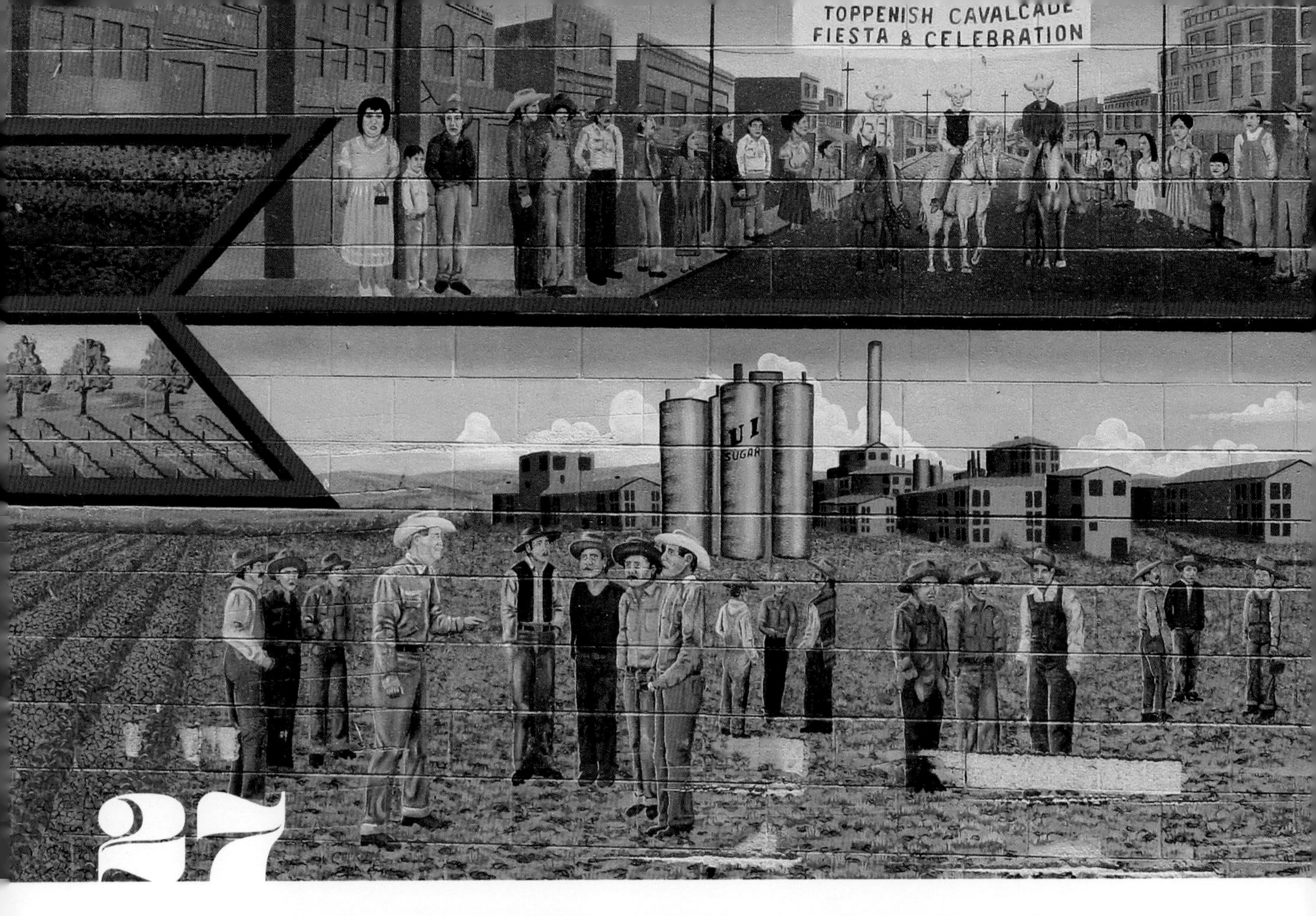

27

FARMWORKER SETTLEMENTS

FIG. 2.43 Bracero mural, Toppenish, Washington. This downtown mural honors the legacy of braceros who arrived in the 1940s. It was painted by Latino artist Daniel DeSiga, born to migrant workers who labored in nearby sugar-beet and asparagus fields.

IN *ILL FARES THE LAND* (1942), SOCIAL CRITIC CAREY MCWILLIAMS described the enduring invisibility of farmworkers: "You do not see them in the fields, on the highways, or in the towns. . . . From the highway, it may look as though a dozen or so hands were stooping over at work in the fields. But get out of your car, go into the fields, and count them. And don't be surprised if you count several hundred" (7–8).

Farmworker settlements are inextricably tied to the West's labor-intensive, industrial-scale agriculture (fig. 2.7; **23–25**). Historically, migrants supplied that labor, including German Russians, Japanese, Filipinos, and those fleeing the Dust Bowl. The number of Mexican workers grew between 1915 and 1940, and labor shortages during World War II prompted the bracero program in which Mexican citizens legally came north to work (fig. 2.43). In places such as southern Arizona (cotton) or Southern California (citrus groves), *colonias* or worker communities (most later razed for suburbanization) were established by growers or informally by workers. The New Deal (**63**) brought more state and federal support (fig. 2.44), and strikes and unionization led by César Chávez in the 1960s and 1970s led to improvements in pay and working conditions (fig. 2.10).

California has more than 650,000 farmworkers, more than one-third of

FIG. 2.44 Arvin Migrant Center near Bakersfield, California. The Arvin Center offers seasonal housing to eighty-eight farmworker families (background). Arvin is also home to historical buildings from the Weedpatch Camp, a migrant worker housing facility, built during the New Deal. The restored library (foreground) was built in the 1930s.

the nation's total. Washington (180,000) and Colorado (50,000) also have large farmworker populations, with significant settlements in Oregon, Idaho, Utah, New Mexico, and Arizona. Changing immigration policies have increased the number of workers who remain year-round in communities, but many stay from April to October.

How do you explore these communities? Farmworker settlements are not advertised, and some workers are undocumented immigrants, factors that limit the legibility of these settlements. In some places, migrant family housing centers provide affordable homes and apartments (fig. 2.44). In California, centers housing 50 to 150 families are located from Kern to Modoc counties on land made available by growers or local governments and financed by state and federal governments and private nonprofits. Elsewhere, you can find clusters of low-cost housing for farmworkers, typically in nearby farm towns (**26**). Workers also live in camps *(cantones),* in tents, shacks, and trailers.

Finally, services are essential to farmworker settlements, and day care centers, schools, and medical facilities might occupy simple storefronts in temporary locations. Churches (often Catholic) also operate help centers (fig. I.14). Neighborhood *carnicerías* (butcher shops) and *taquerías* (taco shops) offer familiar foods to families (**41**). As McWilliams suggests, however, workers are still invisible to many westerners and visitors, revealed less by their physical presence than by the bounty of their labor. ❋

5

LANDSCAPES OF EXTRACTION

FIG. 3.1 Coal unit train near Orin, Wyoming. Fast-moving coal unit trains are a mainstay of the Wyoming landscape, taking valuable low-sulfur thermal coal to power plants in the region and beyond.

IN EAST-CENTRAL WYOMING, NEAR THE ORIN EXIT ON INTERSTATE 25, sixty miles southeast of Casper, there is an overpass that spans the Union Pacific Railroad line. If you stop there, it won't be long before you will see an example of how landscapes of extraction remain crucial elements of the West (fig. 3.1). Coal is king in Wyoming, and untold thousands of coal-filled rail cars have passed beneath the overpass at Orin. The West's larger history of economic development and settlement is deeply rooted in metals mining, energy resources, and commercial forestry, extractive activities that have shaped both the historical landscape and the contemporary scene. To understand these landscapes, it helps to keep five things in mind.

Much of the West's transportation infrastructure and urban geography was created by the extractive economy. The region's great mining, energy, and timber booms depended on major investments in roads, rail lines, and towns (as both production and processing centers), and these early connections and urban centers were formative elements in the West's contemporary human geography. We may not think of Denver and San Francisco as mining towns, but their success is grounded in the mining economy of the late nineteenth century. Dozens of western settlements and much of the region's modern rail and highway infrastructure grew out of the historical imperative to develop extractive resources, and those activities still dominate portions of the West. Wyoming coal accounts for about 40 percent of U.S. production, while other western landscapes are home to gold and copper mines, oil and natural gas wells, coal-bed methane projects, and logging operations. More recently, renewable-energy businesses such as wind and solar farms have also been gaining importance.

The scale of extractive activities in the West is huge. One coal car looks just like the next for the same reason that haul trucks in open-pit gold mining operations are some of the largest wheeled vehicles on the planet (fig. 3.2). There are many examples of "scaling up" on the extractive western landscape: vast

FIG. 3.2 Haul truck, Carlin gold mine near Elko, Nevada. Bigger is better (and more economical) at Newmont Mining Company's Carlin gold mine. These trucks are used at the nearby pit to haul away waste rock and move vast amounts of low-grade gold-bearing ore. *Author photo used with permission of Newmont Mining Corporation.*

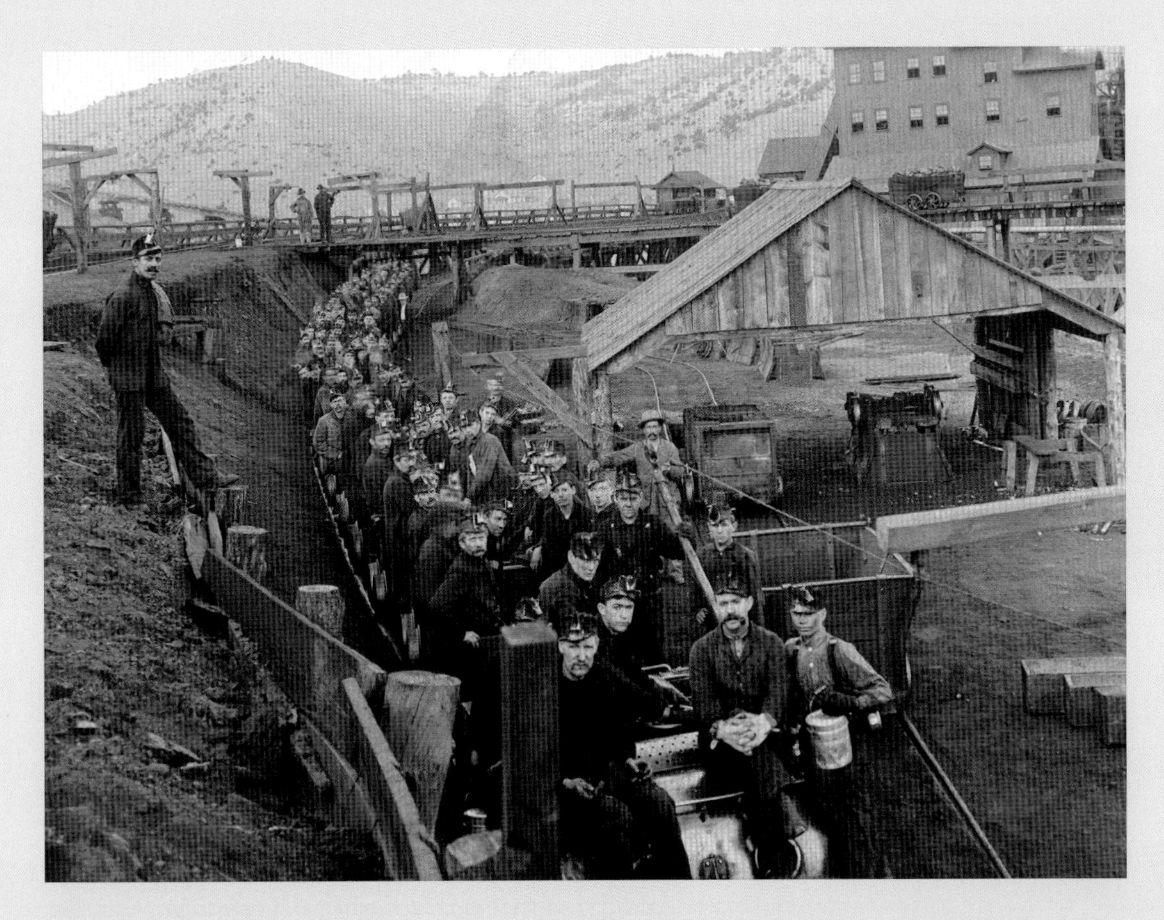

FIG. 3.3 Coal miners, Starkville, Colorado. Southern Colorado's early twentieth-century coal industry featured tough company towns, dangerous underground mines, and large numbers of both skilled and unskilled workers. *Courtesy of History Colorado (Trinidad Collection, image no. 10038297).*

clear-cutting operations, strip mining, huge arrays of wind turbines and solar collectors. This kind of technological gigantism makes economic sense, as larger-scale operations usually succeed in reducing per-unit production costs. Resource extraction is all about the efficient use of capital, labor, and technology and the imposition of a human-crafted environmental system on the landscape.

But that economic logic has important consequences. First, it usually demands a huge scale of capital investment. Today's winners on the extractive landscape are much more likely to be large national and global businesses than small independent producers. Second, traditional labor forces have largely been replaced by equipment and labor-saving technology. Consider the differences between today's coal-filled rail cars (fig. 3.1) and the worker-filled coal mines of the early twentieth century (fig. 3.3). That transformation has huge collateral consequences for the size and number of human settlements associated with extractive activities. The landscape near Trinidad, Colorado, for example, was once filled with coal-mining towns populated by thousands of miners. Today's coal operations in Wyoming's Powder River Basin need only a fraction of the labor force to produce much more coal.

There is also nothing subtle about the visual or environmental impacts of large-scale extractive activities, and their cumulative effects ripple widely and not always predictably across a large area. Massive environmental cleanups in places such as Butte, Montana; Silver Valley, Idaho; and Leadville, Colo-

FIG. 3.4 Temporary housing, Wamsutter, Wyoming. Many male workers in the West's mining and energy sector reside in low-cost housing, sometimes provided by their employers. These RVs, trailers, and mobile homes offer handy, affordable lodging in southern Wyoming's energy patch.

rado, illustrate the consequences of early mining operations. Important as well are the signatures of mitigation and restoration that since the 1950s have attempted to minimize these large-scale impacts. Because of environmental laws and regulations and a growing awareness of the damage that can be done on such landscapes, today's extractive landscapes almost always bear evidence of remedial forces at work.

Extractive activities are typically nonsustainable. Often shaped by boom-and-bust economies, extractive activities create ephemeral landscapes that pass through a predictable life cycle from birth (initial investment, expansion) to death (declining activity and disinvestment). Of course, not all extractive activities follow such a linear path. The mining and energy industries have effectively developed strategies for lengthening the lives of their operations, often by capturing lower-quality or less-accessible resources in ever increasing volumes. Surging prices for a natural resource can resuscitate a mining district, and sustainable logging practices can make a timber stand profitable through the effective use of resources.

When you look at an extractive landscape, think about the story it is telling about the sustainability and maturity of the operation. What phase in the life cycle of the operation is the landscape reflecting? Ghost towns speak volumes about the boom-and-bust nature of the mining business, and it is no accident that many workers in the energy industry live in mobile homes (fig. 3.4), modular housing, or PODs (portable, military-style housing units) in what are known as "man camps."

Extractive activities exist in larger national and global contexts. Historically, the West's extractive landscapes blossomed with nationwide industrialization and economic integration, which was facilitated and supported by the federal government. These landscapes are inevitably embedded in stories of corporate capitalism, changing social and economic geographies of labor, the mass consumption of metals and energy resources, negotiated political power, and environmental impacts that can begin with a hole in the ground and continue with acid rain falling on a pond thousands of miles away. Simply put, connections exist among the coal mines in Wyoming, the modular housing for workers, the network of rail cars, and the constellation of national, and increasingly global, power plants burning the coal for electricity (the most common use of Powder River production).

As you encounter extractive landscapes across the West, consider where they fit into that larger production and consumption process. Are you looking at a site of removal (a clear-cut slope, for example), a place of initial processing and refining (a copper smelting plant), workers' housing (trailers at an open-pit mine), or a landscape where resources are being readied for consumption (a lumberyard on the edge of a suburb)?

Extractive landscapes have accumulated cultural meanings and place identities. Westerners are fascinated by stories of prospectors, lost mines in the desert, and ghost towns. Historically, the challenges of making a living from placer streams, coal mines, lumber camps, and copper towns created a strong sense of place, a shared accumulation of stories and experiences that outlived the economic booms that first attracted people to these landscapes. The legacy of ethnic communities is found in local churches, placenames, and social organizations, and complex social geographies based on wealth, class, and occupation have enduringly shaped localities and neighborhoods across the region.

Even in those places that later experienced economic declines, longtime residents are proud of their history. In towns that survived, you can see how the past has been preserved or reimagined in ways that capture traditional identities (fig. 3.5). Ski towns like Aspen and Telluride market their mining-era origins. In addition, tourist haunts such as Virginia City, Nevada, and Virginia City, Montana, and ghost towns like Tin Cup, Colorado, and Bannack, Montana, have been carefully preserved and interpreted to give visitors a glimpse into the past.

FIG. 3.5 Miner's Hat Realty, Kellogg, Idaho. Just off busy Interstate 90, this real estate office is a reminder of the special role played by silver mining in this northern Idaho community.

Many extractive landscapes in the West were formed between the California gold rush after 1848 and the end of World War I in 1918. A series of precious metals rushes radically refashioned economic geographies, human landscapes, and natural environments in such places as California's Sierra Nevada foothills, central Nevada, central and southwestern Colorado, central Idaho, western Montana, and other places from Republic, Washington, to Elizabethtown, New Mexico. Gritty, industrial-scale copper mining towns often emerged later, but became firmly established by the early twentieth century in Butte, Montana; Bingham Canyon, Utah; and a constellation of mining districts in Arizona that included Bisbee, Globe, Jerome, Miami, and Clifton-Morenci. Politically, the Copper Kings, as the mine owners were called, wielded enormous power, a reminder of connections that still exist today between many of the region's politicians and its extractive industries.

While many of the earlier mining centers have declined, or even ceased to produce metals, other post-1920 mining initiatives continue to shape regional landscapes. Among these are open-pit operations at Utah's Bingham Canyon mine, low-grade gold-mining activities in central Nevada, episodic mining of molybdenum at Climax, Colorado, and the strategic development of one of North America's only platinum mines southwest of Billings, Montana.

The lumber industry in the West began piecemeal, often fueled by the local demands of miners and settlers. California's Sierra Nevada and Colorado's Front Range, for example, saw much of their old-growth timber disappear in initial surges of cutting between 1850 and 1890. In the interior West, stands of commercial timber, such as those found across northern Arizona, spurred production. But ultimately, it was logging in the great forests of the Pacific Northwest that created a more enduring imprint on the land. Between 1870 and 1910, the logging industry transformed the landscape and economy of northwestern California, western Oregon and Washington, and the northern Rockies. National and, later, global connections linked the forests of Douglas fir, spruce, and pine to markets in the Midwest, the East Coast, and across the Pacific. Today the legacy of logging still dominates places such as Eureka, California; Coos Bay, Oregon; and Longview, Washington (fig. 3.6).

You can also see the legacy of the energy economy in the West. In southern Colorado's abandoned coal towns near Trinidad, that older extractive imprint has transformed people and place. Some of that legacy is largely invisible, where oil wells were paved over and suburbanized in Southern California and Boulder County, Colorado. But you can find newer landscapes throughout the West, including oil refineries in California, coal-bed methane wells in New Mexico, natural gas wells in Colorado, strip mines in Wyoming, and tar sands in eastern Utah. There is also an increasing investment in renewable fuels—geothermal projects in California, solar collectors in Arizona, wind turbines in the Columbia Gorge—that integrate the nation's changing energy economy with the landscape of the twenty-first-century West.

Finally, consider the aesthetics of reading extractive landscapes. For many people, clear-cut slopes, wastewater ponds, and piles of mine waste are ugly reminders of the visual and environmental price of extracting resources from the land. Many of these places are the subject of political debate as competing interests see the landscape in very different ways. There is no denying that the long-term environmental price tag of the extractive economy has

been enormous, but these landscapes have also been places of work that produced paychecks and built communities. As you explore these settings, look for all of the stories that extractive landscapes can tell. Their tales of resource discoveries, immigrant workers, vibrant communities, technological innovation, and environmental challenges will remind you how all of our lives remain intimately tied to the earth. ❋

FIG. 3.6 Shipping lumber near Coos Bay, Oregon. Logging trucks converge on this coastal facility in North Bend, where large cranes load debarked and bundled logs onto ships (background) bound for destinations worldwide.

28

SURFACE MINING: GOLD AND COPPER

FIG. 3.7 Dredging landscape, southwestern Montana. Dredge operations extensively reworked gold-bearing gravels and produced an uneven, pond-dotted (right), hummocky (left) landscape that is still a common sight in western mining districts today. This scene is west of Virginia City.

THE WEST'S LARGEST, MOST FAMOUS SITES OF SURFACE MINING CENTER on gold and copper. Surface mining involves the extraction of valuable mineral deposits by removing them from near the Earth's surface. Placer gold rushes transformed California's Sierra Nevada foothills in 1848 and 1849, Colorado's steep mountain canyons in 1859, and the gulches of central Idaho and southwestern Montana from 1862 to 1864 (fig. 3.7). Modern large-scale gold mines continue to alter the landscape, especially in Nevada (about 80 percent of U.S. production) and at Cripple Creek, Colorado. The evidence of copper mining is indelible in Arizona—in Ajo, Bagdad, Clifton-Morenci, Bisbee, Globe, and Jerome—at Utah's Bingham Canyon Mine (one of the world's richest), and in Butte, Montana (**31**).

Some of the fundamentals of surface mining have not changed. Gold and copper mines are still wedded to mineral-forming processes that originally created these metals, and investments remain linked to the complex calculus of weighing development costs against potential revenues. But the scale of surface mining has grown incredibly since the days of the prospector. In the

FIG. 3.8 Lavender Pit, Bisbee, Arizona. The Phelps Dodge Corporation (company buildings on the rim, in distance) extracted copper ore at the pit between 1950 and 1974. Note the remnants of terraces built to provide access to nearby vertical walls (called faces). Nonworking pits can also accumulate toxic water from nearby surface drainage or underground sources.

twenty-first century, a handful of workers can operate a huge mountaintop gold or copper mine, and the impact on the land differs markedly from that of a nineteenth-century placer camp.

Locating sites where materials have been removed is easy where open-pit, strip, or mountaintop removal mining has displaced huge volumes of material (fig. 3.8). Where whole mountains have been consumed or deep pits probe the depths, the metal content of deposits is typically low, but operations are viable thanks to efficiencies in earth-moving machinery and processing technologies (fig. 3.2).

The signatures of classic placer mining are quite different. In these settings, gold pans, rockers, and sluice boxes helped miners separate heavier free gold—that is, gold not embedded in rock—from surrounding material. Look for the ditches and flumes miners used to run water through the material, a practice that produced irregular riparian landscapes of discarded waste rock and recovering vegetation. Follow rivers and streams and watch for altered terrain and old equipment.

Most spectacular are topographies of hydraulic mining, where high-pressure hoses directed water at river gravels to speed up removal and sorting (fig. 3.9). The resulting surges in sediment loads and toxic tailings changed nearby riverine environments forever.

In some areas, earlier placer settings were reworked later in large-scale dredge operations. Miners created shallow lakes along a riverbed and used

FIG. 3.9 Late nineteenth-century hydraulic mining in California, Sierra Nevada foothills. High-pressure water hoses exposed gold-bearing river deposits around the West, reworking local topography. Changing sediment loads in regional drainage basins also transformed riparian habitats. *Courtesy of Denver Public Library, Western History Collection, image no. P-1259.*

flat-bottomed dredge scows to reprocess previous diggings. Look for surviving ponds and an uneven hummocky appearance to the landscape (fig. 3.7).

In general, surface mining is never a delicate affair. Some of the nation's largest Superfund sites, such as Butte's Berkeley Pit, are in historic mining areas. By definition, surface mining reworks terrain, and you can find evidence in the surrounding topography, water management, and vegetation. Watch for piles of removed overburden (waste rock) and related leaching and tailings ponds where materials and waste are processed (**30**). Local drainage patterns may have been extensively reworked, depending on where water was needed in the mining process. Water quality might have been compromised in places where toxic metals ended up in surface and groundwater supplies. Dramatic changes to vegetation are also typical as forests are swallowed up in pit operations or consumed by nearby settlements.

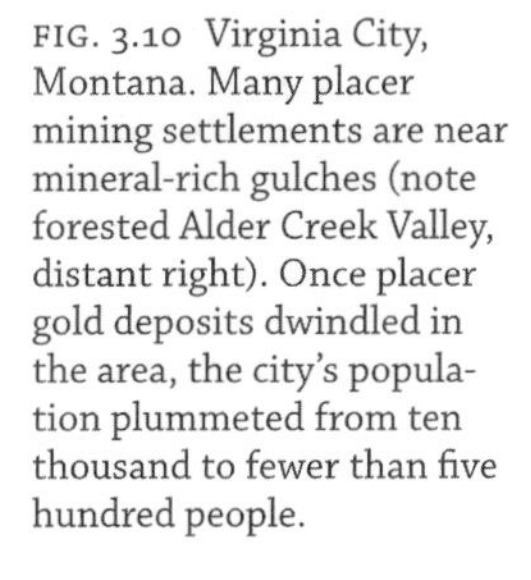

FIG. 3.10 Virginia City, Montana. Many placer mining settlements are near mineral-rich gulches (note forested Alder Creek Valley, distant right). Once placer gold deposits dwindled in the area, the city's population plummeted from ten thousand to fewer than five hundred people.

FIG. 3.11 Dredge Restaurant and Bar, Breckenridge, Colorado. Nine dredges once worked Summit County streambeds in Colorado's high country. This replica of one of the largest dredge boats serves as a local hangout and is a reminder of mining's special legacy on the modern scene.

No mine is an island, so piece together the story of how the whole is greater than the sum of the parts. In other words, appreciate how surface mines depend on labor and settlement (fig. 3.10 and **31**). Also recognize the necessity of a legal landscape of land ownership and environmental regulations; a chain of processing and refining facilities to take raw materials to a purer form (**30**); a supportive transportation network of packhorse trails, toll roads, narrow-gauge railroads (**51**), and highways; and a flow of capital for equipment to dig holes and for paying workers. The timing of mining initiatives also shapes the precise composition and arrangement of this support infrastructure.

Think too about how the history and cultural meaning of the mining experience is reworked on the modern landscape. Not unlike a dredging operation, the contemporary world continues to refashion, rearrange, and reprocess those experiences, forging from them new meanings. It might be a "pan your own gold" business by the side of the road, a souvenir shop selling copper treasures, or a local restaurant that invites customers to relive memories of the old days (fig. 3.11). Excellent museums in Butte, Colorado Springs, and Mariposa, California, can also help you explore how these places have changed over time and how people in these communities have incorporated their mining heritage into the contemporary landscape. ❋

29

UNDERGROUND MINING: GOLD, SILVER, AND COPPER

FIG. 3.12 The subterranean world, Comstock Lode, Nevada. One of the West's richest silver mines sat beneath Virginia City, Nevada, and produced a vast underworld geography that both challenged and rewarded thousands of nineteenth-century mine workers. This engraving by T. L. Dawes appeared in 1876. *Courtesy of Nevada Historical Society.*

WHEN YOU DESCEND INTO AN UNDERGROUND MINE, DAYLIGHT recedes to a pinpoint of light. Subterranean geographies of hard-rock mining are intricate, enveloping—a landscape that both beckons and repels (fig. 3.12). This underground world was often an overheated, underventilated, dangerous place. Still, thousands of men made a living here, helping owners build financial empires even as they exacted a steep human and environmental price.

Hard-rock or lode mining is focused on finding and extracting metals directly from their geological roots. Gold, silver, copper, lead, and zinc are often found in close proximity, and some mining landscapes, such as Butte, Montana, were gold and silver camps before copper and other riches were unearthed. Some metals were easily separated from surrounding rock, but others required elaborate chemical and mechanical processing. Thus, underground mines are inevitably wedded to facilities that mill and concentrate ore into higher-grade material (**30**).

The great gold mining districts included Colorado's Central City and Cripple Creek and such places as Bodie and Grass Valley, California, and Republic, Washington. Silver strikes began with the Comstock lode near Virginia City, Nevada, in 1859; Idaho's Silver Valley near Coeur d'Alene; and later other places in Colorado, such as Leadville, Creede, Aspen, Georgetown, and the San Juans. Underground copper mines often predated open-pit and surface operations (**28**) in places such as Butte, Montana; Bingham Canyon, Utah; or Bisbee, Arizona.

Sizable industrial-scale underground mining operations required large communities to support them (**31**), but the direct surface expressions of underground mining are modest compared to the infrastructure beneath. If you're looking for entrances, you'll see that horizontal tunneling is marked by adits, openings in a mountainside. But it is dangerous to explore this landscape. Inside, collapsed tunnel ceilings, winzes (vertically slanting passages that dive deeper into the mountain), and other hazards lurk.

Many large-scale mines employ shafts—vertical passageways into the mine marked by large wooden or metallic head frames (fig. 3.13). These structures house a hoist that uses pulleys to raise and lower workers, equipment, and material in ore buckets called skips. In a large operation, the shaft beneath the head frame might extend several thousand feet into the earth, accessing miles of tunnels that in later years often flooded when the mine closed and the pumps were turned off. Hoist houses next to the head frame contained the engines and boilers used to run the hoist. In addition, piles of underground waste material often decorate surrounding slopes (fig. 1.8). In many mining landscapes, nearby hills have been stripped of timber or they support second-growth vegetation, evidence that forests were quickly consumed to provide tunnel timbers and fuel. Nearby water supplies are also often tainted from residual metals that have trickled into surface and groundwater systems for decades.

Less visibly, you can hear the stories linked to these industrial centers. Sudden disasters, chronic work-related respiratory diseases, and close-knit communities shaped the bittersweet memories of miners. Inevitably, their lives were interwoven with a landscape too spare in its rewards for those who knew it best. ❋

FIG. 3.13 Head frames on Butte Hill, Montana. Many of Butte's historic underground mines (right) still display both an impressive head frame (flag atop) and an adjacent hoist house (center). In the background, additional mine operations mingle with residences in one of the West's largest copper-mining settlements.

30

METALS MILLING, PROCESSING, AND REFINING

FIG. 3.14 Miami smelters, Arizona. This 1944 view shows Miami's smelter stacks and nearby terraces of slag (waste) that have been contoured and shaped (right) above the town (left). Today, many mining operations in the area have ceased. *Courtesy of Arizona Historical Society/Tucson, image no. PC180F243–0165.*

WHAT DO A PIONEER STAMP MILL (USED FOR CRUSHING GOLD-BEARING ore), an amalgamation vat (used for concentrating silver), and a towering stack above a copper smelter (fig. 3.14) have in common? Each represents a part of what it takes to process raw ore into its refined state. The West is filled with the expression of metals milling, processing, and refining activities, and these industrial pursuits are integral parts of the extractive landscape. They are home to settlements (**31**), technological innovations, and some of the region's most toxic environments.

Gold, silver, and copper all need to be processed to remove waste rock and to concentrate the valuable metallic portion of the ore. Ores are initially processed near the mine (**28** and **29**) to produce refined concentrates. These less bulky materials can then be easily shipped to smelting and refining plants,

FIG. 3.15 Argo Mill, Idaho Springs, Colorado. The Argo Mill west of Denver displays the typical landscape signatures of these processing facilities. Ore was crushed and concentrated as it was gravity-fed through the mill. An associated tunnel (completed in 1910) helped drain nearby mines and transport ore to the mill.

FIG. 3.16 Roasting facility, Newmont Mining Company, Carlin, Nevada. Large, open-pit gold-mining operations often feature on-site processing facilities. These high-tech roasters at Newmont's Carlin mine help separate out minute amounts of gold from surrounding waste rock. *Author photo used with permission of Newmont Mining Corporation.*

often many miles from the mine, where centralized facilities can process them inexpensively and efficiently.

Landscape signatures of this process are amazingly diverse. They include primitive crushing mills in mountain ravines; a golf course carved from reclaimed slag heaps in Anaconda, Montana; and a high-tech refining lab at a gold mine. Some of the most recognizable signatures include mills (watch for long, sloping, stair-stepped rooflines), often built on a hill to feed ore by gravity through the crushing and pulverizing processes necessary to prepare material for separation and concentration (fig. 3.15). Nearby tailings piles and ponds contain waste rock (called gangue) and water. In large-scale copper smelting operations, heat and other materials (fluxes) reduce metal concentrates to blister or anode copper, roughly 97 percent to 99 percent pure. Tall smokestacks help diffuse the toxins into the air, and terraced slag piles of waste, called accretionary features, often sprawl near the smelter (fig. 3.14).

Large-scale gold mining operations are similarly equipped with cyanide heap leach pads and collecting reservoirs to chemically separate and concentrate the pure metal. There are also large facilities, called roasters, that separate gold from waste rock (fig. 3.16). All of these industrial settings are prime turf for environmental despoliation, so watch for altered water, soil, and vegetation and for remediation efforts designed to address such problems. ❋

31 MINING TOWNS

FIG. 3.17 Bisbee, Arizona. This copper-mining town retains a densely built downtown of two- and three-story commercial structures (lower left and center) and distinctive neighborhoods perched on slopes above and tucked into nearby gulches (upper left, far right). Note the hillside letter (**84**).

THERE IS NO MISTAKING THE LOOK AND FEEL OF MINING TOWNS, which are enduring expressions of western identity (fig. 3.17). They are all about labor. Workers needed places to live, eat, drink, raise hell, buy boots, have families, and build communities. What we see today in towns such as Virginia City, Nevada, and Cripple Creek, Colorado, represents the traces left by those lives as well as the story of how the town endured, died, or reimagined itself.

The West's great mining towns (fig. 3.18) exist in places where rich underground lode mines or, later, open-pit operations were particularly long lasting: you can explore California's Sierra Nevada (along State Highway 49 and in nearby Bodie), the Colorado Rockies (Georgetown, Central City, Leadville, Cripple Creek, Aspen, Ouray, and Telluride), central Nevada (Virginia City, Eureka, Austin), and locations in the northern Rockies (Bannack, Virginia City, Idaho City, Wallace). Copper towns such as Jerome or Bisbee, Arizona, remain distinctive western places (fig. 3.17). Coal-mining settlements (**35**) retain a feel all their own, whether they display the legacy of older underground operations in southern Colorado (fig. 3.3) or the more casual layout of newer settlements based on strip mining, such as Wright, Wyoming (fig. 3.19).

Mining towns are never far from mines or processing and smelting facil-

FIG. 3.18 Selected precious-metals mining towns. The West's major precious-metals mining towns cluster in California's Sierra Nevada, the Colorado high country, Nevada, and the northern Rockies.

ities (**28**, **29**, and **30**). In communities like Butte, Montana, you can see how mining operations, processing sites, and residential neighborhoods blended with one another (fig. 3.13). Also, there is usually a strong relation between the size and persistence of the extractive activity and the settlement that supports it. In Colorado's high country, for example, Leadville's sizable downtown (fig. 3.20) suggests its economic importance compared with the smaller business district of nearby Saint Elmo (fig. 3.21).

Many mining towns were created in gulches or on the side of a mountain. Usually an early effort was made to create a formal plat with streets and lots. While most mining towns in the West are organized on some variation of a grid plan, sometimes it was impossible to impose a neat, rectilinear order on a natural landscape characterized by steep slopes and canyons.

Consider Bisbee (fig. 3.17). This classic copper town has deeply embedded itself in the rugged hills of southeastern Arizona, its crowded commercial avenue snaking along the bottom of a gulch and its residential areas creeping up nearby slopes and side canyons. Elsewhere in the West, Jerome, Arizona, is built on the steep side of Mingus Mountain, where stairways are used to navigate the different levels of the town. Central City, Colorado; Virginia City, Nevada; and Butte, Montana (fig. 3.13), also offer examples of unlikely locations for development. In other settings, however, residents have spread their homes across the landscape, choosing privacy and sprawl over the coziness of close neighbors (fig. 3.19).

FIG. 3.19 Wright, Wyoming. Contrast the clustered layout of Bisbee (fig. 3.17) with the peripheral sprawl of houses and mobile homes on the edge of Wright, a newer mining community in Wyoming's coal country.

The main thoroughfares in mining towns, such as Harrison Avenue in Leadville, are vibrant corridors of commercial activity that historically featured a closely packed collection of buildings and businesses, ranging from assay offices and union halls to saloons and boardinghouses (fig. 3.20). Early on, downtowns served largely male populations, but retail services evolved in later years as more women and families became part of the community. False building fronts (fig. 3.21) and high-style architectural flourishes—Leadville's brick and stone building fronts, for example (fig. 3.20)—added a sense of permanence to the town, along with local signposts of civilization such as banks, hotels, and opera houses. You can find the dates of construction on building fronts, and older business signs and advertisements may still be present on the sides of buildings. If you venture down alleyways, you can explore the backside of Main Street's bustle, usually a messy accumulation of architectural and technological eras.

Beyond downtown are residential landscapes, often an eclectic assortment of modest cottages (fig. 3.22), cabins (with front-facing gables most common), and substantial brick or stone houses. Typically, the lot sizes and front yards for working-class miners were small, and architectural designs were simple and functional. You can also spot the multitiered social geographies of laborers, supervisors, and mine owners. The upper crust displayed their good fortune in Victorian, Queen Anne, and Gothic Revival mansions, while the heritage of residents is revealed in street names, local businesses, names

FIG. 3.20 Downtown Leadville, Colorado. This two-mile-high mining town contains a lively central commercial district, which nowadays largely caters to tourists. Older structures date from the late nineteenth century (middle distance, right and left). Their fine brick detailing and Victorian-era embellishments are typical of larger mining towns of that period.

on mailboxes, and meeting places such as churches and fraternal halls. You can also explore cemeteries and learn the names of residents and when they died. Years of high mortality suggest a time of epidemics or mine accidents.

One interesting exception to the untidy nature of most mining settlements is the company town, which bears the signature of a single corporate imperative revealed in a coherent design, uniform architecture, and integrated land use. Designed to maximize efficiency and to offer a productive work environment, company towns (**34** and **35**) symbolized the imposition of large-scale industrial capital on the mining business. Company towns include Copperton, Utah (copper); Morenci, Arizona (copper); Cokedale, Colorado (coal); Climax, Colorado (molybdenum); Tyrone, New Mexico (copper); and Colstrip, Montana (coal). Bagdad, Arizona, a copper-mining town, bears the imprint of corporations that built standardized company housing—about four hundred units—and provided social and recreational services for workers (fig. 3.23).

Finally, think about how time has passed in the mining town. Often, few remnants of the initial settlement survive. Most of the shacks and buildings have fallen down, been cleared away, or been destroyed by fire. Ask some questions: Are there signs of a later, corporate phase in the town's mining history? Is there a surviving population, or has the place reverted to ghost-town status (**32**)? How is the mining era remembered and represented on the landscape? Leadville's streets, for example, are filled with tourist attractions, museums, and souvenir shops (fig. 3.20). Sometimes mining has remained a

FIG. 3.21 Downtown Saint Elmo, Colorado. Modern-day Saint Elmo (near Buena Vista) survives, just barely. The town includes several occupied homes and a general store. False-front architecture and a semblance of sidewalks suggest the urban aspirations of earlier residents.

mainstay of the local economy: Colstrip, Montana, has witnessed the revival of coal; Victor, Colorado, an early gold-mining town, now sits next to an open-pit gold mine; and Utah's Bingham Canyon open-pit mine keeps getting deeper and deeper.

Other mining towns have been transformed and repackaged into ski towns such as Aspen, Colorado, and Red Lodge, Montana (**92**); art colonies such as Bisbee, Arizona, and Madrid, New Mexico (**97**); retirement towns such as Grass Valley, California (**98**); and mountain-biking meccas such as Moab, Utah (**89**). But underneath the New West sheen, you can often see the gritty roots, signatures of an extractive past that both westerners and visitors seem eager to retain. ❋

FIG. 3.22 Miner housing, Leadville, Colorado. In larger mining towns, watch for renovated examples of mining-era housing, complete with Victorian-style embellishments and a fresh coat of paint. Often, retirees and amenity seekers purchase these properties.

FIG. 3.23 Company housing, Bagdad, Arizona. Watch for the standardized architecture (these units offer a pleasant fusion of ranch-house layouts with Southwest-style ornamentation) of the company mining town in selected regional settings. Compare with fig. 3.22.

32 GHOST TOWNS

FIG. 3.24 Hotel Meade, Bannack State Park, Montana. Built in 1875, this impressive two-story brick structure is occupied only by the memories of miners and visitors who once passed through this placer-mining town. The town is preserved as a state park and has been left in a state of arrested decay.

MANY WESTERN MINING TOWNS (**31**) HAVE WITHERED AWAY AND become ghost towns, with a population of zero or close to it. As the process of decline unfolded across the West, phrases such as "dead camps" and "ghost cities" were used to describe a dying settlement. Sometime between 1920 and 1930, the term *ghost town* came to be used in the popular press, and guidebooks, Hollywood movies, and television westerns only added to their mythic quality.

We can learn several things from ghost towns. First, they are revealing signatures of industrial capitalism, of how the West's boom-and-bust economy led to the widespread abandonment of many localities in favor of more promising opportunities elsewhere. Second, the collections of remaining foundations, buildings, and objects that remain today, usually in decaying, weather-beaten condition, are material artifacts of western history. And ghost towns are an enduring part of popular culture through the myths and landscapes that capture the legacy of the Wild West. They incorporate ideas of freedom, independence, violence, greed, and risk-taking, all characteristics of the West in the national imagination.

Most ghost-town landscapes fall into one of three basic types. In the abandoned settlement, no one is around and surviving buildings are rapidly decaying, reclaimed by the surrounding environment. Dozens of such small mining camps can be found in remote mountain and desert locations, such as Masonic, California; Rochester, Montana; and Eureka, Colorado. These towns are now often on private land, so be respectful, talk softly, and take nothing with you. Also be careful of rotting floorboards and nearby mine shafts.

Other ghost towns, by contrast, are in a state of arrested decay, with structures exposed to the elements but protected from complete deterioration. These ghost towns are under some form of protection, either on public land (often as state parks) or under private ownership. State parks include Bodie, California; Berlin, Nevada; and Bannack, Montana (fig. 3.24). Privately owned sites include Cerro Gordo, California; Elkhorn, Montana; and Saint Elmo, Colorado (fig. 3.21). For a twist, look at Uravan, Colorado, and Jeffrey City, Wyoming, mid-twentieth-century uranium boom towns (**65**) that later saw their fortunes fade (fig. 3.25).

Still other ghost towns are full-fledged tourist attractions, with saloons, gold panning, train rides, staged gunfights, and cemeteries. Tombstone, Arizona (the Helldorado Days festival began in 1929), and Virginia City, Nevada (where you can visit the Bucket of Blood saloon), were early participants in this reimagining of the Old West town experience. Most commercialized ghost towns—including the Bovey family's Virginia City and Nevada City, Montana, and the Knott family's Calico and Knott's Berry Farm, California—feature a mix of original, moved, restored, and reconstructed buildings. Such settings may not be entirely authentic, but they both capture and craft our social memories of one of the West's most iconic landscapes. ❋

FIG. 3.25 Abandoned Firestone tire store, Jeffrey City, Wyoming. When the uranium mining boom went bust in the 1980s, so did the central Wyoming town of Jeffrey City. Weeds now have the upper hand.

55 LOGGING

FIG. 3.26 Logging in western Oregon, near Grand Ronde. This patchwork of clear-cut and unlogged land is a common sight across the Northwest. It represents a complex assortment of cutting strategies and management practices, both on private and on public lands.

IF YOU FLY OVER THE WESTERN FORESTS ON A CLEAR DAY, ESPECIALLY IN the Pacific Northwest, you are likely to see signs of commercial logging, a patchwork of mature trees, cut-over land, and replanted forest (fig. 3.26). The landscape is part of a long story of human modification. Native Americans burned forest understories to drive game and increase desirable habitats. Large-scale cutting increased with nineteenth-century Euro-American settlement as miners, farmers, and city-dwellers needed lumber. As the railroads arrived (**50** and **51**), huge tie drives floated down western rivers, reshaping stream ecologies (**7**) to send thousands of logs downstream to the mill. When unmanaged cutting accelerated, many parts of the West witnessed the creation of stump forests.

By the 1880s, the growing demand for lumber brought an entirely new scale of industrial logging, with more efficient saws and dragging and hauling technologies. Corporate operations consumed the coastal forests of the Pacific Northwest. As national forests (**67**) were created, private lumbering was allowed on public lands, subject to management strategies that evolved over time. After World War II, demands for new housing (**76**) ushered in another surge in logging, especially across the northern Rockies and the Pacific Northwest.

Today's megalogged landscapes are concentrated in Northern California, western and central Oregon and Washington, northern Idaho, and northwestern Montana. In these settings, the complex ecosystems of virgin old-growth forests have largely been replaced with managed crops of single-age commercial timber. More broadly, logging is sometimes used as an environmental management tool to control disease and insect infestations (fig. 3.27), manage wildfire threats (**16**), and shape viewsheds. A ground reconnaissance can reveal examples of clear-cutting, where virtually all trees are removed from an area of the forest (fig. 3.26). In other settings, you can see evidence of selective cutting, where trees of a particular size, age, or species are removed. You will also see unpaved logging roads, abandoned railroad tracks and skid roads (paths along which logs are dragged), temporary logging camps, piles of felled trees (**34**), and logging trucks on the highway. Once an area is logged, you can spot slash piles of remaining material (fig. 3.27), which are often burned on-site, and blocks of even-aged trees, which reflect reforestation initiatives and a modern commitment to sustained-yield logging practices that are designed to generate more commercial logging opportunities in the future (fig. 3.28). ❋

FIG. 3.27 Slash piles near Grand Lake, Colorado. Massive pine-beetle infestations across central Colorado have prompted logging efforts to manage wildfire hazards and improve viewsheds. These slash piles are found near the resort community of Grand Lake close to Rocky Mountain National Park.

FIG. 3.28 Reforestation east of Eugene, Oregon. Large stands of even-aged trees (contrast left and right) suggest the area has been repeatedly logged and reforested. Sustainable-yield forestry practices produce a landscape mosaic of different age-stand trees that can be harvested in sequence over time as they mature.

34

LUMBER MILLING AND PROCESSING

FIG. 3.29 Lumber mill near Flagstaff, Arizona. This early twentieth-century aerial view of a large lumber mill in northern Arizona reveals a sprawling landscape of milling facilities (center right), finished products (upper left), and worker homes (lower left). *Courtesy of Arizona Historical Society/Tucson, image no. PC196F368–04.*

JUST AS MINERAL ORE REQUIRES SMELTING AND REFINING, RAW LUMBER needs to be milled into marketable products. In some cases, timber is processed close to where it is harvested (**33**), but in other operations logs are milled in distant places (fig. 3.6). You can see larger western sawmilling operations in northwestern California, Oregon, Washington, Idaho, and western Montana. Markets for their products are global, with Asian consumers' demand increasing significantly since 1980.

The scale of lumber milling has increased greatly over the past century, and mills usually cover a large area (figs. 3.29 and 3.30). You can see piles of freshly arrived logs ready for decking, a process by which logs of different size, species, or intended uses (lumber, plywood, chips) are sorted before milling. Then logs are run through a variety of operations that might include sawing, edging, trimming, planing, and drying. Finally, finished products are stacked for later shipment.

Older mills might reveal signs of conical-shaped sawdust burners, but most modern facilities use mill waste and sawdust to produce fiberboard

(used in industrial products) or wood pellets (used for heating). Other related facilities include boutique mills that specialize in log-home building kits. Western Montana's Bitterroot Valley, south of Missoula, is a major regional producer in the log-home industry (fig. 3.30). Large, industrial-scale pulp and paper mills, which also consume forest products, are concentrated in California, Oregon, and Washington. It's easy to see their high emission stacks and, if the wind is right, to smell their characteristic odor.

Traditionally, these processing facilities employed many workers. In communities located near the mill, you can spot the legacy of company towns or company housing (fig. 3.29). In Bonner, Montana, company-built housing early in the twentieth century survives even though a nearby mill periodically closes (fig. 3.31). Elsewhere you can find signs of other former company towns in places such as Samoa, California (near Eureka); Gilchrist, Oregon (south of Bend); Potlatch, Idaho (north of Moscow); and Longview, Washington (south of Olympia). ❋

FIG. 3.30 Log-home sawmill near Hamilton, Montana. Western Montana's Bitterroot Valley has become a mecca for high-end log-home building companies, which mill the wood and package kits for regional and national customers.

FIG. 3.31 Company housing, Bonner, Montana. The mill town of Bonner (near Missoula) was established in the 1880s, and many company-built cottages date from the early twentieth century (compare with fig. 3.23). Today, many of these structures remain tidy, well-kept homes.

33

COAL

FIG. 3.32 Mining coal: Navajo Mine south of Farmington, New Mexico. The massive scale of coal strip mining is apparent as huge dragline excavators remove overburden (waste rock) to expose underlying coal seams.

WHEN YOU FLIP ON A LIGHT SWITCH, COAL IS THE LAST THING ON your mind, but it does much of the everyday work of our economy, mostly by producing relatively inexpensive electricity. Despite its attractions, coal exacts a price, and the human and environmental costs of coal mining have left a bittersweet legacy in the West. Global-scale implications also loom as many scientists warn of its role in sending carbon into the atmosphere, which contributes to the warming of the Earth's climate.

A carbon-rich fossil fuel, coal comes in varied forms, defined by heat-producing capacity and chemical content. Most western production is low-sulfur, cleaner-burning, bituminous and subbituminous coal (used for electricity production), not hotter-burning anthracite coal (used in metals processing). Wyoming, particularly the Powder River Basin, dominates western coal mining, but Montana, Colorado, New Mexico, and Utah are also key producers.

Coal's signature across the West can be traced in three different ways. First, there are places shaped by coal mining. Some of the region's earliest operations developed across Colorado, and the area around Trinidad is especially worth exploring (see fig. 3.3 and **31**). Nearby, Crested Butte and Redstone, Colorado, had high-quality anthracite mines. In the twenty-first century, most of the West's largest coal mines are surface operations that feature terraced, strip-mined topography designed to remove both the coal and the overburden (waste rock) efficiently. Under the impetus of recent environmental regulations, reclamation efforts have restored these areas to varying degrees. Colstrip, Montana; Gillette and Wright, Wyoming (fig. 3.19); and Farmington, New Mexico, are all good base camps for exploring coal mining in the interior West (fig. 3.32). Nearby, watch for coal preparation plants where waste rock is removed and the coal is cleaned.

Second, watch for coal on the move once it has been extracted from a pit, open seam, or underground mine. Over short distances, giant trucks remove coal and waste from the mine and conveyor-belt systems and loading towers fill waiting rail cars. Unit trains, endless lines of identical metal cars, are filled to the brim with the raven-hued rock (fig. 3.1). The West has also been home to slurry pipelines, in which coal and water are mixed and sent long distances. The large Black Mesa mine in northeastern Arizona once sent coal to Laughlin, Nevada (a journey of more than 270 miles) via a controversial pipeline, but environmental and cost issues shut down operations. Growing Asian demand has also created new markets and a greater need for controversial West Coast loading facilities.

Finally, the West is liberally sprinkled with points of coal consumption, mostly in the form of coal-fired power plants. Some of these are close to points of coal production, while others are near urban electricity markets. Watch for converging rail lines, large boiler furnaces, and tall stacks where heat is generated to turn the turbines (fig. 3.33). From there, electrical transmission lines reach out to your light switch and the world beyond (**56**). ❋

FIG. 3.33 Burning coal: San Juan Generating Station west of Farmington, New Mexico. This power plant burns locally mined coal to make steam-generated electricity that serves more than two million customers across the Southwest.

56

OIL AND NATURAL GAS

FIG. 3.34 Oil well, Coalinga, California. The Iron Zoo, created by local artist Jean Dakessian, animates an otherwise bleak landscape in the barren, pump-dotted hills near Coalinga, southwest of Fresno, California.

PEOPLE STILL COME TO SEE JEAN DAKESSIAN'S IRON ZOO. THE PAINT has faded on the animals and fanciful creatures that adorn oil pumps near Coalinga, California, but the local artist's work, dating from the early 1970s, continues to celebrate the role of petroleum in the town's history (fig. 3.34). Many western towns have a legacy based in the extraction of oil and natural gas and other organic, carbon-based products. These resources date from a warmer and wetter geologic past such as the Cretaceous period (about 60 million to 140 million years ago). Decaying plant and animal life accumulated in sedimentary basins (**4**) near seas, rivers, and marshes, were subjected to changes in temperature and pressure, and transformed into the fossil fuels we now covet.

While early settlers made use of oil springs and seeps, the first large-scale boom in the West's energy economy occurred between 1890 and 1920. Southern California's rich basins were plumbed with abandon; more selectively, interior landscapes from southeastern New Mexico to north-central Montana became part of the new economy. Accelerating national demand for energy contributed to cyclical expansion between 1920 and 1970, and periodic energy crises since then have spurred new booms. Oil and gas production in the twenty-first century is a response to rising energy prices and growing global demands for fuel and electricity.

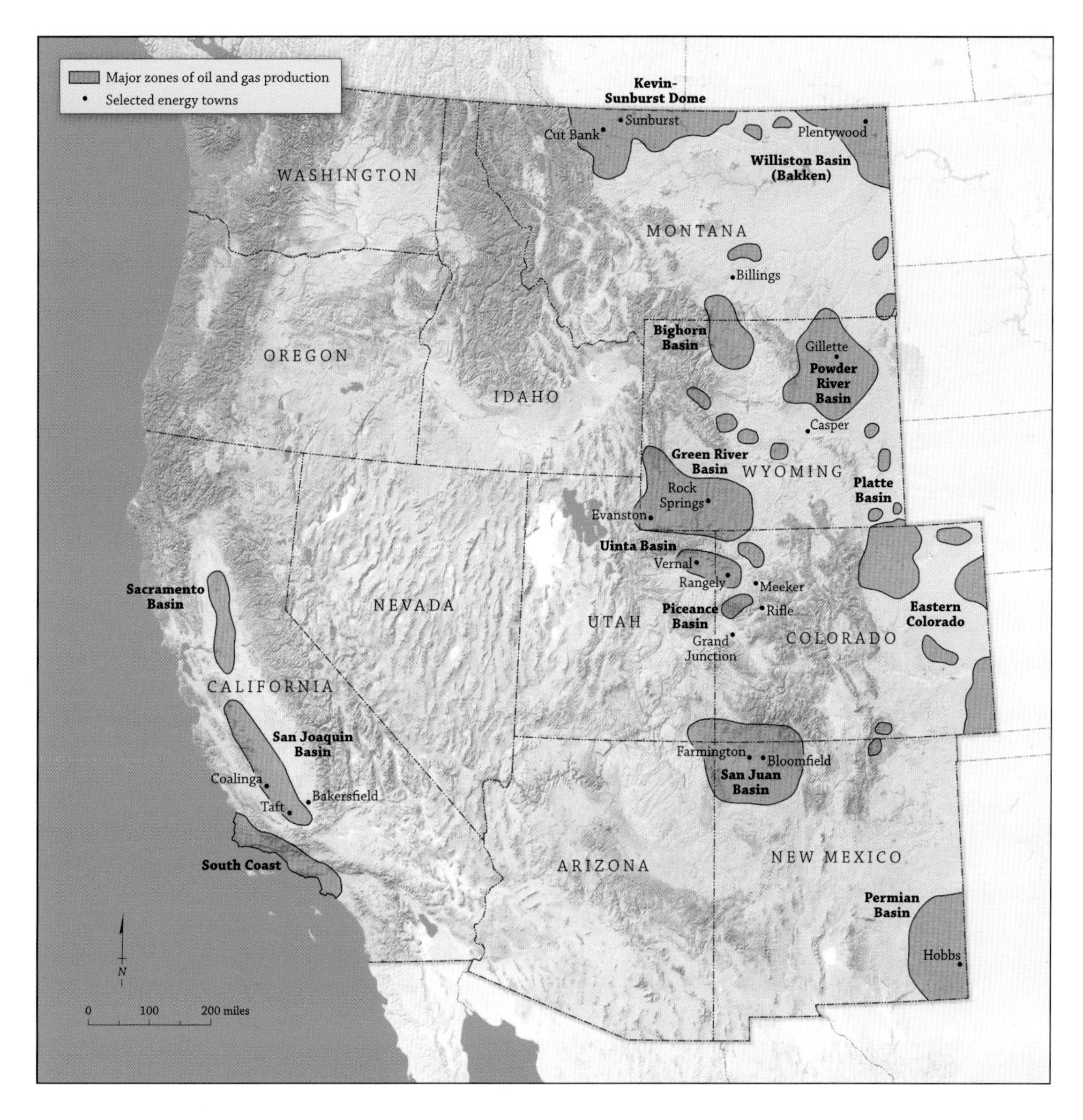

FIG. 3.35 Major zones of oil and natural gas production. California crude remains a vital part of the West's energy budget, but varied interior locales from southeastern New Mexico to northern Montana also contribute substantial amounts of petroleum and natural gas to the regional economy.

The map of major production zones (fig. 3.35) suggests three geographical patterns. First, California continues to dominate in oil, accounting for more than 10 percent of U.S. production (13 percent of proved reserves). In Kern County, often called California's Texas, and other parts of the San Joaquin Valley near towns like Coalinga, you can see some of the nation's largest oil fields (Belridge South near Bakersfield is one of the highest-producing oil fields in the lower forty-eight states). Elsewhere, explore the Santa Maria basin or Southern California's Wilmington oil field, where offshore oil platforms around Long Beach Harbor have been cleverly disguised.

Second, the interior West's dynamic oil patch is the source for another 10 percent of national production (11 percent of reserves). New Mexico (the Permian and San Juan basins) and Wyoming (the Powder River and Big Horn basins) dominate interior output, but you can also find major production

FIG. 3.36 Natural gas well near La Plata, New Mexico. In production zones, farms and fields (left) are punctuated with reminders of the energy economy. In this patch, gas is brought to the surface by a small pump (center left) and passed through a gas separation device (far right) that has a cellular antenna for monitoring well activity. Natural gas liquids are stored in the tank (center right, painted green to blend in).

zones in Colorado (east of Denver), Utah (the Uinta Basin near Vernal), and Montana (the Williston Basin and the Kevin–Sunburst Dome near Cut Bank).

Third, the interior West also produces natural gas, with Wyoming (especially the Green River and Washakie basins between Evanston and Wamsutter), Colorado (Weld County and the Western Slope's Piceance Basin), New Mexico (near Farmington or Hobbs), and Utah (the Uinta Basin) accounting for more than 25 percent of U.S. production (30 percent of reserves). Drillers increase production through hydraulic fracturing—or fracking—which injects a mix of water, sand, and chemicals underground to break apart sedimentary rocks and release gas. Critics of the process believe fracking pollutes groundwater and creates hazardous environmental conditions as well as earthquakes.

The West also leads the nation in the development of unconventional fossil fuel resources. Coal-bed methane drilling (70 percent of national production) is a controversial process in which natural gas is harvested from coal seams. Often, gas can be released only after these formations are partially dewatered, a process in which underground water is pumped to the surface. Critics argue that the process affects water tables and brings polluted water to the surface. Shale gas and shale oil projects in northeastern Montana (Bakken shale) and southwestern California (Monterey shale) or tar-sand developments in eastern Utah (on the East Tavaputs Plateau in the Uinta Basin) require large amounts of water, chemicals, and energy to unlock useful hydrocarbons but can also produce environmental hazards or threaten activities such as ranching.

Landscapes of oil and gas production are varied. Site-specific impacts focus on small pumping platforms and pads that thickly dot major producing districts (fig. 3.34). These signatures are often interwoven into agricultural settings, where farms and fields are juxtaposed with pumping outfits and storage tanks (fig. 3.36).

But this is only the beginning: watch for "upstream" elements such as drilling rigs, road construction projects, and oil-field service and geological

FIG. 3.37 Oil storage tank farm near Casper, Wyoming. Large storage tanks dot the western landscape and hold unrefined crude oil as well as refined gasoline and other petroleum-related liquids.

exploration companies. Also consider what happens "downstream" from these activities: wastewater and oil-field fluid storage facilities, holding and evaporation ponds, and waste pits. In Kern County, the oil industry generates more than twice the volume of wastewater (often infused with salts, metals, and petroleum residue) produced by the City of San Francisco. Follow the trucks, rail cars, and pipelines that move resources through several stages of storage and processing (fig. 3.37). Initial gas collection and oil refining (watch for a skyline of tanks, pipes, fume vents, and towers) occur near production sites to reduce costs, but there are also coastal facilities (see large tank farms from El Segundo, California, to Anacortes, Washington) that refine foreign oil for domestic consumption.

In oil and gas towns in California (Coalinga and Taft), New Mexico (Hobbs, Bloomfield, and Farmington), Colorado (Grand Junction, Rifle, and Meeker), and Wyoming (Green River, Rock Springs, Casper, Evansville, and Gillette), you can see the effects of a boom-and-bust energy economy. You can see part of that history in their place identities—school mascots, business and street names, derrick and oil pump statues, local museums. In boom times, these towns are places of long work shifts, traffic jams, tight housing markets (fig. 3.4), and stressed schools and social services. In leaner times, neighborhoods are dotted with "For Sale" signs, the shopping mall feels empty, U-Haul trailers are on the road, and there is too much time for drinking at the local bar. The region's geological wealth and global patterns of demand suggest that these landscapes will endure, supported by drilling and processing technologies designed to harvest new riches from the subterranean world. ❋

37

WIND AND SOLAR

FIG. 3.38 Wind farm, Tehachapi Pass, California. Joshua trees and rotating turbines populate this breezy, high-desert landscape west of Mojave, California.

THE WEST IS HOME TO MANY OF NORTH AMERICA'S MOST AMBITIOUS and visible wind and solar energy initiatives. Many westerners are attracted to the enduring availability, lower environmental impacts, and potentially lower costs of these technologies. In addition to fluctuating federal subsidies, most western states (led by California) offer tax incentives for investing in renewable energy production. The result is a regional landscape increasingly punctuated by signatures of energy sustainability.

Traditional windmills, typically used for pumping water to the surface, made wind energy a long-standing part of the western American landscape (**20**; figs. 1.7 and 2.20). Since the early 1980s, however, a new wind revolution has swept the region as industrial-scale wind farms, sometimes with hundreds of one-hundred- to three-hundred-foot-high turbines (three-bladed propellers are most common), have been built across breeze-prone landscapes (fig. 3.38). You can see these wind farms at Altamont Pass, Tehachapi Pass, and San Gorgonio Pass in California; the Columbia River Gorge in Washington and Oregon; the northeastern plains of Colorado; the eastern plains of New Mexico; and at Judith Gap and east of Glacier National Park in Montana. Southern Wyoming may have some of the continent's best wind energy potential.

FIG. 3.39 Solar thermal power plant near Kramer Junction, California. The Solar Electric Generating System is one of the largest of its kind in the world. Curved mirrors concentrate the sun's heat to warm a thermal liquid within the pipes. The heated liquid produces steam, generating clean electricity.

Proponents point to the sustainable, low-cost benefits of wind energy, while critics cite problems with unpredictable power (**56**), noise, and viewshed pollution. The turbines also are responsible for killing birds, especially raptors, and some critics refer to them as "Cuisinarts of the sky."

In similar fashion, the West is well situated to produce solar energy. Much of the Southwest boasts more than 2,800 hours of annual sunshine, twice as much as New England. Encouraged by federal and state incentives, California, Nevada, Colorado, and Arizona have taken a leading role in solar energy production. You can see several varieties on the landscape. Passive solar design elements on buildings capture heat and light through the use of dark surfaces or south-facing windows. At southwestern Arizona's Agua Caliente plant, you can see rows of solar panels that use photovoltaic cells to convert sunlight to electricity.

Solar thermal technology concentrates energy from sunlight to heat a thermal liquid that produces steam to power a turbine that generates electricity. One of the world's largest examples is SEGS, the Solar Electric Generating System (begun in the 1980s), in California's Mojave Desert (fig. 3.39). From a distance you can see the metallic parabolic mirrors that capture and concentrate heat. A related approach focuses sunlight on a central receiver system in which thousands of flat mirrors (heliostats) direct sunlight at a tower where energy is collected to produce electricity. These projects produce their own variety of pollution and waste (especially in the manufacturing of solar panels) and they also take up a good deal of open space and wildlife habitat. Some environmentalists worry that solar sprawl is already a threat to animals such as the desert tortoise (**18**). But the number of new projects suggests that this renewable energy source will be an ever more visible part of the regional landscape. ❋

PLACES OF SPECIAL CULTURAL IDENTITY

FIG. 4.1 Road sign, Flathead Indian Reservation, Montana. With the Mission Mountains in the background, this road sign along U.S. 93 near Saint Ignatius celebrates the rich linguistic legacy of the Salish-Kootenai nation.

NORTH OF MISSOULA, MONTANA, U.S. 93 ENTERS THE FLATHEAD Indian Reservation, bends northeast, and then gently descends into the lovely Mission Valley at Saint Ignatius. With the dramatic Mission Range in the background, you cross Sabine Creek and encounter a bilingual landscape that demonstrates how the Salish-Kootenai have used a process called ethnic reidentification to shape the way people look at the land (fig. 4.1).

In southeastern Arizona, north of the border town of Douglas, State Route 80 takes you along the backside of the Chiricahua Mountains. There is a small roadside cross alongside the highway (fig. 4.2). Roadside memorials, which are found throughout the West, are called *descansos* in the Southwest. They designate a "resting place" that commemorates the last place a person was alive before dying in a roadside accident. Often decorated with flowers and personal items, *descansos* are subtle reminders of the Southwest's deep connection with Hispanic traditions.

Monterey Park's Garfield Avenue is a busy Southern California thoroughfare just off the San Bernardino Freeway in the Greater Los Angeles Area. As you walk down the street, you can smell the pungent Chinese and Vietnamese dishes cooking at local restaurants, hear a Chinese American couple in conversation, and ponder the diversity of non-English signage that marks nearby strip malls (fig. 4.3). Monterey Park has become one of North America's largest Asian American communities, where people from many countries have established a sprawling ethnic suburb (sometimes termed an *ethnoburb*). Many portions of the West have become increasingly globalized since 1970, particularly as more Latin American and Asian immigrants have settled in the region. Their transformative impact on the West is especially notable in urban areas. From Seattle and Yakima to San Jose and Denver, immigrant communities have reshaped neighborhoods and produced new cultural landscapes. In the case of Los Angeles County, about 35 percent of residents in 2012 were foreign-born (up from 11 percent in 1970). The scene on Garfield Avenue is a reminder that these recent immigrants have added even further cultural variety to the area's already diverse population.

FIG. 4.2 Roadside *descanso* (resting place) northeast of Douglas, Arizona. Quiet reminders of a lost loved one, these crosses and monuments mark the location of a deadly auto accident; many derive from Hispanic traditions of resting on the long, sad journey between church and cemetery.

FIG. 4.3 Asian strip mall along Garfield Avenue, Monterey Park, California. This largely Chinese ethnoburb caters to the diverse needs of the neighborhood's Asian American population and also effectively links the area to the global economy.

These snapshots suggest how people have expressed their cultural identities and shaped landscapes across the West. As you explore the region's cultural geography, it is useful to keep in mind a few basic definitions. Geographers Paul Knox and Sallie Marston offer some guidance in their study, *Human Geography: Places and Regions in Global Context. Culture,* they remind us, is a "shared set of meanings that is lived through the material and symbolic practices of everyday life." In other words, culture is not inherent to a place; people create culture in a place through practices, symbols, and identity. A *culture region* is an area "where certain cultural practices, beliefs, or values are more or less practiced by the majority of its inhabitants." Furthermore, a "geographic focus on *ethnicity* is an attempt to understand how it shapes and is shaped by space, and how ethnic groups use space with respect to mainstream culture." A *cultural landscape* is seen as "a characteristic and tangible outcome of the complex interactions between a human group and a natural environment."

Places of special cultural identity can be explored in a variety of ways. Material elements include tangible expressions of cultural history, such as the *descanso* on State Route 80 in Arizona. Nonmaterial elements—such as spiritual beliefs, language, and social practices—also give places their character. Identifying key landscape features can tell you a great deal, but you can also learn about a place by listening to the rhythms of language in conversations. You can visit coffee bars and taverns and tune in to local radio stations, from Navajo news broadcasts in the Four Corners country to Latina hip-hop in Albuquerque. You can watch how people relate to one another and consider how cultural differences are revealed and negotiated among groups. And, of course, you can sample the local food.

Social media and the Internet are also playing a role in reshaping the cultural geography of the West. The connectivity made possible by the virtual world can be a powerful force that contributes to globalization and cultural convergence, a process through which local cultural differences—revealed in languages and landscapes, for example—diminish in favor of homogenizing

FIG. 4.4 Hopi cornfield near Walpi, Arizona. Traditional cultivation of maize still supplements the Hopi economy and sustains agricultural traditions within the semiarid region of northeastern Arizona.

influences. Navajo, Latino, and Korean teens may use the same Web sites to purchase music, check out the latest news, or buy concert tickets.

At the same time, these technologies open the door to new opportunities for cultural differentiation, preserving and enhancing local cultural and ethnic values through the preservation of traditional languages, the celebration of shared traditions, and a reinvigorated sense of cultural homeland and place-based identity. For example, geographers have created a software program that uses Google Earth to enable Arapaho children in Wyoming schools to learn local place-names in their native language. Facebook, Internet radio and newspaper sites, and Web-based cultural organizations make it easier for immigrant members of Latin American and various Asian communities (Chinese, Indian, Vietnamese, Filipino, and others) to stay in touch with one another and with people in their home countries. Members of American Indian nations can now connect as never before with indigenous peoples in Australia, northeastern Siberia, and South America and in the process forge a new sense of their own place in a larger world.

The West's diverse cultural geography is the product of its settlement history and regional and global migration patterns. Native Americans have a long history of movement and encounters with many North American settings, dating from fifteen thousand to thirty-five thousand years ago. By the time Europeans began further complicating the region's cultural geography in the sixteenth century, Native people had already carved out place-defined relationships with one another and with the environment (fig. 4.4). In the early twenty-first century, these relationships rooted in place remain evident in everything from ancient mountaintop medicine wheels to contemporary reservation communities.

Similarly, Hispanic and Latin American people have had an enduring influence on western landscapes, creating one of North America's best examples of an ethnic province or cultural homeland—a region where a large geographical area has been shaped by a particular culture over several centuries. That influence can be seen in the design elements of a Mediterranean-style home

in the California suburbs (fig. 4.5) and in Latino street art on South Atlantic Boulevard in East Los Angeles (fig. 4.6). You can explore the special character of Hispano plaza towns, which have their roots in Spain's colonization of the upper Rio Grande Valley in the 1590s; the impact of the Spanish Colonial Revival architectural style, a popular early twentieth-century movement in California that draws on Spanish and Italian traditions; and the vitality of modern Latino communities, which reflects the influx of new immigrants from countries such as Mexico, Nicaragua, and Guatemala.

On a more local scale, there are ethnic enclaves in urban settings and ethnic islands in rural settings, smaller areas or neighborhoods that are dominated by a cultural group with characteristic social networks, landscapes, and practices. For example, European neighborhoods that began as clustered immigrant settlements persist through shared religion, remnants of language, and ethnic identity. Well-established communities of Chinese and Japanese Americans also have shaped the region over the past century, particularly in West Coast cities. A number of African American neighborhoods, also primarily in the urban West, were established between 1890 and 1950. Many of these communities, often revealing powerful forces of racial segregation by Anglo residents, were settled by southern blacks migrating westward for a better life. In addition, a growing number of western

FIG. 4.5 Mediterranean-style home, Glendale, California. This early twentieth-century home in Southern California's San Fernando Valley celebrates elements of the Spanish Colonial Revival and the region's long affiliation with Mediterranean cultural influences.

FIG. 4.6 Latino street art, East Los Angeles, California. Atlantic Boulevard's commercial landscape reflects the vibrancy of Southern California's largest Latino community.

FIG. 4.7 Mormon temple, Manti, Utah. Completed in 1888, the Gothic- and French Second Empire–style temple overlooks central Utah's productive Sanpete Valley and marks the enduring influence of the Mormon Church in the region.

localities (both urban and rural) became home to newer neighborhoods of various Asian immigrants who have arrived since 1970. Examples of these important enclaves include the immigrants in Monterey Park, Vietnamese communities in San Jose, and Hmong farmers in western Montana's Bitterroot Valley.

The American West has also been fertile ground for social experimentation, utopian dreams, and movements that flowered in relative isolation. An early example was the creation of Mormon country, which had its beginnings in the mid-nineteenth century when thousands of members of the Church of Jesus Christ of Latter-Day Saints fled persecution in the Midwest to establish a homeland in Utah (fig. 4.7). In a somewhat different way, the West—particularly in its more progressive cities such as San Francisco and Seattle—has been a formative setting for gay and lesbian communities, which have imprinted their character and diversity on regional landscapes (fig. 4.8). Countercultural impulses include rural utopian colonies anchored in sequestered corners of the West in places like the deserts of Southern California or the outback of central Oregon. More diffusely, the West also is a congenial home to New Age spiritual beliefs, many of which have strong ties to place.

All of these settings are found within a larger Anglo cultural and political landscape that established dominance in the West after 1840. History matters. Racialized landscapes of white privilege were violently imposed on Native peoples, African Americans, Latinos, Asians, and others across the

Fig. 4.8 The Palms Bar, West Hollywood, California. An unassuming visual landscape can represent a legendary cultural landmark. This West Hollywood dance club has long served the lesbian community, revealing its enduring role in shaping the city's cultural landscape.

West in the last half of the nineteenth century. This cultural imperialism was part of a larger transformation that included the geopolitical expansion of American power into the region and the imposition of a capitalistic economic system that forever redefined both the value of the natural environment in the West and the class relationships of the people who settled and worked there. Any twenty-first-century journey to the West's culturally distinctive places must be undertaken with that rather sober preamble. Latino barrios, African American suburbs, and Native American reservations are all expressions of that history and of all that has happened since to rework, redefine, reinforce, reverse, or erase imperatives forged in an earlier time. Western landscapes—including the remains of Wyoming's Heart Mountain internment camp for World War II–era Japanese Americans, Southern California's impressive Asian Garden Mall (a Vietnamese American business center in Orange County), and western Oregon's Native American–owned Spirit Mountain Casino—all bear tangible witness to that legacy. ❋

INDIAN COUNTRY

FIG. 4.9 Isleta Resort and Casino, New Mexico. A reimagined native landscape—complete with pueblo-style architecture and engineered stream and pond—welcomes visitors to this fashionable New Mexico resort near Albuquerque.

INDIAN COUNTRY FEATURES A CONSTELLATION OF DISTINCTIVE LANDscapes across the interior West (figs. 4.9–4.14). The Southwest is often thought of as the center of Indian country (the Automobile Club of Southern California has long published a map of the region by that name). In percentage terms, New Mexico has the highest Native population in the West (10 percent), and Arizona has the largest reservation acreage in the region (the Navajo nation alone contains more than seventeen million acres). Reservations are found throughout the West, particularly in Montana, Washington, Oregon, western Nevada, eastern Utah, and central Wyoming. California has the largest number of self-identified Native Americans of any state in the country (more than 350,000), the vast majority of whom live in cities such as Los Angeles, which has the nation's largest urban Indian population.

The archaeological record suggests that much of the Native West was initially peopled from northeastern Asia between 15,000 and 35,000 B.P. (before the present), and many of the region's archaeological sites date from 9,000 to 12,000 B.P. Maize agriculture arrived in the Southwest from Mesoamerica more than four thousand years ago. By the time of the first contact between Native people and Europeans, around 1500, there were many distinctive Native languages spoken across the region's varied ecological niches.

FIG. 4.10 Native American housing, San Carlos Indian Reservation, Arizona. This Apache community in southeastern Arizona remains one of the poorest in the state. A mix of institutional and vernacular building styles marks the scene.

The traditional lifeways of many western tribes still reflect these broad patterns of cultural and environmental diversity. The interior Southwest was home to both crop-growing Pueblo peoples—the Hopi (fig. 4.4) and Zuni, for example—and to ancestors of the Navajo and Apache, more recent arrivals from a western Canadian homeland. The Great Plains area was populated by bison-hunting peoples—the ancestors of the Blackfoot and Kiowa—and agriculturalists who grew crops along major rivers. The arid Great Basin stretching across the interior West was the homeland of small, mobile bands—the Paiute, the Goshute, and the Shoshone—who hunted and gathered within a challenging natural setting. The coastal California region—homeland of the Chumash, the Pomo, the Yurok, and others—was densely settled, an area where a complex environment offered a varied diet of nuts, seeds, animals, and marine resources. On the northwestern coast, the Siuslaw, the Chinook, and many other groups were connected to the ocean and rivers and worked the region's fisheries, especially to harvest salmon as a major part of their diet, which also included local game animals and edible plants. Similarly, in the interior plateau zone, ancestors of the Yakama, the Umatilla, the Nez Perce, and others supplemented their fish diets with hunting and gathering.

European contact brought disease, destruction, and disaster to most western tribes, whose populations were decimated by smallpox, measles, malaria, and other diseases for which they had no immunity. After years of conflict with whites who intruded on Native homelands to settle or to exploit the region's rich natural resources, by the end of the nineteenth century most of those who survived had been sequestered on tribal reservations—often far away from their homelands. The U.S. government forced many Native people to become farmers, prohibited traditional religious and spiritual practices, and sent children to boarding or reservation schools. Nevertheless, many practices and lifeways survived, as people worked to speak their languages, retain their social relations, and teach their children in traditional ways.

Among a long list of laws and regulations that shaped Indian country, several are particularly important in understanding the modern landscape. The

FIG. 4.11 Sky City, Acoma Pueblo, New Mexico. For almost a thousand years, Native peoples have called Acoma Pueblo home. Today, some residents still live atop the mesa, surrounded by a landscape constructed long before the arrival of pickup trucks.

Dawes Act, passed in 1887, divided many reservation lands into individual plots, prescribed strategies for Indian farming, and opened unallotted land to white settlement. As a result, significant parts of some reservations, such as the Flathead in Montana, are owned by non-Indians. The Indian Reorganization Act of 1934 recognized that tribes were sovereign nations and gave Native Americans who were members of those nations more control over their economic resources. That process was reinforced by the Indian Self-Determination and Education Assistance Act, passed in 1975. The Indian Gaming Regulatory Act of 1988 made it possible for tribes to own and operate casinos, transforming the lives of tribal members through new jobs and improved social services. New Indian casinos have also refashioned many reservation landscapes around the West (compare figs. 4.9 and 4.10).

Particularly around large reservations, the landscape can offer clues that address some important questions:

What cultural elements remain of the pre-European legacy? Local Native languages are evident in place-names, road signs (fig. 4.1), radio broadcasts, and conversations. Thanks to the Native language revitalization movement, many tribes have worked to document their language and teach it to their children. For example, Montana's Blackfeet tribe founded the Piegan Institute in 1987 in order to preserve the Blackfeet language and cultivate its use within the community.

The layouts of towns and neighborhoods, roads, and irrigation systems may also reflect very old patterns on the land. Hopi farming practices, for example (fig. 4.4), represent methods that long predate the agricultural innovations introduced by Europeans. In more subtle ways, because many Native people used burning for hunting and agricultural purposes, ecologists argue that much of the West's subsequent plant and animal life represents a fire-modified landscape that has been extensively changed by the long human presence in the region.

Preserved archaeological features include petroglyphs (rock engravings), pictographs (rock drawings), remains of settlements, and buffalo jumps, many of which are protected from tourists. Many sites located off reservations are in national parks and monuments (**66**). For example, western New Mexico's Petroglyph National Monument protects one of the largest petroglyph sites in North America and Dinosaur National Monument (in Utah and Colorado) contains accessible examples of both petroglyphs and pictographs. Native religions and beliefs are often closely connected with such settings and, more broadly, with the landscape itself—the importance of sky and earth, the significance of cardinal directions, and the role played by elements of the natural environment, including plants, animals, rivers, and mountain peaks (**6**).

How and where do people live on or near a reservation? More than 35 percent of Native Americans live on reservations. These settings range from tiny tribal reserves in Southern California to sprawling rural tracts that cover large areas of the western interior from eastern Washington to the Navajo Nation. Traditional houses such as hogans, wickiups, and tepees can still be seen on reservations, but they are rarely used as primary residences. At Acoma Pueblo in western New Mexico, a few dozen tribal members still live in Sky City, a thousand-year-old mesa-top settlement (fig. 4.11). For many, how-

FIG. 4.12 New housing subdivision, near Chilchinbito, Arizona. Almost ready for new families, this neighborhood in northeastern Arizona's Navajo Reservation exemplifies the mass-produced housing common across Indian country today.

ever, recently constructed federally subsidized housing creates a very different residential experience (fig. 4.12). In some places, mobile homes, trailers, and modular housing provide affordable options. Overall, reservation residents survive on household incomes that are 30 percent below the American average (fig. 4.10).

What are the expressions of tribal and federal authority? Both tribal and federal bureaucracies play a large role in Indian country. The Bureau of Indian Affairs (BIA) in the U.S. Department of the Interior is charged with administering and managing Indian trust lands. While tribes are considered sovereign nations (and tribal members are citizens of two countries), federal agencies within the BIA and the Indian Health Service (part of the U.S. Department of Health and Human Services) provide many essential resources to the Native American community and often have local offices on reservation lands. In addition, tribal law enforcement agencies, tribal government offices, and sometimes a tribal college or vocational school are common institutional elements of the landscape.

In many ways, Native and non-Native people also share common lives across the West. Near reservations are so-called border towns—such as Hardin, Montana, and Gallup, New Mexico—places where Indians and non-Indians interact on a daily basis. Local high schools feature a mix of students from Native and non-Native communities, tribal members often work and shop in these settlements, and tourist businesses, both Native- and non-Native-owned, take advantage of their proximity to reservation lands. On the reservation, signs of cross-cultural interactions include Catholic and Protestant churches, bilingual road signs (fig. 4.1), trading posts or roadside shops and stands selling Native arts and crafts, and a growing number of tribally run museums, hotels, and casinos (fig. 4.9). Visitors to reservations may also be able to take a guided tour of traditional settlements (fig. 4.11) or attend powwows, rug auctions (fig. 4.13), dance performances, and other activities that attract locals and tourists.

More generally, many non-Indians have used elements of Native culture

FIG. 4.13 Rug auction, Crownpoint, New Mexico. Local residents and visitors mingle at the monthly rug auction at the Crownpoint Elementary School in the Navajo nation east of Gallup.

FIG. 4.14 IHOP, Southwest style, Santa Fe, New Mexico. At this popular eatery on Santa Fe's commercial strip, earth-toned pueblo-style architecture with Native American design elements blends with the rocky simplicity of high-desert landscaping.

for other purposes, often mingling symbols, traditions, architecture, and music from disparate tribal groups to promote everything from franchise restaurants (fig. 4.14) to New Age spiritual practices (**48**). Navajo rugs, turquoise jewelry, representations of tepees and hogans, salmon iconography, flute-playing Kokopellis, and modern sweat-lodge cures are popular in different parts of the West, all part of a phenomenon known as ethnic commodification in which these traditional cultural symbols and practices are exploited and assigned commercial value.

The cultural landscapes of Indian Country tell complex stories that reflect rich traditions and complicated relationships with non-Native peoples. For all Indian country's cultural richness and vitality, many people there continue to struggle. While gaming revenues and natural resource industries have stimulated some tribal economies, most reservations are places of substandard housing, high unemployment, and enduring poverty. High teen suicide rates, alcohol and drug abuse, and other health issues (especially obesity and diabetes) are persistent concerns. Still, Indian country is changing, drawing strength from several factors that include increased economic opportunities, improved health care, a diversifying economy, forward-looking tribal leadership, and the increasingly powerful legal status of Indian nations. ❋

HISPANO PLAZA TOWNS

FIG. 4.15 Sangre de Cristo Catholic Church, San Luis, Colorado. Established in 1851, San Luis retains many characteristics of the Hispano plaza town, including this graceful Catholic church, which sits near the center of the small settlement.

SPANIARDS BEGAN TO SETTLE THE UPPER RIO GRANDE VALLEY IN 1598. They initiated a distinctive regional settlement pattern that included towns, missions, presidios (military forts), farming villages, and ranches. Hispano plaza towns, clustered settlements oriented around a central open space and a Catholic church, are signatures of that era (fig. 4.15). In New Mexico, Santa Fe (1610), the colonial capital of northern New Spain, and Old Town Albuquerque (1706) were larger plaza towns (called *villas*), but dozens of smaller settlements also appeared as the Hispano homeland expanded and became part of Mexico in 1821 and the United States in 1848.

The precise arrangement of plaza towns varies. Many feature a Catholic church and a collection of houses and shops that reflect grid, irregular, or linear layouts (figs. I.24, 4.15, and 4.16). Vernacular, earth-toned adobe and brick structures predominate, interspersed with metal sheds, trailers, and modular housing. There are also surviving traditions of log building in high-elevation Hispano settings, and some homes have adobe *hornos*, beehive-shape cooking ovens. Picket-style vertical fences and street-facing walls emphasize privacy and inward-looking residential landscapes.

Near these towns, you can often find long, narrow, irregular strips of farmland, fenced or lined with vegetation, that run between irrigation ditches

FIG. 4.16 Potrero Plaza, Chimayó, New Mexico. This settlement northeast of Santa Fe has an irregular layout, with housing clustered around a central Catholic church (El Santuario de Chimayó attracts three hundred thousand pilgrims annually) and nearby pastures (beyond the village) watered by the tree-lined acequia (irrigation works) near the base of the distant hills.

(acequias) and streams (**23**; fig. 4.17). These so-called long lots are allocated to village residents, though later patterns of land subdivision can complicate the scene.

Plaza towns are located close to the Rio Grande between Las Cruces and Albuquerque (Mesilla and Socorro, for example); to the east, along the Pecos River (Puerto de Luna, Dilia, Mora, Villanueva, and Anton Chico, as well as Las Vegas, where there is a classic plaza, fig. 4.17); northeast of Albuquerque (Chimayó, fig. 4.16, and Las Trampas); and San Luis Valley in southern Colorado (fig. 4.15). Santa Fe, which has been reinvented as a tourist town and artist colony (**97**), is a carefully manicured example of the Hispano plaza town.

Modern life has intruded into these towns as older farming practices waned and newly arrived Anglo residents (who often build upscale houses) bring a different set of values into these settings. In addition, new Latin American immigrants (**27** and **41**) can also come into conflict with traditional elements of the Hispanic community as they compete for jobs and assert their own cultural and political identities in these locales. ❋

FIG. 4.17 Highland Hispano long lots, Villanueva, New Mexico. The tree-lined Pecos River in the distance forms the southern boundary of the long, narrow irrigated fields located just west of the village.

40 SPANISH COLONIAL REVIVAL ARCHITECTURE

FIG. 4.18 Catholic Church of the Immaculate Conception, Old Town San Diego, California. Completed in 1917, this structure celebrates the Spanish Colonial Revival tradition. The church tower (left) houses one of the original bells from nearby Mission San Diego de Alcalá (1769).

THE TWO EVENTS WERE SEEMINGLY UNRELATED. IN 1884, HELEN HUNT Jackson published the novel *Ramona,* a nostalgic look back at California's Spanish missions and quaint Mexican adobes. Three years later, Florida developer Henry Flagler completed his grand Ponce de León Hotel in Saint Augustine, a resort designed in high Spanish Renaissance style to recall Florida's Iberian glory. Both reimaginings of Spain's legacy in the United States resonated with the public. They were harbingers of a flowering of Spanish Colonial Revival architecture, most designed by Anglos, which shaped many coastal California communities, particularly from San Diego to San Francisco. The revival never recognized the region's contemporary Mexican culture, preferring to identify instead with a more remote, mostly fictional past.

The elements of the revival crystallized at San Diego's Panama-California Exposition, held between 1915 and 1917. Inspired by the exposition, the next generation of Southern and central California's public buildings, railroad depots, churches, commercial structures, and homes combined characteristics of Spanish Baroque, Mission, and Moorish Revival styles, along

with vernacular pueblo and rancho elements (fig. 4.18). The legacy of those decades is in place-names (including a plethora of Ramonas) and in a blending of architectural traditions that includes white stucco and earth-toned structures; low-pitched clay-tile roofs; arched windows, walkways, doors, and bell towers; circular Moorish-style domes; decorative balconies and iron grillwork; and a landscape of Mediterranean plantings (such as cypress trees and Mediterranean dwarf palms), fountains, decorative tile and red brick, outdoor patios, and arcaded courtyards (figs. 4.5 and 4.18–4.20).

Stylish archways and tile roofs dominate Santa Barbara, California, where city fathers decreed, after an earthquake in 1925, that new downtown construction would reflect that look. Many subsequent redevelopment efforts in the city, including the Paseo Nuevo midtown commercial district in the 1980s, reinforced the enduring visual power of Revival architecture. You can also see signs of this tradition in San Diego's Balboa Park, the site of the exposition; in the Mission Inn Hotel and Spa in Riverside, California; at Stanford University and many other college campuses; and in the Pima County Courthouse in Tucson. Based on the style's popularity in exclusive California communities such as Carmel, Montecito, San Clemente, and Avalon (on Santa Catalina Island), the Spanish Colonial Revival's architectural appeal was embraced in dozens of middle-class communities from Glendale to Fresno (fig. 4.5). Decades later, the Revival is evident in upscale shopping centers such as Tlaquepaque in Sedona, Arizona, in master-planned housing developments (**79**; fig. 4.20), and in many retirement communities (**98**). ❋

FIG. 4.19 Inner courtyard, San Francisco Art Institute. Tiled roofs, arched walkways and windows, bricked courtyard, Moorish-tiled fountain, Mediterranean plantings, and a nearby mural (Diego Rivera artwork inside) make this 1920s-era Bakewell and Brown–designed building a classic celebration of the Spanish Colonial Revival.

FIG. 4.20 Nouveau Mediterranean-style home, Verrado, Arizona. This post-2000 Spanish Colonial Revival home, built in a planned subdivision west of Phoenix, suggests that this architectural impulse remains alive and well in the twenty-first-century West.

41

LATINO COMMUNITIES

FIG. 4.21 Latino market, Toppenish, Washington. The hybrid wording in Mi Favorita Market suggests how this central Washington farm town is changing and how the West's Latinos have reshaped the regional landscape.

MI FAVORITA MARKET IS A NEIGHBORHOOD GROCERY STORE (FIG. 4.21) in the farm town of Toppenish, Washington, far from Latin America. Or is it? Since about 1960, the growth of Latino communities has been one of the most transformative forces in the West. Latinos account for 29 percent of the West's population, making it the most latinized region of the United States (fig. 4.22). California (38 percent) and New Mexico (46 percent) are the most latinized parts of the West, but populations in both Arizona (30 percent) and Nevada (27 percent) have had spectacular increases since 1990. Even Washington (11 percent) and Idaho (11 percent)—far removed from the Mexico-U.S. border—are home to large Latino populations. Numerically, the Latino heartland is in Southern California, where more than 4.7 million Latinos live in Los Angeles County alone, more than the entire population of Costa Rica.

The roots of communities of Spanish speakers in the West extend deep into the past (**39**), and their legacy remains indelibly etched across the landscape, particularly in the Southwest. Between 1890 and 1930, however, newer immigrants, many of them from Mexico, began arriving in the West, drawn by opportunities in mining, agriculture, and industry (**27** and **31**). During World War II, the bracero program also welcomed workers from Mexico to the

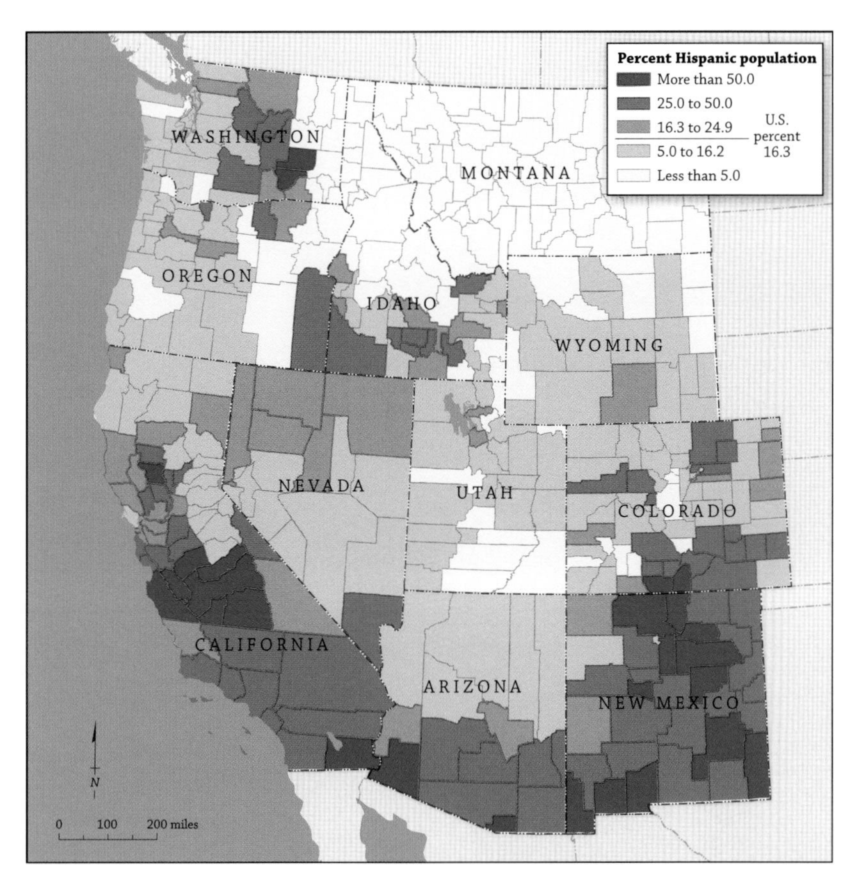

FIG. 4.22 Hispanic or Latino population as a percent of total population, by county, 2010. The latinization of the West is clearly shown, both within the border states and beyond. *Source: U.S. Census Bureau, 2010 Census.*

United States to help fill demand for manual laborers. Workers from Central and South America also migrated northward to find jobs. The numbers of people moving north accelerated after the Immigration and Nationality Act of 1965 reduced barriers to non-European immigration.

Their destinations included agricultural zones such as California's Imperial and Central valleys, western Oregon, central Washington, southern Idaho, southern New Mexico, and eastern Colorado. An even greater number of people were drawn to cities. East Los Angeles (more than 90 percent Latino, according to the U.S. Census in 2010) and nearby areas now make up the nation's largest Latino community, a diverse, sprawling landscape of several hundred square miles and more than two million people. You can find other examples of large and vibrant Latino communities in Phoenix, Tucson, Denver, Albuquerque, Las Vegas, San Diego, Fresno, and San Francisco.

Some Latino neighborhoods (also called barrios) began as early twentieth-century worker settlements, known as *colonias*. Over time, the neighborhoods that were not razed to make room for Anglo suburbs and other development attracted additional migrants in search of low-cost housing. Latino communities in Pacoima and Santa Ana, California, are examples. Other barrios are the result of what social scientists call barrioization, the external and often discriminatory pressures of housing markets, capital investment, and urban development, along with the blatant anti-Mexican racism that isolated,

LOTUS
NIGHTCLUB & VIP ULTRALOUNGE
211 GOLD AVENUE
(505) 243-0955
WWW.LOTUSABQ.COM
FREE EXERCISE OF RELIGION FREEDOM OF SPEECH AND OF THE PRESS; THE RIGHT TO ASSEMBLE AND TO PETITION THE GOVERNMENT

MEXICA-TESSEN

marginalized, and spatially concentrated Latino populations. As Latino communities grew, many nearby white homeowners moved elsewhere (a process known as residential succession), allowing the barrio to expand. This occurred, for example, in the older industrial suburbs of southeastern Los Angeles. Finally, the place identity of Latino communities, particularly since the 1960s, has been better defined and enhanced by the notion of barrio identity, which cultivates a sense of pride and ownership in these communities. Cultural geographer Lawrence Herzog argues in his book *From Aztec to High Tech* that this is an era when "we find an increasing shift in consciousness by the Latino population toward viewing their neighborhoods as a kind of valued cultural and social space" (p. 124). A good example is the flowering of urban muralism, a traditional practice that celebrates Latino art and identity in public places (figs. I.16, 2.10, 2.43, and 4.23).

FIG. 4.23 *Frutos de la Expresión*, downtown Albuquerque. This example of urban muralism (completed in 2000) celebrates the liberating impact on the region's population of the First, Fifteenth, and Nineteenth amendments to the Constitution. Next door (left), note the Asian presence of the Lotus Nightclub.

Barrios are dynamic and diverse. In Los Angeles, a five-minute drive along South Atlantic Boulevard will take you from the heart of a Chinese ethnoburb (**46**) to a Latino community. Not far away in Van Nuys, well-defined communities of Mexican, Salvadoran, and Guatemalan immigrants live on different streets and in separate apartment blocks. To the north, Mexican immigration has transformed the ethnic makeup and cultural landscapes of many Central Valley farming towns, a process paralleled in locales such as the westside suburbs of Phoenix, southside Tucson, and South Tucson. In San Francisco's Mission district (fig. 4.24), an early Mexican community was settled by European immigrants (Italians, Irish, Poles, and others) in the late nineteenth century, and then a new Latino barrio took shape there after World War II.

Brightly painted houses line many residential streets in older Latino communities, where a preference for color and display are evidence of the residents' roots in Mexican and older indigenous and Iberian cultural traditions. Latinos now live in the Victorian, bungalow, or ranch-style housing originally constructed for Anglos (**75** and **76**). Some of these newer residents have also added low walls, grillwork, and fencing that enclose front-yard space but permit interaction with the sidewalk and street (fig. 4.25). The practice reflects cultural preferences for courtyards and enclosed yet accessible space and is seen across the West in traditional plaza towns and in high-style Colonial Revival designs (fig. 4.19; **39** and **40**). Yard and porch spaces in these neighborhoods are highly personalized, often decorated with religious shrines, plants, and furniture.

Many businesses cater to the Spanish-speaking population. *Taquerías* offer a quick street-front meal, nail and beauty salons cater to current Latina fashion, and *carnicerías* and *fruterías* serve as meat shops, produce markets, and grocery stores. These shops are also meeting places where community news and gossip are exchanged. Informal retailing is another characteristic element of the Latino landscape: street vendors, food carts and trucks (fig. I.11), and flea markets are all grounded in Latin American cultural traditions of the public market, or *mercado*. Day laborers often head to the street as well, using a designated place where they can connect with potential employers. Cultural signatures also include traditional Catholic churches, storefront religious outlets (fig. 4.26), cultural centers, and *botánicas* (or *yerberías*), which sell religious statuary and herbal remedies. The landscape is often decorated with more casual graffiti and street art, which can include gang tagging and

FIG. 4.24 Cultural mixing in the Mission district, San Francisco, California. Once a working-class neighborhood for immigrant Europeans, the Mission district is now home to San Francisco's largest Latino community. La Palma Mexica-tessen is found along 24th Street. Murals adorn nearby walls.

FIG. 4.25 Front yards, South Tucson, Arizona. Low fences, walls, and grillwork define the boundaries of small front yards on Tucson's south side. Residents may interact with people on nearby sidewalks and streets, yet they shape the personalized spaces of their distinctive yards and porches.

political protests on public signs, utility boxes, and trash bins as well as whimsical works of creative expression (fig. 4.6). There are also signatures of poverty, poor urban planning, and what are called "disamenities." Latino communities are often close to freeways, airports, industrial and toxic-waste sites, and a host of other land uses that wealthier members of the urban population avoid.

Latino neighborhoods are transnational communities. Many residents maintain close ties to their villages and neighborhoods in their home countries of Mexico, El Salvador, Colombia, or elsewhere. They travel between their homes in the American West and their hometowns and they also send money (known as remittances) to their families back home. Many shops advertise that they perform money transfers, or *envíos de dinero,* and some businesses also sell phone cards, change currency, cash checks, and offer travel services, all of which help maintain connections between countries. Shops display Latin American national flags and colors (Mexico's is white, green, and red) as well as use Latin American place-names (such as the Mexican states of Sinaloa, Michoacán, and so on) in business names and advertisements.

Finally, these landscapes are parts of transcultural communities in the West, where Latino cultural elements mix with other cultural traditions and influences. The borderlands region in the Southwest has been described by geographers as a setting for such cultural transactions, and you can see evidence in the blending of architectural traditions (fig. 4.25); the use of Spanglish (a combination of English and Spanish, figs. 4.21 and 4.24); regional music, festivals, and cooking styles; and the daily flow of money, labor, and products across one of the world's busiest international borders (**59**). These dynamic interrelationships—often displayed boldly on the landscape—are reshaping both the Latino community and the cultural geography of the entire American West. ❋

FIG. 4.26 Iglesia La Luz del Mundo, Douglas, Arizona. This storefront church in the southern Arizona border town of Douglas represents a growing religious movement within the Latino community. Based in Guadalajara, Mexico, the controversial Light of the World Church is a charismatic offshoot of Catholicism with many members north of the border.

42

MORMON COUNTRY

FIG. 4.27 LDS church and community complex, Manassa, Colorado. Established around 1880, this southern Colorado Mormon community boasts a handsome modernist-style ward chapel that remains the center of the community.

WHILE THE CHURCH OF JESUS CHRIST OF LATTER-DAY SAINTS (LDS) has changed greatly since it was founded in 1830 by Joseph Smith, the location of Mormons in the Great Basin has created a distinctive cultural landscape still apparent today (fig. 4.27). In July 1847, Mormon leader Brigham Young guided about 150 people to the fertile, well-watered base of northern Utah's Wasatch Mountains. Refugees from persecution (Smith was murdered by a mob in jail in Carthage, Illinois, in 1844), Young and his followers domesticated the semiarid landscape, laid out towns, dug irrigation ditches, and planted crops. Salt Lake City, the colony's primary settlement, was organized around central squares and had wide, cardinally oriented streets (**70**) based partly on the model of the City of Zion sketched out by Smith.

The Mormons envisioned a State of Deseret that would reach from the Wasatch Mountains to the Pacific coast. That vision spawned a pattern of settlement expansion that sent Mormons to many corners of the West, both north and south of Salt Lake City and as far away as Southern California and western Nevada. Although the State of Deseret was never established—the smaller Utah Territory created in 1850 was admitted as a state in 1896—the Mormon expansion was enduring, particularly across Utah and in eastern Idaho (fig. 4.28).

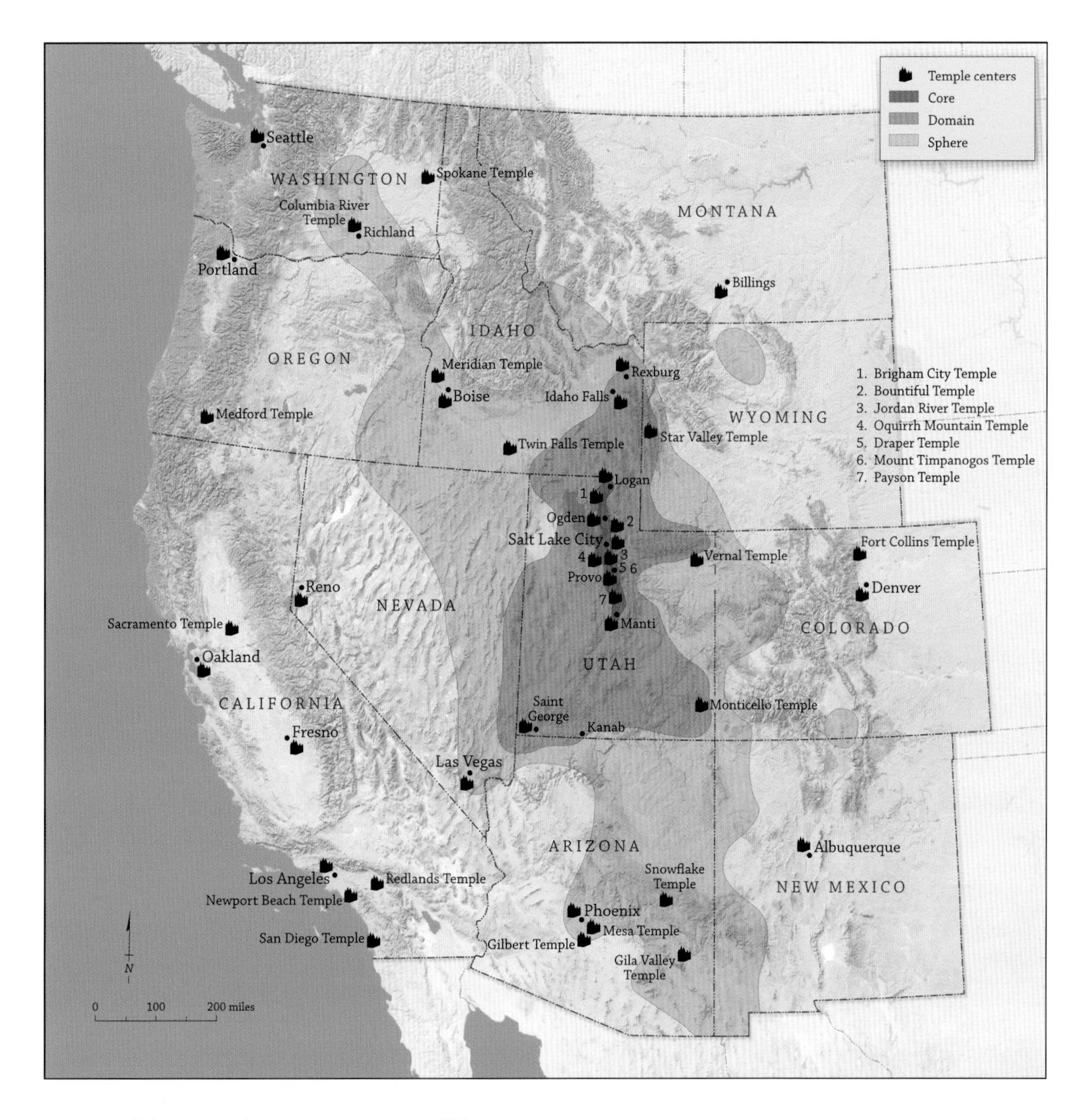

FIG. 4.28 Mormon country. While Utah and southeastern Idaho remain primary areas of Mormon influence, growing numbers of adherents live in major metropolitan regions. U.S. highway 89 transects Utah from south (Kanab) to north (Logan). *Sources: D. W. Meinig, "The Mormon Culture Region: Strategies and Patterns in the Geography of the American West, 1847–1964."* Annals of the Association of American Geographers *55 (1965): 191–219; Paul F. Starrs, "Meetinghouses in the Mormon Mind: Ideology, Architecture, and Turbulent Streams of an Expanding Church,"* Geographical Review *99 (2009): 323–55; and ldschurchtemples.com/maps.*

One of the most obvious signatures of Mormon country is the large regional temple. Salt Lake City, still the spiritual core of Mormonism and headquarters of the church, remains home to the impressive Salt Lake Temple. Highly visible temples are also found not only in traditional Mormon strongholds such as Manti (fig. 4.7) or Logan, Utah, but also in Las Vegas and Los Angeles.

The best way to see the traditional Mormon landscape is to explore small towns and rural settings, focusing on areas that are still predominantly LDS. One excellent Utah transect is Highway 89 between Kanab and Logan (fig. 4.28). Along

FIG. 4.29 Corner of 200 East 200 South, Spring City, Utah. This quiet Mormon village in central Utah includes a standard grid and street-numbering system, large in-town lots, many brick houses, and unpainted wood outbuildings.

this route, you can find Mormon influences in Panguitch, Circleville, Manti (fig. 4.7), Spring City (fig. 4.29), and Mount Pleasant (fig. 2.41).

You will find many Mormon-inspired place names—Zion, Nephi, Deseret, Manti, Moroni—and symbols—the beehive, representations of the Salt Lake City Temple, the Angel Moroni, and the Book of Mormon.

A number of standard elements appear in and around Mormon towns. A ward chapel, or meetinghouse, usually occupies a large, impeccably landscaped lot. The architecture of early chapels varies, but meetinghouses built since the 1950s are vaguely colonial Georgian or modernist, often brick, with a standardized school-like layout (often wings with a separate chapel complex, classrooms, gymnasium, and meeting rooms), and a simple steeple or spire (fig. 4.27). A rectangular grid—with many local variations—with square blocks, large in-town lots, and wide streets tends to characterize Mormon towns. Standardized numeric street-naming systems are also common (fig. 4.29), similar to the one in Salt Lake City.

Rural and urban landscape elements intertwine in these settings, based on the church's colonization strategy of providing to each family an in-town lot large enough for a home, barn, garden, and livestock, which were also pastured on land outside town. In small Mormon towns, you will see barns (typically unpainted), fencing, sheds, even farm animals that are ordinarily associated with dispersed farmsteads in the country (fig. 4.29). While most Mormon farmers have relocated to isolated farmsteads (**19**), these in-town relic elements provide an open, low-density feel to traditional Mormon towns. In older parts of town, the residential architectural landscape emphasizes a more widespread use of brick and stone (and even adobe) than in most western towns (fig. 4.29), early symmetrical colonial-style Federalist and central-hall houses with end chimneys, as well as more traditional homes with Greek, Gothic Revival, and Victorian elements (fig. I.12).

If you widen the landscape lens, you can see an intricate network of irrigation canals and ditches both in towns and around dispersed farmsteads (**23**; fig. 4.30). The Mormons cooperatively constructed these systems, mindful of the scarcity of water in the Great Basin. Many earthen in-town ditches have been altered or abandoned in favor of more modern technologies. Relic rural landscapes include often-abandoned Scandinavian-style hay derricks (pole booms used for stacking hay), untidy picket fences, unpainted barns, obsolete farm equipment, and planted rows of Lombardy poplars.

You can also see the ways in which Mormon country is changing. Vacant houses and businesses suggest a landscape of abandonment and economic decline. In favored farming districts, more intensive, specialized forms of livestock and crop production have largely replaced traditional, diversified agricultural practices. Non-Mormon retirees, tourists, and amenity seekers are changing places like Saint George and Cedar City in southern Utah, where you can see a mix of Mormon, Californian, and southwestern influences. Just as some elements of the LDS church are becoming global (its rapid ascent, for example, in Latin America), changes in the West appear likely to dilute and redefine the visible legacy of Mormon country in its Utah heartland. ❋

FIG. 4.30 Mormon ditch system, Rockville, Utah. In smaller Mormon towns, watch for remnants of community-managed irrigation systems. In Rockville (near Saint George), neighbors share water for flood-irrigating lawns and gardens through a system of street-lined ditches and gates (foreground) that allow users their allotted amount of water.

PERSISTING EUROPEAN COMMUNITIES

FIG. 4.31 German Russian house, Fort Collins, Colorado. German Russian immigrants came in sizable numbers to northeastern Colorado after 1890 This German-Russian-style home, complete with gingerbread porch, central gable, and front balcony, is on the city's north side, not far from the former site of a sugar-beet factory.

SOMETIMES THE SIGNS ARE SUBTLE, AS IN THE UNASSUMING SIGNAtures of German Russian immigrants (German dissenters who lived in Russia after the 1760s and immigrated to the United States starting in the 1870s) in older neighborhoods of Fort Collins, Colorado (fig. 4.31), or in eastern Washington on a thriving colony of Hutterite (German Anabaptist) farmers engaged in twenty-first-century agribusiness (fig. 4.32). More visibly, street murals in Elko, Nevada, celebrate that city's Basque community (fig. 4.33).

These European communities persist in both urban enclaves and rural neighborhoods. Unlike Latinos and Mormons, they lay no claim to larger regional identities but are islands of self-identified ethnicity within contemporary American culture. Some communities represent a long presence in a place, while others have been shaped by more recent immigrants. European communities have two interrelated imprints: their material expression is revealed in place-names, structures, festivals, and visible reminders of heritage; and their identification with place secures for them a meaningful cultural connection with particular localities.

FIG. 4.32 Hutterite farm near Ritzville, Washington. This large, successful eastern Washington Hutterite farm includes elaborate irrigation works (center and right foreground), outbuildings for equipment storage (left), and colony complex and school (distant center and right).

You can see the European West in western cities and mining towns (**31**). It is visible in synagogues; Russian- and Greek-Orthodox churches; and some Catholic parishes, bars, and restaurant districts (such as Little Italy in San Diego and San Francisco). Communities of Russians live in West Sacramento and San Francisco's Richmond district; Butte, Montana, has Irish bars and Cornish pasty shops; Slovenes, Greeks, and Italians live in Pueblo, Colorado; Los Angeles has communities of European Jews, Russians, and Armenians; and Seattle's Ballard district still reflects its Scandinavian heritage.

Enduring rural communities include Scandinavians in eastern Montana, where you will find Lutheran churches and Norwegian lutefisk suppers. You will also see surviving German Russian communities in the South Platte (eastern Colorado) and Yellowstone (Montana) valleys. Elsewhere, Hutterite farms and rural settlements in Montana, eastern Washington, and Oregon often produce premium-quality agricultural goods for commercial sale. In California's Central Valley (explore Hanford and Gustine), Azorean Portuguese dairymen remain important parts of local communities. Also watch for expressions of Basque settlement, such as hotels and restaurants, in Boise, Elko, Reno, and Bakersfield. In the high country of the Great Basin, you can also see Basque rock cairns and tree carvings.

FIG. 4.33 Basque Mural, Elko. This northeastern Nevada town celebrates its Basque heritage with public art, ethnic eateries, and an annual Sheepherder's Ball.

European ethnicity is celebrated across the West at festivals and gatherings through foods, dances, and games. Red Lodge, Montana, has a Festival of Nations; Solvang, California, celebrates Danish Days; and dozens of parades take place in the West on St. Patrick's Day and Columbus Day. You will also see synthetic ethnic landscapes, contrived places that have no longstanding ethnic identity. In Leavenworth, Washington, for example, the town uniformly adopted Bavarian architecture in the 1960s to attract tourists. ❋

44

AFRICAN AMERICAN NEIGHBORHOODS

FIG. 4.34 African American business, Seattle, Washington. The long, narrow shotgun-style house that is home to Awa's African Braids Express and Beauty Supply reflects a building type common in the South, source-region for many of Seattle's post-1900 African American migrants.

African American neighborhoods offer another reminder of the diverse cultural geography of the West and of how large-scale forces produce enduring local patterns and place identities. The subregional geographies are striking: Montana, Idaho, and Wyoming remain among the states with the smallest number of African American residents, while California is home to the largest number in the region, more than 2.2 million. Other western states with sizable African American populations include Colorado (Denver), Washington (Puget Sound), Arizona (Phoenix), and Nevada (Las Vegas).

While there is a long and rich history of African Americans on the western frontier—for example, as forty-niners in California, military scouts, cowboys, Buffalo soldiers, and sodbusters—the large-scale migration of southern blacks to the West occurred mostly in the twentieth century. African American communities coalesced in urban areas between 1890 and 1930, often displacing older European immigrant populations (**43**). But World War II was the great watershed, when more than two hundred thousand African Americans came west, many from southern states such as Texas and Louisiana.

FIG. 4.35 Beebe Memorial Christian Methodist Episcopal (C.M.E.) Church, Oakland, California. This large cathedral was built in 1966 and reflects enduring connections between the C.M.E. Church and the city's African American community.

Thousands worked in shipyards and at other industrial jobs and established neighborhoods in South Los Angeles; Oakland, Richmond, and Hunters Point in the Bay Area; and the Central district in Seattle (figs. 4.34 and 4.35). Large interregional migrations between the South and the West continued between 1945 and 1965, a period when the number of African Americans in the West more than tripled.

Prior to 1970, white racism, police repression, legal restrictions such as race-based housing covenants, and de facto segregation practices by real estate agents and bank lenders combined with the lower income levels of black migrants to severely limit their residential options. Los Angeles, which had the sixth largest African American population in the United States in 1960, was one of the most segregated cities in the nation. The Watts riots in Los Angeles in 1965 and the creation of the Black Panther Party for Self-Defense in Oakland in 1966 are reminders of roles played by the region's African American communities during the turbulent 1960s.

After 1970, changes in federal law—including the Civil Rights Act (1964), the Voting Rights Act (1965), and the Equal Employment Opportunity Act (1972)—created new opportunities for employment, housing, and mobility. In large metropolitan areas, African Americans left older inner city neighborhoods for the suburbs, moving, for example, from South Los Angeles to Chino, Tustin, Rialto, Palmdale, and Victorville. Bay Area residents in Hunt-

FIG. 4.36 Medgar Evers Pool, Central district, Seattle, Washington. This area, near Garfield High School and the Garfield Community Center, serves as a hub for Seattle's African American community. During the 1950s and early 1960s, Evers (left) was a pivotal civil rights activist in the South.

ers Point and Oakland left for the suburbs or more distant destinations, and Seattle inner-city residents migrated to the suburban Rainier Valley. The dispersal of people is also regional, with African American communities in cities such as West Las Vegas, Fresno, and West Phoenix receiving new residents. Growing Latino (**41**) and Asian American (**46**) populations have also accelerated the exodus of African Americans from some communities. In addition, whites who want to live in central-city neighborhoods have priced some African Americans out of what had been affordable housing.

African American geographies in the West are changing as nonblack residents, especially Latinos, move into established neighborhoods and newer black communities remain relatively small. Nevertheless, the landscapes of enduring African American neighborhoods have some common characteristics. The central role of religion in the community is visible in the prominence of churches, including long-established Protestant African American denominations (fig. 4.35) as well as charismatic and Pentecostal organizations that may operate out of storefronts. The Nation of Islam and other Muslim groups have also built mosques in these neighborhoods. Many of these institutions have extended their reach beyond the pulpit, operating community development corporations to fund local businesses, providing food and clothing to those in need, and offering family support services.

Public celebrations of African American identity are visible in street names (consider how many cities have named streets and schools after Martin Luther King Jr.), public art (fig. 4.36), museums and cultural centers, and events, parades, and celebrations that recognize black history, music, and culture.

The enduring and transcultural appeal of African American food and music is legible in restaurants, clubs, and performance venues. Restaurants reflect traditional ties to the South—such as the longtime popularity of Bertha's Soul Food in Los Angeles and Ezell's Famous Chicken in Seattle—and other businesses reveal connections with newer African immigrant communities, such as Little Ethiopia along Fairfax Avenue in Los Angeles. In many forms

FIG. 4.37 Gateway Supermarket, San Pablo Avenue, Oakland, California. A local institution on Oakland's north side, the Gateway serves the area's African American community.

of music, African American heritage has also deeply influenced mainstream American culture. You can hear jazz, rhythm and blues, and soul music in clubs and on the radio, along with hip-hop, rap, and Afro-Caribbean music.

A vernacular landscape, mainly visible to local residents, helps define the character of neighborhoods. Grocery stores (fig. 4.37), nail parlors, and beauty salons (fig. 4.34) not only offer basic goods and services but are also places where people meet and exchange news. The evidence of gang territories, often visible in distinctive graffiti and street art, is another component of social geography that is a reality in many African American communities.

A racialized landscape of poverty, particularly in larger, older neighborhoods, is evident in substandard housing, poor access to health care and educational services, and the toxic environmental effects of nearby industrial facilities (active or decommissioned), all of them reinforced by a set of social relations that produce high rates of African American unemployment, crime, and drug abuse (**74**).

The legibility of African American neighborhoods will no doubt change with the times, but the communities that endure will reflect national priorities as well as the identity of people who have shared a common cultural history in some of the West's most challenging and troubled places. What is also clear, however, is that these regional settings have been key crucibles of national political change and that they are likely to continue to play a vital role in defining the larger multicultural character of both the American West and the country as a whole. ❋

45 JAPANESE INTERNMENT CAMPS

FIG. 4.38 Remains of the hospital heating plant, Heart Mountain Relocation Center, Wyoming. Only wind in the sagebrush breaks the silence at Heart Mountain (butte, distant left). In the early 1940s, ten thousand Japanese Americans lived at the camp.

ANGULAR HEART MOUNTAIN OVERLOOKS THE WINDSWEPT SHOSHONE Valley (fig. 4.38). Surrounded by ranching country, the nearby community of Heart Mountain was once Wyoming's third largest city with 10,767 residents, most of them Japanese Americans who were imprisoned there. From August 1942 to November 1945, Heart Mountain was one of eight major Japanese internment camps in the West.

The removal of Japanese Americans to internment camps from what was called the excluded zone along the Pacific Coast was set in motion by Executive Order 9066 in February 1942. Seen as a security threat after the Japanese attack on Pearl Harbor on December 7, 1941, about 110,000 people—most of them second-generation American citizens, or Nisei, from California—were forced to report to assembly centers at racetracks, fairgrounds, and migrant labor camps before being shipped to internment camps in the interior West (fig. 4.39). Forced to sell or abandon their homes and businesses, thousands of people left their neighborhoods in Seattle, suburban Los Angeles, and California's agricultural Central Valley. In addition to Heart Mountain, they were

FIG. 4.39 Japanese American internment in the West. This map shows the Japanese American exclusion zone, initial assembly centers near areas of settlement, and the relocation centers or internment camps located in the interior.

sent to isolated camps in western Arizona (Poston and Gila River), the Great Basin (Manzanar, Tule Lake, Minidoka, and Topaz), and southeastern Colorado (Granada). The camps were laid out on a grid plan (**70**), with schools, hospitals, stores, farms, and crude barracks-style housing. Most camps operated until late 1945.

After the camps were closed, the land was returned to federal agencies or private owners. Most of the buildings were reused, moved, or made available to veterans, and the remaining structures were left to weather away (fig. 4.38). In the late 1980s, surviving internees received reparations and an apology from the U.S. government. As surviving artifacts on the western landscape, the camps at Heart Mountain, Manzanar, Minidoka, Tule Lake, and Topaz constitute an important chapter in U.S. history and in the region's Asian American experience (**46**).

Beyond foundations and structures, there are also landscapes of remembrance and representation at the camps. At the Manzanar National Historic Site near Lone Pine, California, for example, the National Park Service offers interpretive exhibits and auto tours. The camp serves as a reunion site for internees and their families (fig. 4.40). Other camps have plaques, exhibits, and restored buildings, creating a landscape and collective memory designed to heal wounds inflicted at a time when fear trumped understanding. ❋

FIG. 4.40 Monument to console the souls of the dead, Manzanar, California. Located near the camp's western edge at the small cemetery (Sierra Nevada in background), the monument at Manzanar National Historic Site is visited by surviving internees and their families.

46

EMERGENT ASIAN MOSAIC

FIG. 4.41 Morning meditation, Seattle's international district. A small park plays host to practitioners of Falun Dafa (also called Falun Gong), a Chinese meditation tradition. The city's mostly Asian American international district is just southeast of downtown.

THE WORK AND LIFE OF BLACK EYED PEAS ARTIST APL.DE.AP (ALLEN Pineda Lindo) illustrate one aspect of the intimate twenty-first-century connections between Asia and the American West. Born in Barrio Sapang Bato, a suburb of Angeles City in the Republic of the Philippines, Lindo moved to Los Angeles, California, when he was fourteen years old. His hip-hop lyrics reflect the hard life he lived on the streets of the city, but he persevered and adopted California as his home. His story is part of an emergent Asian mosaic that includes everything from kung fu movies to new sources of entrepreneurial capital.

The modern Asian imprint on the West is strikingly focused. About one-third of the nation's Asian Americans—about five million people—live in California, with 70 percent in the Los Angeles metropolitan area or the Bay Area. There are also sizable and diverse Asian American communities in Seattle, Portland, Sacramento, San Diego, and Denver (fig. 4.41). Newer pieces of the mosaic can be seen in Phoenix, Fresno, Las Vegas, and Albuquerque.

A quick history lesson: It was primarily Chinese laborers, living in conditions of extreme hardship and discrimination, who built the railroads and mined many of the region's streams and hillsides in the nineteenth-century West. By 1880, more than a hundred thousand Asian immigrants lived in the region. Legal barriers such as the 1882 Chinese Exclusion Act slowed the influx, and the National Origins Act of 1924 severely limited Asian migration, reflecting an attitude of suspicion that culminated in the wartime internment of Japanese and Japanese Americans who lived along the coast (**45**).

The tide turned dramatically in 1965 with the passage of the Immigration and Nationality Act, which led to greater legal immigration from Asia, especially by professionals, skilled workers, and people with family already in the United States. The Refugee Act of 1980 amended the Immigration and Nationality Act, making it easier for refugees from Vietnam, Cambodia, and elsewhere to begin new lives in this country. The U.S. Immigrant Act of 1990 further increased limits on legal immigration to the United States, opening the way for more high-tech engineers from Bangalore, India; budding millionaires from Taiwan; and relatives of Filipinos like Allen Pineda Lindo.

Once Asian immigrants arrived and settled in the West, the powerful force of chain migration encouraged them to settle in neighborhoods near relatives and friends, where they could find social and economic support. As in Latino communities (**41**), people maintained close connections with their families and hometowns in Asia through social networking, visits, and remittances.

Part of the region's Asian American landscape can be found in traditional central-city agglomerations such as San Francisco's Chinatown and Japantown, though many Chinese now live in the Richmond and Sunset districts of the city and many Japanese live in nearby San Mateo. Central Los Angeles has its own Chinatown, Little Tokyo, and Filipinotown in the Echo Park district. In Koreatown, west of downtown Los Angeles, several thousand businesses owned by Korean Americans (fig. I.10) cater to more than a hundred thousand Korean residents and other (mainly Latino) populations. Seattle's International district may enjoy the greatest variety of Asian Americans, including people with families from China, Vietnam, Thailand, Laos, Japan, Korea, and elsewhere (fig. 4.41).

Asian Americans have also established large, economically successful suburban communities far beyond the central cities, typically aligned along major

FIG. 4.42 Little India, Artesia, California. Strip malls along busy Pioneer Boulevard are the commercial focus for Southern California's largest South Asian community.

FIG. 4.43 Asian Garden Mall, Westminster, California. Part of Little Saigon, this large, enclosed shopping mall and business center in Southern California is home to noodle (phở) shops, music stores featuring recordings of popular Asian singers, and restful altars reserved for religious meditation.

commercial boulevards and anchored by shopping centers, grocery stores, and strip malls (fig. 4.3). In Southern California, these communities include Little India in Artesia (fig. 4.42); Little Saigon in Orange County, where about 10 percent of the nation's Vietnamese Americans live (fig. 4.43); North Hollywood's Wat Thai complex (the largest Thai Buddhist temple in the United States); and across the San Gabriel Valley, from Monterey Park eastward to Rowland Heights and Diamond Bar. To the north, Chinese, Indian, and Vietnamese American neighborhoods dot the Silicon Valley (**82**) and nearby communities. Similar suburban expressions of different Asian cultures shape westside Las Vegas: a trip out Spring Mountain Road west of town can lead to dim sum, Korean barbecue, and Vietnamese noodle restaurants. Most sizable western American cities offer similar examples.

Distinctive signs and conversations on the street add character to an Asian-American neighborhood. You can sample the restaurants from Thai Town in West Los Angeles and Cambodia Town in Long Beach to Little Arabia in Anaheim and Little Saigon along Sacramento's Stockton Boulevard. Other commercial and financial services also cater to Asian populations, including grocery stores, whether they be mom-and-pop operations, large chains like Shun Fat Supermarket or 99 Ranch Market, or Seattle's wondrous Uwajimaya marketplace.

Every Asian American community has annual parades and festivals as well as cultural centers, monuments, temples, churches, and parks. Also be aware of more subtle clues. There are driving ranges in places like Koreatown in Los Angeles (golf is popular in the Korean community), and Indian-owned motels in every major city in the region (a common business choice for many South Asian immigrants). In addition, there are Asian architectural touches added to structures and gardens, and in affluent residential districts, some wealthier Asian Americans have torn down old houses and erected high-end mini-mansions on small lots.

The West also reveals many examples of how diverse Asian influences have blended with one another and with other cultural elements within the region. For example, suburban apartment buildings in Monterey Park have been reinvented with subtle Chinese signatures (fig. 4.44). Nearby, in Koreatown, you can buy Korean-style tacos. Also catering to changing American tastes in food, Laotian immigrants in California's Central Valley have opened Thai and Vietnamese restaurants in Visalia. Throughout the West, Asian fusion cooking combines spices and food-preparation methods from countries as far-flung as China, Indonesia, Mexico, and Morocco.

Asian American influences in the West are not limited to its cities. Rural signatures include Buddhist monks living in the Colorado outback, Southeast Asian bok choy farmers in California's Central Valley (more than 20 percent of Fresno County farmers are Asian), and Laotian gardeners (many are displaced Hmong) who have unlocked the fertility of western Montana's Bitterroot Valley.

By 2050, analysts with the U.S. Census Bureau predict, about 10 percent of the nation's population will be Asian American. New migrants—many from Asia's growing cities like the Philippines' Barrio Sapang Bato—will make the journey, heading east to arrive in the West. ❋

FIG. 4.44 Apartment building, Monterey Park, California. This Monterey Park apartment building illustrates how a structure may be reinvented to reflect newer ethnic landscape tastes. Note the bamboo plantings, colorful tiling, decorative columns, and ideographic lettering ("a peaceful, auspicious home and garden").

47

GAY AND LESBIAN NEIGHBORHOODS

FIG. 4.45 Castro district, San Francisco. Since the 1970s, San Francisco's Castro district (anchored by the Castro Theatre) has retained its character as one of the nation's best-known gay communities.

A FIXTURE IN WESTERN AMERICAN CITIES DURING MUCH OF THE TWENtieth century, the region's well-established gay and lesbian neighborhoods—sometimes called "gayborhoods"—are changing. While there were earlier visual expressions of gay and lesbian place identity, especially in San Francisco, clearly recognizable neighborhoods flowered across the United States starting in the 1970s. Because of broader social changes, greater acceptance, and economic incentives—that is, gay and lesbian residents seeking less expensive, older urban neighborhoods—places like San Francisco's Castro district, West Hollywood, Denver's Capitol Hill, and Seattle's Capitol Hill grew into some of the country's largest LGBT neighborhoods. The Castro (fig. 4.45) in particular assumed national prominence in gay political activism, punctuated by the 1977 election of Harvey Milk, an openly gay politician, to the city's Board of Supervisors, and his assassination just one year later.

Gay- and lesbian-friendly neighborhoods also became more legible in Portland, Phoenix, Boulder, Salt Lake City, Albuquerque, San Diego, and Sacramento. Smaller towns such as Palm Springs and Guerneville, California,

FIG. 4.46 Strip mall, Hollywood, California. Free HIV testing and the Out of the Closet Thrift Store are part of the local retailing landscape at this Hollywood strip mall.

as well as Ashland, Oregon, also developed reputations for being welcoming vacation destinations for gays and lesbians.

The semiotics (signs and symbols) of these localities sometimes plays on vampy or kitsch stereotypes but is often more subtle. The symbolic landscape includes the rainbow flag, shades of lavender and purple, and the use of "Pride" or "LGBT" on bumper stickers, yard signs, and businesses. Repositioned gender symbols, tongue-in-cheek wordplay, and suggestive place-names—from the unambiguous Moby Dick bar in the Castro to the more nuanced Here dance club in West Hollywood—are parts of a sometimes whimsical, yet purposeful built environment. In neighborhoods, you can find renovated housing, strip malls, cafés, and hotels that cater to gay and lesbian patrons (fig. 4.46). Along with stereotyped landscapes of leather bars, gyms, and dance clubs (fig. 4.8), there are annual parades, public art, and social services that include AIDS counseling, gay and lesbian service centers, and same-sex adoption counselors.

Twenty-first-century transformations are changing the West's gay and lesbian communities. Some traditional gay neighborhoods are being gentrified, diluting their identity within larger culturally diverse urban settings. Many gay couples also raise families in socially diverse neighborhoods throughout the West, especially in larger urban and suburban settings. Elsewhere, the graying of gay America has spawned a growing number of specialized retirement communities (**98**) in smaller places such as Santa Fe, New Mexico; Santa Rosa, California; and Gresham, Oregon. ❋

48 COUNTERCULTURAL IMPULSES

FIG. 4.47 Housing complex, former Big Muddy Ranch commune, near Antelope, Oregon. Now a Christian youth camp, from 1981 to 1985 this isolated central Oregon complex housed several thousand devotees of the controversial Bhagwan Shree Rajneesh, who advocated group sex and attempted to take control of the local government.

THE AMERICAN WEST HAS LONG BEEN A PLACE FOR UTOPIAN EXPERIMENTS. The region has been the setting for the mingling of ideas from many traditions, and a tolerance for and interest in everything from Montana's armed militias to California's Esalen Institute. Place figures importantly in these stories, and perhaps the West's open spaces allow for the unusual and the unexpected. In exploring these countercultural landscapes, there are several variants to keep in mind.

Some countercultural groups are geographical isolates that are eager to be off the grid, free from interference from state and federal authority. The Home of Truth, for example, was a utopian community established in a remote area of southeastern Utah in the 1930s; the Llano Del Rio commune was a socialist experiment started near Lancaster, California, in 1914; and southern New Mexico's Shalam colony began in 1884 as a utopian community for children. Recent iterations include eastern Oregon's orgiastic Big Muddy Ranch commune of the Bhagwan Shree Rajneesh (fig. 4.47, now a Christian youth camp); the still-thriving Tassajara Zen Mountain Center in central California; and

the well-armed Church Universal and Triumphant in southwestern Montana, just miles from the entrance to Yellowstone National Park.

In some settings, supportive cultural milieus provide fertile ground for such communities. Bohemian lifestyles in California, from Berkeley and Haight-Ashbury to Carmel, were reinforced in the 1960s and 1970s by the hippie movement, drug culture, and radical politics. You can find other countercultural impulses in Southern California (fig. 4.48); on Puget Sound in Washington; in Oregon's Willamette Valley; in Boulder, Colorado, and other high-country incubates; in Santa Fe and Taos, New Mexico; and in places as disparate as Missoula, Montana, and Arcata, California. There are frequently connections between these countercultural hotbeds and broader regional arts communities (**97**). Often, you can also see how these settings incorporate trendy versions of traditional Asian practices (**46**), which might include everything from shiatsu massage and Kundalini yoga to Zen meditation and Krishna consciousness.

Countercultural currents often swirl around places that are considered sacred. Some people, for example, believe in the healing power of vortexes (fig. 4.49), focal points of cosmic energy that attract modern-day pilgrims. Some of the region's best-known vortexes are near Sedona, Arizona, and Mount Shasta, California. More broadly, setting has been an active agent in many countercultural localities such as California's Big Sur or Mendocino. And part of the countercultural cachet of states such as Montana (the "Last Best Place") and New Mexico (the "Land of Enchantment") is defined by the appealing environments these settings have within the popular imagination. They are seen as special places that offer attractive opportunities to practice New Age healing, cleansing, spiritual exploration, vision quests, and other promised paths to self-awareness. ❋

FIG. 4.48 Hazy Moon Zen Center, West Los Angeles. Housed in a fine English-style Victorian house, the Hazy Moon Zen Center is a reminder that many Eastern-inspired practices of meditation and education have blossomed in Southern California's diverse cultural setting.

FIG. 4.49 Vortex parking lot, Sedona, Arizona. This parking lot near Sedona is busy as visitors hike to the site of the famed vortex in the nearby hills. Thousands of New Age pilgrims are drawn annually to this beautiful red-rock country.

CONNECTIONS

FIG. 5.1 Connections along the Columbia River near Wallula, Washington. Road, rail, and river are reminders that the Columbia is the region's greatest natural gateway into the western interior.

MUCH OF THE WESTERN AMERICAN LANDSCAPE IS DEVOTED TO CONnecting one place to another. Whether it be the maze of interstate highways—the roads, on-ramps and off-ramps, rest areas, service centers, motels—or the infrastructure of railroad lines—tracks, depots, water towers, grain elevators, maintenance shops, and railroad towns—the West is interlaced with connections that link the once-isolated region with the rest of the world.

Place identity in the West also has been constructed around many of these linkages. Much of the region's legacy and character is rooted in the stories, artwork, and symbols of movement—the Lewis and Clark Expedition, the Oregon Trail, the Donner Party, the Pony Express, the transcontinental railroad, the cross-country road trip. These shared stories are integral parts of American culture and defining elements of the western landscape.

Between 1790 and 1940, a short 150 years, the West witnessed a transformative spatial reorganization. In 1790, the best way to get around in the region was to follow waterways and the trails that linked them. The great drainage basins of the West were its interior highways (fig. 1.27). For centuries, Native people had aligned their seasonal movements—for trade, warfare, hunting, and gathering—along these natural thoroughfares, following the Missouri, the Platte, and the Columbia rivers (fig. 5.1). An intricate geography of overland trails crisscrossed between watersheds long before Lewis and Clark or other Euro-American explorers asked Indians for their help in negotiating the region's complex terrain. Spanish, English, Russian, and American expeditions mapped out the coastal geography, and by 1790 maritime charts were bringing into focus features such as San Francisco Bay and the mouth of the Columbia River. Outside of the Spanish-controlled Southwest, Natives held most of the region's geographical cards.

By 1860, on the eve of the Civil War, two generations of American exploration and a surge of American settlement had produced a great deal of new geographical knowledge, changed the balance of power in the West, and reconfigured the regional landscape. New connections, both for migration and for commerce, linked the West with the rest of the country. Thousands of

FIG. 5.2 Toll road above Ophir, Colorado. Southwestern Colorado's rugged San Juan Mountains proved challenging for man, beast, and freight wagon. This nineteenth-century view (near modern Telluride) shows the tenuous nature of early mountain roads. Remnants of pioneer toll roads remain visible today. *Courtesy of Denver Public Library, Western History Collection, image no. C-139.*

FIG. 5.3 Railroads in the West, circa 1910. By the end of the nineteenth century, the West was well connected to the rest of the country via a growing number of east-west trunk lines. Within the region, numerous feeder lines added density to the network.

travelers had experienced the drudgery of the Santa Fe, Oregon, California, and Mormon trails. In places like central Wyoming and northeastern Oregon, remnants of these pioneer trails are still visible, either as ruts in the soil or as routes followed by modern roads and highways.

In addition to immigrant trails, the federal government designated overland mail and stage routes (the Pony Express began in April 1860) and wagon roads, all designed to promote development and provide essential connections. A network of mountain roads probed the high country, particularly where there was mineral wealth, and private entrepreneurs in Colorado built toll roads (fig. 5.2), convinced that such alpine connections could pay. Few turned a profit.

To identify the best routes through the West, the federal government undertook the Pacific Railroad Surveys, which were completed in the 1850s. The Pacific Railroad Act followed in 1862, leading to the completion of the

FIG. 5.4 Navigating the Arrowhead Trail, southwestern Utah, late 1920s. Unpredictable road conditions, poorly marked highways, flat tires, and overheated radiators challenged many a cross-country motorist. This was the main road between Los Angeles and Salt Lake City in the late 1920s. *Courtesy of Utah State Archives, Utah Road Collection, Washington County, image no. 117.*

nation's first transcontinental line in 1869. More rail building followed, creating a trackside landscape that included more cross-country connections and a network of narrow-gauge lines that penetrated the mountains.

By the late nineteenth century, the key transport connections in the West were essentially defined by this new technology (fig. 5.3). The railroads also produced their own peculiar, booster-oriented image of the West as a promised land of opportunity for tourists, businessmen, and farmers. Railroads reworked basic notions of time and distance, reoriented landscapes, created new economic relationships, sparked the creation of hundreds of towns (many of them later abandoned), and facilitated and guided settlement. Subsequent connections built around commercial trade, communications lines, and modern highways more often than not simply followed the tracks.

By 1940, while railroads continued to dominate the movement of freight, trucks and automobiles were taking advantage of a growing network of highways. The highway system had come together in piecemeal fashion from local initiatives. Between 1912 and 1925, privately promoted "auto trails" created the first long-distance routes in the West, including the Lincoln Highway, the National Old Trails Road, the Pacific Highway, the National Park-to-Park Highway, and the Yellowstone Trail. But road surfaces were unpredictable, federal aid to highways was in its infancy (the Federal Aid Road Act was passed in 1916), and the entire system was confusing (fig. 5.4). Finally, between 1925 and 1930, the nation moved toward a broader federal and state network of standard numbered routes and traffic signs. By the beginning of World War II, the massive efforts to improve and pave these roads had brought people in the West closer together.

Postwar efforts to build an integrated national highway system produced the standardized interstate landscapes we see today (fig. 5.5). Many of the older routes became bypassed highways that are still used by commuters or, like U.S. Route 66, attract their own brand of leisure traveler. Americans loved their automobiles and had an equal zeal for the attractions of the open road, particularly in the West. In urban areas, automobiles redefined suburban space, and the region led the nation in the sprawling expansion of metropolitan areas.

After 1940, commercial air travel blossomed across the West, further reducing the travel time between cities. Along the Pacific, the improvement of coastal connections, much of it facilitated and shaped by government funding and in response to new regulations, created a predictable landscape of wharves and jetties, ferry boats and lighthouses.

If we take a broader view of connections, energy pipelines and irrigation and water projects also make modern life possible in the West. Just as essential is the electrical grid that powers millions of western localities. Newer networks of fiber-optic cables (most of them buried) and other communications technologies (increasingly wireless) also connect the West, enabling everything from daily cell-phone conversations to banking transactions.

The West, with its abundance of open space, has even been congenial ground for exploring intergalactic connections. For decades, an area of New Mexico west of Socorro has been home to the VLA (Very Large Array) astronomical radio observatory, which explores deep space for radio waves (fig. 5.6). Elsewhere in the state, Roswell is claimed to be the site of a 1947 UFO crash, and the Spaceport America initiative chose a landscape north of Las Cruces to build the world's first commercial gateway to outer space. You can also journey along eastern Nevada's Extraterrestrial Highway or visit Wyoming's Devils Tower National Monument, home to Hollywood's famed landing spot in *Close Encounters of the Third Kind*. ❋

FIG. 5.5 Interstate 82 north of Yakima, Washington. Product of the nation's ambitious interstate highway system, this soaring bridge spans a deep canyon in central Washington's rugged volcanic tablelands and makes travel easy between Yakima and Seattle.

FIG. 5.6 Out-of-this-world connections: Very Large Array, New Mexico. Clusters of radio antennas dot the landscape west of Socorro at the National Radio Astronomy Observatory. They make up one of earth's most sophisticated listening devices. Sparsely populated New Mexico is ideal for connecting with worlds far beyond the solar system.

49 HISTORIC TRAILS

FIG. 5.7 Santa Fe Trail marker near Wagon Mound, New Mexico. The National Historic Trails network is a federal initiative that identifies, interprets, and preserves many pioneer-era connections across the West. Here, the Santa Fe Trail crosses highway 120 in this quiet, windswept corner of northern New Mexico.

SOMETIMES HISTORY IS A JOURNEY, AND HISTORIC TRAILS IN THE WEST offer a wonderful way to capture a glimpse of the past and to reflect on why these early expressions of the region's modern musculature still matter. Trails were key corridors of commerce and immigration, and they often paved the way for railroad routes (**50** and **51**) and interstate highways (**55**). Most important, the myths and romance associated with historic trails shaped the West's regional identity. The trails became the focus of stories captured in artwork, books, and film, as well as dozens of museums, interpretive centers, and trailside exhibits and landmarks (fig. 5.7).

About a mile south of Guernsey in eastern Wyoming, there is a hill dotted with stunted pine trees. The Oregon Trail runs through this location, and you can see where the wagons of 350,000 to 400,000 travelers wore deep ruts into the sandstone ridge (fig. 5.8). This is one of the best localities in the West to see a tangible trail landscape.

Historically, different types of trails crisscrossed the West. Settlers used trails of immigration to move west. Between 1843, when John Charles Frémont popularized the route, and 1869, when the transcontinental railroad was completed, the Oregon Trail was the interstate of its day. Large parts the route also served as integral segments of the California Trail and the Mormon Trail (**42**; fig. 5.9).

FIG. 5.8 Oregon Trail ruts, Guernsey, Wyoming. This spot at Guernsey State Park remains one of the best places in the West to see the remains of pioneer trails on the twenty-first-century landscape.

Travelers left Independence, Missouri, in the spring and headed northwest to the Platte River near present-day Kearney, Nebraska. Following the North Platte River, they passed landmarks such as Chimney Rock and Scotts Bluff before arriving at Fort Laramie in early summer. With luck and good planning, they made it past Independence Rock west of Casper, Wyoming, by July 4—you can climb the rock to read their inscriptions in the stones—before rolling toward South Pass, a low and open traverse of the Continental Divide. From there, branches angled northwest across the Snake River plain and Blue Mountains toward Oregon or bent southwest through the Salt Lake Valley, across the Great Basin, and over the Sierra Nevada to California.

Some trails are routes of exploration. The Corps of Discovery, led by Meriwether Lewis and William Clark, made one of the West's best-known exploratory traverses between 1804 and 1806, traveling from Saint Louis to the

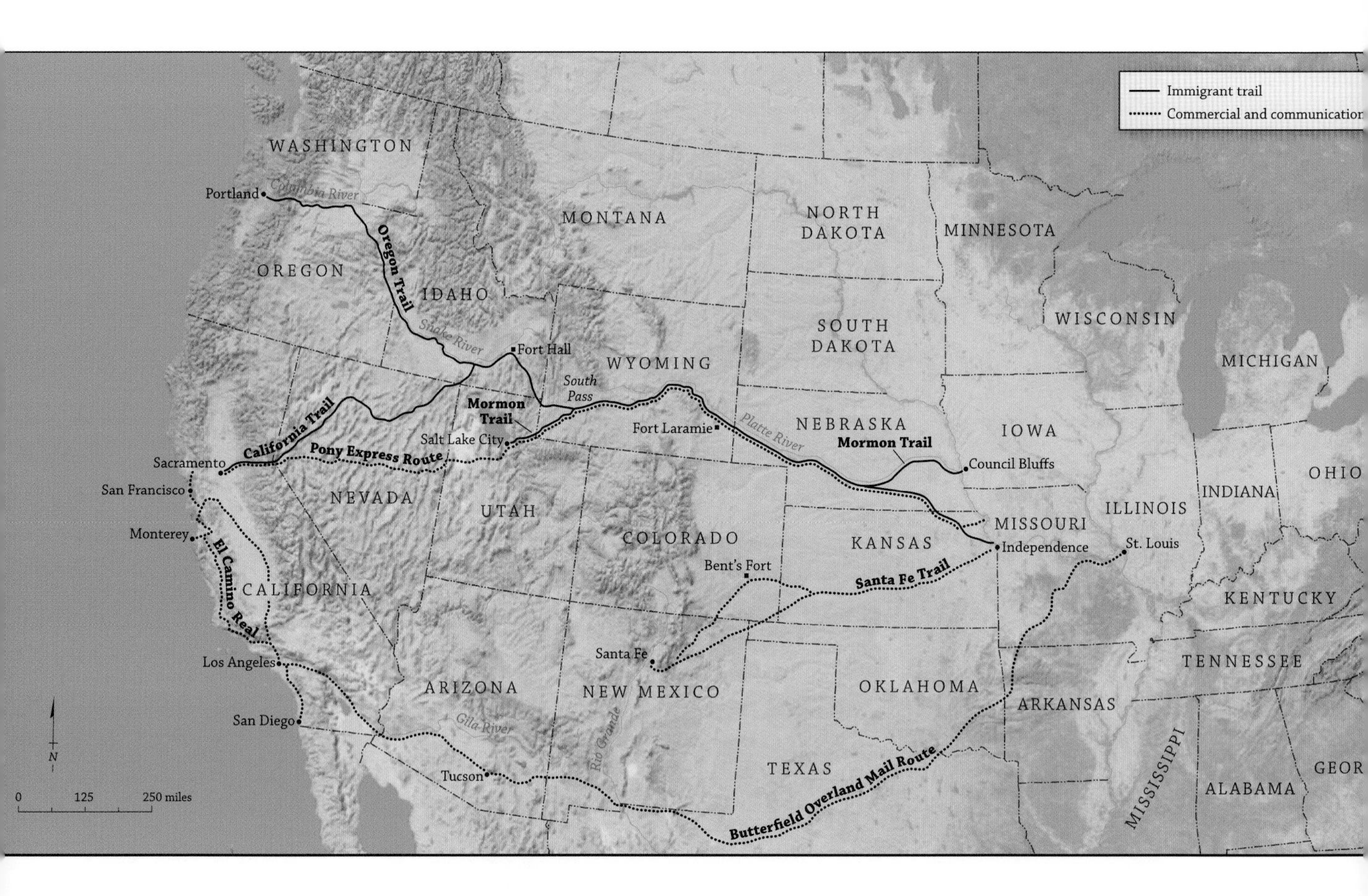

FIG. 5.9 Selected western trails. Several of the West's key historical connections are shown on this map. Note the central role played by the Platte Valley and South Pass (in Wyoming), particularly for migrants headed to Utah, Oregon, and California.

Pacific coast. The journey is described at interpretive centers on the Missouri River (including one near Great Falls, Montana), at Fort Clatsop on the Oregon coast, and at Montana's Pompeys Pillar National Monument, where you can see Clark's hand-carved signature on a rock near the Yellowstone River. In the interior Southwest, the journeys of Spanish explorers from Coronado (1540–1542) to Domínguez and Escalante (1776–1777) are described in museums, on landmarks, and in books.

Trails of commerce and communication were the pivotal commercial conduits in the West, and the trade they generated helped forge enduring economic relationships. Often, the first linkages were later reinforced by railroad lines and modern highways. Roadside signs and monuments identify these trails, and their significance is described in local museums, state and national park units, and interpretive material.

The Santa Fe Trail, for example, linked the Midwest with northern New Mexico between the early 1820s, when an independent Mexico fostered American trade, and the 1870s, when regional railroads captured the trade (fig. 5.9). In southern Colorado, Bent's Old Fort National Historic Site, near the Arkansas River, is a good place to get a sense of the trail's commercial importance. Farther west, the Old Spanish Trail carried nineteenth-century trade goods to Southern California. California's El Camino Real, much of which runs north-south along U.S. 101 (figs. 5.9 and 5.10), linked Spanish settlements and missions from San Diego to San Francisco. Curiously, the styl-

ized mission bells that mark the trail were part of a promotional campaign of mostly Anglo women who wanted to celebrate the Spanish Colonial Revival (**40**) during the state's early automobile era.

More fleeting, but still retaining great regional cachet, are the ephemeral mail routes used to send communications westward, especially to California. In the Southwest, the Butterfield Trail blossomed briefly in the 1850s, taking overland mail along a southern route before the Civil War disrupted the connection. Farther north, you can trace the Pony Express Route (fig. 5.9), a mail route followed in 1860 and 1861 by fast-moving relay riders on horseback. Express riders could take a letter from Saint Joseph, Missouri, to Sacramento, California, in ten days. Remnants of Pony Express stations, originally spaced about ten miles apart, are found in many places across the West. The service ended in 1861 with the introduction of the telegraph. ❋

FIG. 5.10 Mission bell, El Camino Real, Morgan Hill, California. Southern California women's clubs campaigned to preserve portions of the historic "King's Highway." By 1915, more than 150 bells marked the route that once connected Spanish settlements and missions. Many survive today.

50

TRACKSIDE

FIG. 5.11 Depot and tracks, San Luis Obispo, California. For hundreds of western towns, the railroad depot was the main connection to metropolitan America. This Spanish Colonial Revival structure (built in 1942) still serves Amtrak passengers along California's scenic Pacific Surfliner and Coast Starlight lines.

THE ECHO OF WESTERN RAILROADS CAN BE PICKED UP IN MANY WAYS. The depot was the hub of railroad movement, commerce, and travel (fig. 5.11), connecting communities to the rest of the world. The depot represented a new kind of western space, an efficient, rational, urbanized world of linkages that historian John Stilgoe refers to as the "metropolitan corridor." These corridors produced new economic geographies, and they knitted the West together with the rest of the nation.

It is not always easy to find the depot in a town. Many have been abandoned or torn down, and others have been moved or now perform other functions (a practice referred to as adaptive reuse) such as serving as chambers of commerce, museums, and restaurants. The depot's size, construction materials, and architectural style reveal a great deal about the history of both the companies that owned them and the residents who used them. A large brick-and-stone structure suggests lofty expectations for a town, while a smaller wood-frame building sufficed for a more modest trackside burg. Many Santa Fe Railroad depots in the Southwest featured Mission Revival flourishes (**40**), while larger depots along the Milwaukee Road boasted impressive brick structures with Romanesque detailing and distinctive clock towers. Elsewhere, companies built their depots using standardized formats.

FIG. 5.12 Aerial view of Casa Grande, Arizona, circa 1930. The town's layout parallels the track's angled alignment (bottom) and sets it off from the surrounding cardinal orientation (top) of the township-and-range survey system. The trackside depot (lower center) and water tower (far right) are visible with downtown just across the street. *Courtesy of Arizona Historical Society/Tucson, image no. PC196F351–01.*

Water towers still survive near the tracks (fig. 5.12), along with grain elevators (**22**) and other storage and processing facilities. Railroad hotels typically were built near depots. Those that survive, such as La Posada Hotel and Gardens in Winslow, Arizona, are a reminder of the role railroads played in promoting the West as a national playground.

To take a wider view, explore the town and its relation to the depot and tracks. Railroad towns, frequently platted by the company or its corporate interests, are typically oriented to the rail line, sometimes interrupting the geometry of the surrounding, cardinally oriented township-and-range survey system (**58**). In some settings, commercial business lots form a downtown district parallel to and across the street from the tracks (fig. 5.12). In other towns, a *T*-town alignment was used, with a primary commercial street running perpendicular to the rail line (fig. 2.40). Regardless, these basic patterns of street orientation and lot subdivision often organized the landscape in ways that long outlived the commercial viability of the rail line itself.

Also think beyond the depot and the town to more strategic regional and national alignments. The tracks and depot are local expressions of continental-scale geographical imperatives. Each town and depot, each main line and spur, formed a part of an elaborate corporate strategy. Railroad networks were the complex expressions of perceived economic risks and opportunities. This was a carefully engineered world: everything from grades, crossings, tunnels, and trestles to more abstract estimates of regional profits and losses was planned and calculated, creating a railroad landscape and a larger financial calculus that attempted to impose modernity and order upon the world.

How did the logic of that larger geography evolve? Ceremonies at northern Utah's Promontory Summit (today the Golden Spike National Historic Site) on May 10, 1869, sealed the marriage between the West and the railroad. That first transcontinental line, built by the Union Pacific and Central Pacific railways with federal land subsidies, linked Chicago, Omaha, and San Francisco. Competing east-west alignments followed, including the Southern Pacific and Santa Fe lines to California (1882, 1885), the Northern Pacific to Tacoma (1883), the Great Northern to Seattle (1893), the Milwaukee Road to the Puget Sound (1909), and the Western Pacific to Oakland (1909). These great east-west links became key components of the region's railroad network (fig. 5.3), representing capital investments that opened regional markets to economic development and settlement. Narrow-gauge lines, capable of penetrating smaller markets and more varied terrain, also connected remote landscapes to these larger regional systems (**51**).

The way these strategic alignments worked can be understood by following a surviving railroad corridor. Montana's Hi-Line (U.S. 2) parallels the Great Northern line; U.S. 40 follows part of the original Union Pacific system between Denver and Cheyenne Wells; and U.S. 350 and U.S. 50 run alongside the Santa Fe line between Trinidad and Lamar, Colorado. Older trackside settlements were designed to service the trains (steam locomotives required frequent water and fuel); to sell railroad-owned land at appreciated prices; and, most important, to stimulate commercial use of the line. The West's settlement geography reveals many of these linear town alignments, and the

FIG. 5.13 Abandoned rail line north of Condon, Oregon. Can you see the old track bed curving gently toward you from the upper left? Between 1905 and 1993, a spur line (part of the Union Pacific system) once ran from the Columbia River to Condon.

decline of many railroad lines has occurred as farming areas and farm towns have lost population (**26**). Thousands of miles of rail line have also been abandoned, a function of shifting economic realities (fig. 5.13).

Watch for the trains themselves—along transcontinental routes or in coal country—and experience the visceral thrill of the approaching whistle and the earth rumbling as the cars hurtle by. You can also take advantage of heritage railroads, where public and private partnerships have refurbished trains and tracks for tourists (fig. 5.14). The railroad's power to shape place identity survives, even as the economic geographies that created it have been realigned. ❋

FIG. 5.14 The Virginia and Truckee Railroad, Virginia City, Nevada. This famous short line in western Nevada has enjoyed a recent renaissance and is now a popular heritage railroad for tourists near Carson City.

51 NARROW-GAUGE RAILROADS

FIG. 5.15 Colorado's Georgetown Loop, west of Denver. This early twentieth-century view highlights one of the West's narrow-gauge wonders. Today, the engineering marvels of the Loop can still be enjoyed near Georgetown, Colorado. *Courtesy of Denver Public Library, Western History Collection, image no. MCC635.*

IT'S A QUINTESSENTIAL WESTERN EXPERIENCE: CLIMB ABOARD A narrow-gauge railroad car, listen to the steam whistle, and feel the measured jerk of the cars as the train heads up the track. Such rail lines have played a special role in linking western places and in defining their identity.

The track width of a narrow-gauge line is less than the standard gauge of 4 feet 8½ inches (**50**), with many smaller lines using a 3-foot gauge. Narrow-gauge systems were less expensive to build than standard-gauge lines, and they worked better in mountainous terrain where the curves were sharper and the grades steeper (fig. 5.15). In Colorado's high country, William Jackson Palmer's Denver and Rio Grande Railroad became the West's most elaborate narrow-gauge network after 1870. These Colorado lines, mostly oriented around mining, seemed to perform engineering miracles as they snaked and tunneled themselves into what were once some of the West's most isolated alpine settings. Places such as Creede, Crested Butte, Aspen, and Silverton suddenly became connected with the rest of the world, all thanks to these narrow-gauge railroads. Other mining (**28** and **29**) and timber (**33** and **34**) industries were drawn to the technology, from Arizona to California and the Pacific Northwest. The Carson and Colorado Rail Company, for example, probed mineral-rich desert areas in western Nevada and

eastern California (fig. 5.16), and the Oregon, California and Eastern Railroad operated a line out of Klamath Falls, Oregon (fig. 5.17).

Many lines had short life spans or converted to standard-gauge tracks. Most were abandoned, although the right-of-way sometimes survives in different form, thanks to programs such as the Rails-to-Trails Conservancy, which have reworked selected lines for recreational purposes (fig. 5.17).

No narrow-gauge line enjoys greater fame than the Durango and Silverton Narrow Gauge Railroad, which is part of the original Denver and Rio Grande system. The line runs for forty-five miles between Durango and Silverton through the San Juan Mountains, much of it along the Animas River. Nearer Denver, the Georgetown Loop Railroad (fig. 5.15) has also been revived, as have the Cumbres and Toltec Scenic Railroad in southern Colorado and northern New Mexico, the Sumpter Valley Railway in Oregon, and the Roaring Camp and Big Trees Railroad near Santa Cruz, California. ❋

FIG. 5.16 Engine no. 9, Laws Railroad Museum, Laws, California. Note the narrow track width (compare with fig. 5.14) along the remnants of the Carson and Colorado Rail Company line. The line ran from Mound House, Nevada (near Carson City), to the mining town of Keeler, California (near Lone Pine) between 1883 and 1960.

FIG. 5.17 The Oregon, California and Eastern Woods Line State Trail, Klamath Falls, Oregon. The hundred-mile-long linear park is Oregon's longest, following the course of the abandoned narrow-gauge line from Klamath Falls to beyond Bly, Oregon.

52

THE OPEN ROAD

FIG. 5.18 Cruising the open road on the Old Lincoln Highway (Nevada State Route 722) between Carroll Summit and Austin. This highway, once a main thoroughfare across the Silver State, is now an uncrowded stretch of western bliss. Enjoy the ride!

FOR MANY AMERICANS, THE OPEN ROAD BEST CAPTURES THE ESSENtial character of the West—unfinished, open-ended, a marriage of the human psyche with the earth, sky, and highway. The genius of the open road lies in its simplicity, reducing to a bare geometry of space and form all the possibilities that a full tank of gas can offer (fig. 5.18).

The modern aesthetic of the open road—with its celebration of individualism and freedom—blossomed between 1915 and 1935 as Americans learned to love their automobiles and motorcycles. Open roads drew Jack Kerouac on his treks between Denver and San Francisco. They defined the blue highways that so delighted William Least Heat-Moon in the West. The open road inspired Bobby Troup, who wrote "Get Your Kicks on Route 66" in 1946; Chevron Oil's 1950s "See Your West" campaign; and modern muralists (fig. 5.19) drawn to a seemingly endless linear landscape.

Learn to appreciate the subtleties of the open road. Ponder the space it produces (fig. 5.20). An open road creates its own shapes, which shift with distance and time. The ribbon of highway is an invitation of dirt or asphalt defined by long stretches of straight road with an occasional angle or bend. There are variations on the surface—washboard, bits of gravel, painted lines, skid marks, road kill. Embankments, shoulders, fences, utility poles, and

signs sharpen its definition. The larger ambient environment shapes the experience—the topography, the vegetative mosaic, the isolated placement of human artifacts, the horizon, the color and clarity of sky and clouds (**1** and **11**). Open roads are often best encountered in solitude, defined by individual experiences and expectations.

Some of the best open roads are in New Mexico, both eastern and western, in the Four Corners area and through Navajo and Hopi landscapes. Open roads also stretch through the Great Basin, especially in central Nevada, western Utah, and southeastern Oregon, as well as through California's Coast Ranges, from Cuyama north to Hollister. Finally, you can drive the open road through the prairies, from Montana to the outback of southeastern Colorado. And try to keep it under eighty. ❋

FIG. 5.19 Street mural, San Francisco, California. Painted on the side of a building in one of California's busiest urban areas, this archetypal scene captures all the joy and abandon the open road can bring.

FIG. 5.20 Making time between Chester and Fort Benton, Montana. The appearance and shape of the road itself become central players in the landscape as the highway disappears in the distance.

55

MOUNTAIN ROADS

FIG. 5.21 Switchbacks on Sitgreaves Pass, old U.S. 66, near Oatman, Arizona, 1933. Many early mountain roads were built around steep, short grades with tight switchbacks and hairpin turns. This stretch of U.S. 66 west of Kingman is still open to traffic. *Courtesy of Arizona Historical Society/Tucson, image no. PC180B8-F64-001.*

> The surface of a country, the relations of the hills and valleys which go to make up what is called the topography, profoundly affects the cost of roads and requires a peculiar skill in planning the line which the way is to follow. . . . Any slope beyond that required for the removal of the water is a hindrance to transportation which increases at a very rapid rate with the steepening of the declivities.
>
> —Nathaniel Southgate Shaler, *American Highways* (1896, 33–34)

HARVARD GEOLOGIST NATHANIEL SOUTHGATE SHALER WROTE THIS passage for *American Highways* in 1896. His assessment of the challenges of mountain roads is particularly true in the West, where engineering a road over rugged mountains was a complex and costly undertaking. It took a real talent to plan an affordable road over the Rockies, the Cascades, or the Sierra Nevada.

Consider all of the ways a road can climb a mountain, the engineering challenges of building grades, working with the uncertainties of rivers and streams (culverts, drains, and bridges), constructing switchbacks (figs. 5.21 and 5.22), and coping with unpredictable mountain geology ("Watch for Rocks") and weather ("Chain Up Area"). Look at the logic of the landscape

FIG. 5.22 Independence Pass, Colorado. As highway construction budgets grew along with motorist demands for fewer hairpin curves, later mountain roads featured smoother turns and less-tortuous switchbacks. This view shows the east-side approach to the summit along Colorado Route 82.

and think about why mountain-road builders chose one canyon or side hill over another. Mountain roads often follow low-lying valleys as far as possible before making a sharp turn and beginning a vigorous, often twisting ascent. Once they reach a divide, the process is repeated on the opposite side of the pass. The results are roads that travelers find both frightening and exhilarating (**6**). While New England and Appalachia have their share of hilly transects, it is the West that has the country's great mountain highways, including the tortuous road up Pikes Peak, whose 14,115-foot summit was first achieved by a Locomobile in 1901, or the graceful S-shape curves of an interstate highway over the Continental Divide (fig. 5.23).

Modern mountain-road building in the West is an extension of early trail construction through the region (**49**) combined with the engineering expertise learned from building railroad grades over the mountains (**50** and **51**). Many of the great transmontane traverses grew from these early efforts: over Donner Pass, California; through Oregon's Blue Mountains; over Snoqualmie Pass, Washington, and Monarch Pass, Colorado. Other mountain roads began as the product of early federal investment, starting in the 1850s. The civilian Wagon Road Program and the U.S. Army's Mullan Road across the northern Rockies to Walla Walla are early examples. Western states and private toll-road companies (fig. 5.2) also built mountain roads, especially after 1860.

The first Federal Aid Road Act was signed in 1916. Federal participation in road-building expanded greatly nationwide after World War II, especially

FIG. 5.23 Homestake Pass, Interstate 90 near Butte, Montana. Modern high-speed mountain interstates have made high-country travel even easier, but watch out for slow trucks in the right lane. Contrast engineering strategies with figs. 5.21 and 5.22.

FIG. 5.24 Snow fence, MacDonald Pass, Montana. This location along U.S. 12 is on the Continental Divide west of Helena. As is typical of many open, exposed mountain areas, winter wind and snow can combine to build drifts across the highway. Watch for varied snow-fence designs that share the goal of interrupting the wind and keeping the road open.

with the interstate highway system beginning in the 1950s (**55**). In the West, the U.S. Forest Service (**67**) constructed thousands of miles of mountain roads for fire control, logging, mining, grazing, and recreation. The government also recognized that automobile-related recreation was key to attracting visitors to national parks, and the region's most spectacular mountain roads (**66**) include Going-to-the-Sun Road in Glacier National Park, the Zion–Mount Carmel Highway in Zion National Park, the Hurricane Ridge Road in Olympic National Park, the Paradise Valley Road in Mount Rainier National Park, and the Trail Ridge Road in Rocky Mountain National Park.

Many other mountain highways are also notable. Interstate 70 tunnels beneath the crest of the Colorado Rockies at Loveland Pass west of Denver (the Eisenhower Memorial Tunnel was completed in 1973) and then clings to the canyons above the Colorado River near Glenwood Springs. Similar engi-

neering marvels can be found along Interstate 90 across Homestake Pass (fig. 5.23), Lookout Pass, and Snoqualmie Pass and along Interstate 80 over Donner Pass. At high elevations, snow fences keep snow from drifting onto roads (fig. 5.24), and signs for steep grades and runaway truck ramps warn of the special challenges for big rigs (fig. 5.25).

FIG. 5.25 Runaway truck ramp, U.S. 6 near Keystone, Colorado. A steep descent from Loveland Pass (11,990 feet) can broil the brakes of an eighteen-wheeler. Ramps like this, usually graded and filled with sand, are situated on long downhill stretches across the West.

You can find smaller paved ascents in Colorado (summit road to Mount Evans, roads through the Independence, Red Mountain, and Wolf Creek passes), California (Angeles Crest Highway, summit roads to mounts Diablo and Tamalpais, and Tioga, Sonora, and Monitor passes), and Montana and Idaho (the Beartooth, Lolo, and Galena summits). Circular ascents up Steptoe Butte in Washington and Capulin Mountain in New Mexico are also worthwhile options, and unpaved Forest Service roads (for high-clearance vehicles only) can take you to a lookout tower, an out-of-the way campground, or a wilderness trailhead (**69**). ❋

54

BYPASSED HIGHWAYS

FIG. 5.26 San Juan grade between San Juan and Salinas, Old Pacific Highway. This thin, twisting ribbon of patched concrete was part of the main road between San Francisco and Los Angeles in 1918. The bypassed highway (today called San Juan Grade Road) offers a trip back in time and is a great way to motor through the Coast Ranges.

FOUR WORDS OF ADVICE: *TAKE THE OLD ROAD*—ROUTE 66, THE YELLOWstone Trail, the Old Ridge Route, the Columbia River Highway. Bypassed highways offer a different experience of western topography and landscapes, and they help you understand how western places were linked together in the past and why those connections changed over time. Many routes have also been preserved amid nostalgia for the open road (**52**) and an earlier automobile culture.

Economic changes and technology destroy and refashion landscapes. Roads across the West needed to be replaced, switchbacks were smoothed and straightened, and two-lane roads were improved with four-lane highways. In the West, the early (1912–1925) named highways—the Lincoln Highway was the most famous (fig. 5.18)—were often little more than dirt roads, but they shaped the lineaments of the federally numbered road system that was to become the national norm by the late 1920s. Over time, early roads have been swept aside for shorter, faster alternatives. Winding portions of the Pacific Highway, for example, were left to weather beneath the California sun when U.S. 101 offered a faster trip between San Francisco and Los Angeles (fig. 5.26). Similarly, while U.S. 10 followed the general route of the Yellowstone Trail—you can still spot black and yellow trail signs and markings across Montana—the improved highway meandered less and was more efficient.

In 1953, an entire portion of U.S. 66 in western Arizona, near Oatman and Goldroad, was bypassed by a segment that avoided the more mountainous route (fig. 5.21). The old road remains one of the West's best bypassed highways. In other settings, downtown main streets that were once part of the numbered highway system were replaced by roads along the urban periphery. Many old U.S. highways became frontage roads near interstates (fig. 5.27) or interstate business routes through urban areas. Denver's east-west-oriented Colfax Avenue, for example, is old U.S. 40.

On bypassed roads, watch for three stories on the landscape: submersion and incorporation, abandonment, and adaptive reuse. Some older roads have been paved over and fully integrated into a newer highway. Miles of interstate make use of former U.S. highway rights-of-way, and older two-lane routes have become parts of newer four-lane alignments. By contrast, if a route segment has been completely abandoned, the route surface erodes and the edges of the road blur, the grass grows, and opportunistic plants return (fig. 5.28). Sometimes, encroaching vegetation can provide linear clues to former roads (fig. 5.29). Third, bypassed roads can survive through adaptive reuse: it is common for a long-distance route to be repurposed as a still-maintained local road (figs. 5.26 and 5.27), typically with little fanfare.

Increasingly, bypassed roads—and their associated commercial landscapes of gas stations, cafés, tourist attractions (fig. 5.30), and motels—are being preserved (**80**). Brown-toned interpretive signs indicate some of the surviving segments of the U.S. highway system in the West. Instead of taking Interstate 15 from Southern California to northern Montana, you can travel

FIG. 5.27 Former U.S. 87 (left) and Interstate 25 (right) near Glendo, Wyoming. This portion of the old road (today the South Glendo Highway) remains open to local traffic, but most travelers prefer the adjacent interstate.

FIG. 5.28 Bypassed freeway off-ramp, Irvine, California. Even off-ramps can go to seed as decorative ice plant (an exotic succulent) slowly occupies once-congested ground adjacent to the San Diego Freeway. A newer exit has been built nearby.

FIG. 5.29 Abandoned road west of Dillon, Montana. A linear swath of sagebrush agreeably occupies this former roadbed (right). As disturbed ground, bypassed highways are often colonized by distinctive vegetation that can reveal the now-vanished roadbed.

FIG. 5.30 Wigwam Motel, old U.S. 66, Holbrook, Arizona. Fascination with an earlier era of motoring has led to preservation of the West's bypassed highways and tourist services. Watch for vintage diners, motels, and gas stations. An example of duck architecture (a building that visually advertises its product or service), this spot remains a classic stop along the old highway.

on U.S. 91, driving over the Beaver Dam Mountains in southwestern Utah or taking the scenic route past Mesa Falls northeast of Ashton, Idaho. Or you can pick up fragments of U.S. 80 between San Diego and El Paso and avoid long stretches on Interstates 8 and 10.

The creation of organizations such as the Route 66 Corridor Preservation Program, managed by the National Park Service, and New Mexico's Heritage Preservation Alliance, which is concerned with cultivating interest in older roadside motels, suggests that people are making connections between bypassed roads and a fondly remembered past. That place identity has emerged along most of the West's early east-west highways, especially U.S. 2 (the former Theodore Roosevelt International Highway), U.S. 10 (the Yellowstone Trail), U.S. 30 (portions of the Lincoln Highway in Wyoming and the Columbia River Highway in Oregon), U.S. 40, U.S. 66, and U.S. 80. On the Pacific coast, roads attracting similar interest include U.S. 101 (El Camino Real) and U.S. 99 (and the nearby Ridge Route northwest of Los Angeles). Along these routes, new social memories have been produced and in that process new landscape signatures, meanings, and experiences have been created that seem likely to endure and to encourage us, when time allows, to take the old road. ❋

55

INTERSTATE LANDSCAPES

FIG. 5.31 A good day on the 405, Irvine, California. Urban interstates move huge volumes of traffic from one suburb to another within the West's largest metropolitan areas. On some days, however, even these eight lanes (each way) can grind to a halt.

MOST ROAD TRIPS ARE DOMINATED BY INTERSTATE LANDSCAPES. Like the rest of America, the West has been crisscrossed by routes on which one stretch of highway looks just like the next and the same gas stations, fast food restaurants, and motels seen in Arizona can be found in Idaho and Illinois. The franchised, cookie-cutter landscapes of the interstate have been criticized by travelers who want more variety and connection with the surrounding countryside. But who can criticize the speed, relative safety, and convenience of the interstate? And in the West, these modern routes can give travelers a different kind of open-road experience (**52**).

The West's interstate landscapes include vast stretches of highway where rural exits are separated by miles of thinly settled country (fig. 5.23). At eighty miles per hour, large areas of eastern Montana (Interstate 94), northern Nevada (Interstate 80), and central Utah (Interstate 15) pass by in a surreal panorama of far horizons and occasional eighteen-wheelers. Other western interstates, such as Southern California's infamous 405 (surely the birthplace of road rage), are crowded urban arteries choked with commuter traffic (fig. 5.31).

Much of the nation's interstate system was built between 1956 and 1980. A federal system had been discussed and debated since the military sketched out the Pershing Plan of national highways in 1922. President Franklin D. Roosevelt toyed with the idea of a national network of superhighways in the 1930s, both to bolster the nation's defensive capability and to spur its economic integration. It was not until 1956 that President Dwight D. Eisenhower signed the Federal Aid Highway Act, which dramatically increased Washington's role in constructing a nationwide system of limited-access, high-speed highways.

Many rural westerners just shook their heads in disbelief as the new roads were built and hundreds of small towns were bypassed (**54**) by the new system. Miles of four-lane boulevards dramatically reworked open stretches of the West, and the lack of on-ramps and off-ramps did little to serve local needs. But an adrenaline surge of federal cash from the new interstate funding formula revolutionized once-small state highway budgets. Utah's annual allocation, for example, leaped from six million to sixty million dollars after passage of the 1956 legislation. In the end, interstates attracted people as well, allowing for longer commutes from new suburbs and the development of more regional playgrounds within a weekend drive of metropolitan centers.

Meanwhile, hundreds of the now-familiar diamond- and cloverleaf-shape interchanges became destinations for investors. The function and design of new service centers built on older traditions of western travel (fig. 5.32), when long-established local businesses served travelers (fig. 5.33). In the new iterations, interstate service centers are almost wholly made up of national chains and franchises. Further innovations to these commercial strips appeared in the 1980s and 1990s, including the towering signs and other signals that food, a bathroom stop, and an air-conditioned motel are only an off-ramp away (fig. 5.34; **80**).

FIG. 5.32 Two Guns Service Station, old U.S. 66, Arizona. An early version of the modern interstate travel plaza, this famous stop along Arizona's U.S. 66 east of Flagstaff offered high-octane fuel, cold soda, hamburgers, and even a roadside zoo. *Courtesy of Arizona Historical Society/Tucson, image no. PC180F67-4-1.*

CLINES CORNERS
BIENVENIDOS
BELTS
CAPS
MUGS
BUFFET

Chevron
McDonald's
IHOP

Truck stops, which date from the 1920s, are also part of interstate travel; there commercial drivers can secure a shower, a hot meal, fuel, a motel room, and a bit of conversation (fig. 5.35). Standardized signage on the interstate indicates speed limits, distance to the next town (along with white or green roadside markers that indicate the miles from one state line to the next), approaching services (with the familiar, easy-to-see symbols of national chains), and other points of interest. Rest stops for travelers in both cars and trucks include large parking lots, a scattering of picnic tables, flushing toilets, perhaps a soft-drink machine, and a place to walk the dog. ❋

FIG. 5.33 Clines Corners, New Mexico, near Interstate 40. Roy Cline built several roadside businesses along U.S. 66 before World War II and eventually relocated to this still-popular stopping place east of Albuquerque.

FIG. 5.34 A corporatized landscape along Interstate 5 near Eugene, Oregon. This landscape reveals the impact of global chains, towering signage visible from the highway, familiar logos, and standard fare for food, fuel, and lodging.

FIG. 5.35 Truck stop on the west side of Albuquerque, New Mexico. Big rigs have created their own roadside landscape along many busy interstates. Trucks await fueling in this massive high-volume operation. Adjacent eateries and motels cater to those who can afford a longer stop.

56 ELECTRICAL GRID

FIG. 5.36 Producing the juice: Grand Coulee Dam, Washington. Completed along the Columbia River in 1942, this facility is one of the largest electric-power-producing plants in the world.

ALL BUT INVISIBLE YET UBIQUITOUS, THE WEST'S ELECTRICAL GRID creates a landscape all its own. If you walk beneath a high-voltage line, you can hear the crackle and pop of the electrified air above.

This method of long-distance power delivery became possible only after the adoption of alternating current (AC) circuit technology in the 1890s. Thanks to these innovations, many of the West's mountainous, well-watered districts became major producers of hydroelectricity, a system whose potential was revealed in Oakland, California, in 1901 when streetlights glowed from power generated along the Yuba River, more than 140 miles away.

You can watch for three expressions of the electrical grid on the landscape: power production, the transmission grid, and electricity consumption. Nationally, about two-thirds of electricity comes from fossil fuels (mostly coal and natural gas), and thus important parts of the grid begin with generating facilities near these natural resources (**35** and **36**). Nuclear power plants (**65**), including facilities in California, Arizona, and Washington, also generate a significant share of the nation's electricity. The remainder comes from renewable sources, including hydropower (**64**) and geothermal, wind, and solar technologies (**37**). Some of the region's most impressive human engineering is evident in hydroelectric dams, especially on the Columbia and Colorado rivers (fig. 5.36).

The transmission grid is designed to take electricity from where it is produced to where it is consumed. Regional trunk lines of the huge Western Grid—with a capacity of 240,000 megawatts and more than 120,000 miles of high-voltage lines—connect sites where power is produced and consumed. Tower sizes and shapes vary widely, a function of demand, line capacity (high-voltage lines transmit 110 kilovolts and above), and local terrain. Look for varying heights, base configurations, and arrays of high-voltage wires (conductors), and for the insulators that block and control electrical current (fig. 5.37). Off the Pacific Coast, submarine power lines are less-visible components of the grid.

Ultimately, these connections guarantee delivery to consumers, and the result is particularly apparent on the landscape in large urban areas (fig. 5.38). Poles, wires, transformers, and local generating and switching substations ensure that power is available where it is needed.

Some experts warn that the intricately interconnected Western Grid is increasingly vulnerable to shorting out, especially as a warming climate creates higher peak loads. Keep a flashlight handy. ❋

FIG. 5.37 Moving the juice: central Washington. Watch for transmission lines and towers dramatically marching across the western landscape, bringing electricity to the region's centers of population. These high-voltage lines are east of Grand Coulee Dam.

FIG. 5.38 Consuming the juice: San Francisco, California. Within sight of Coit Tower (right), this high-density residential district requires a spidery web of wires to keep the lights on.

57

COASTAL CONNECTIONS

FIG. 5.39 Newport harbor, Oregon. Fishing boats and recreational craft dominate this central Oregon harbor. Locals skillfully navigate these inner harbor channels within Yaquina Bay before exiting via a narrow jetty-defined passage to the open sea.

ENDURING COASTAL CONNECTIONS MAKE LIFE ALONGSHORE POSSIBLE and contribute to place identities never far from high tide. The West's spectacular coastline and harbors (**10**), recreational seaside activities (**87**), and global seaborne trade have produced cultural landscapes from the Strait of Juan de Fuca to San Diego Bay (fig. 5.39). Lighthouses were among the earliest structures on the coast (fig. 5.40). Along with published coastal charts (the U.S. Coast Survey dates from 1807), Pacific lighthouses (most of them built between 1855 and 1915) kept vessels off the rocks. Early lighthouses included a tower for the lantern room, where the lens signals its identifiable flash, and a lighthouse keeper's residence. Many lighthouses are still in operation, most with automated systems. Some Pacific lighthouses (explore Point Cabrillo, California, or Heceta Head, Oregon) are home to modern tourist facilities, including opportunities for overnight stays. They have also been used as an icon of place, appearing on everything from postcards to coffee cups.

Pacific coast harbors, which are jointly controlled by public and private agencies, are hybrid landscapes, part natural, part manmade. North America's largest container cargo ports—Los Angeles (number 1) and Long Beach (number 2)—sprawl along more than fifty miles of waterfront. Both the Bay

Area (fig. 5.41) and the Port of Seattle are also home to a huge commercial coastal infrastructure, much of it oriented to Asia.

Harbors are complex, intricate places. Have you ever captained an ocean freighter into San Francisco Bay, helmed a ferry in Puget Sound, or navigated a fishing boat out of Newport, Oregon's busy harbor? Novices, beware: none is a casual activity. Mastering them requires an ability to carefully read an intricate set of shoreline signposts that permit safe passage.

Stone, gravel, and concrete jetties and breakwaters help define entrance channels and protect the harbor from waves and the open sea. Buoys—moored floating objects that have particular colors, markings, or lights—signal channel boundaries and hazards. The docks are often as regulated as a city street (fig. 5.39). For commercial traffic, large wharves house fixed facilities for mooring ships and storing cargo. Specialized examples include oil terminals (huge tank farms), container facilities (stacked container units, container ships, and cargo cranes), and cruise-ship docks. ❋

FIG. 5.40 Yaquina Head Lighthouse, central Oregon coast. Dating from 1873, the lighthouse (Oregon's tallest) was automated in 1966 but still uses a French-made lens manufactured in 1868. The exposed, cliff-top locality is part of the federally managed Yaquina Head Outstanding Natural Area. Bring a windbreaker.

FIG. 5.41 Embarcadero and San Francisco Bay, California. The historic Embarcadero was once one of the busiest parts of the Port of San Francisco. While still offering some dockside services, the area now has shops, restaurants, tourist attractions, and great views of the bay, including the Bay Bridge and Treasure Island (center).

6

LANDSCAPES OF FEDERAL LARGESSE

FIG. 6.1 Band shell, Lingle, Wyoming. Thousands of public structures owe their origins to New Deal programs designed to jump-start the Depression-era economy through federal intervention. This band shell is in Lingle, a small farm and ranch town in eastern Wyoming.

THERE ARE SEVERAL VERSIONS OF A GAME CALLED UNCLE SAM. ONE version that can be played for this portion of the field guide awards a point for every sign of federal evidence you see on the landscape. Post offices; red, white, and blue signs; eagles; flags; federal prisons; veterans' hospitals; military cemeteries; and army recruiting stations are obvious places to begin. But you can also find Uncle Sam watering the lawn, using water from a federal dam project; helping someone get a mortgage with a federally guaranteed loan; and determining what shoppers buy at a grocery store through agricultural policies and import regulations. Although you can play Uncle Sam in every state, the West is particularly fertile ground for racking up high scores.

Westerners have a love-hate relationship with Uncle Sam. Many western myths are grounded in stories of how individuals have struggled on their own against a tough regional environment—the independence of the free trapper, the lone cowboy in the saddle, the hardscrabble farmer or rancher fighting the elements—and there is an enduring skepticism about big government. But the landscape suggests otherwise, in multiple layers of federal influence, some quiet and omnipresent, others never far from controversy. Westerners struggle with the bargains they have struck with the government since the mid nineteenth century. One hand reaches out for what Uncle Sam can offer while the other shakes a fist at interference from Washington, D.C.

Lingle, a small town in eastern Wyoming, is situated near the North Platte River, not far from where the Oregon Trail passed through. It is a good place to play Uncle Sam. The town has a fine city park at the junction of U.S. 26 and U.S. 85, with shade trees and places for kids and dogs to play (fig. 6.1). The patriotic painting on the band shell, boasting that the town is "Small but Proud," certainly evokes Uncle Sam, flags unfurling, with the added resonance of the Liberty Bell and the stars above. But there is more. The band shell was built during the Great Depression, one of thousands of structures across the West funded through the federal relief efforts of Franklin D. Roosevelt's New Deal. The electric lights and speakers are powered by Wyrulec, an energy cooperative created when Congress passed the Rural Electrification Act in 1936. The water that keeps the lawn green and the drinking fountains flowing is carefully regulated by the Environmental Protection Agency, and local officials are mandated to test its purity according to a set of federal standards.

In a very different part of the West, you can look across the suburbs on Tucson's north side from the foothills of the Santa Catalina Mountains (fig. 6.2). While much of what you see is a result of private capital investment and the efforts of individual homeowners, many residents in these suburbs receive Social Security checks that have made their retirement in Arizona easier. Because of federal air-quality standards, you can see the Santa Rita Mountains (left) near the Mexican border on the horizon to the south. Both legal and undocumented immigrants are also an integral part of this community, and their lives are shaped by a maze of federal laws, regulations, and barbed-wire fences. Much of the Tucson area depends on both groundwater and water from the Colorado River brought in by the Central Arizona Project, created by President Lyndon B. Johnson in 1968 and one of the West's most ambitious and costly federal water initiatives. Finally, much of southern and western Arizona is federal land, a complex mix of Indian trust lands, Bureau

FIG. 6.2 Federal largesse, Tucson. Much of the regional landscape owes its origins to policies and expenditures of the federal government. Land surveys, water initiatives, Social Security payments, immigration policies, and economic stimulus programs are among the many ways in which Uncle Sam has transformed Arizona's landscape.

of Land Management (BLM) holdings, national forests, military spaces, and units of the national park system.

As both Lingle and Tucson suggest, federal largesse is ingrained across the West, but that was not always the case. Until the Civil War, the impress of federal power across the West was intermittent, mainly focused on extending and then asserting national authority over a continent-size nation. Coincident with that westward expansion were federally sponsored explorations—military, scientific, and commercial—designed to foster economic development. Programs to dispose of the region's vast public lands were also put in place.

But after 1870, the emphases on initial exploration and encouraging private development began to change. Over the next fifty years, the federal government became a more active manager of its remaining public lands. Fences were built, dams were constructed, and federal agencies of central importance to the West, such as the U.S. Forest Service and the National Park Service, were established. Federal largesse in the West reflected the shift toward bigger government during the Progressive Era and a growing confidence in government's ability to manage both civil society and natural resources efficiently and scientifically. The West, historian Richard White argues, was a pivotal proving ground for a larger "Bureaucratic Revolution" that modernized and professionalized the national government.

The Depression, the New Deal, and World War II brought an even larger, more visible federal impress across the West. In the postwar atomic age, the region's economy grew more consistently and rapidly than the national norm, and the region received more than its share of federal spending in defense, infrastructure, and urban development programs. The government also played the central role in trade legislation and global development initiatives that secured for the West a pivotal place in the international economy.

How does one make sense of this accumulation on the landscape? There are four broad categories of connections between Uncle Sam and the West. First, watch for signs of federal influence that are found everywhere in the

United States, but have particular regional expression in the West. For example, national policies have long regulated farming and ranching in America, especially since the Great Depression. Federal legislation has shaped working conditions for farmworkers, the availability of water for irrigation, and the regulation of pesticides. Similarly, environmental regulations have shaped extractive landscapes, from coal mines to forest clear-cuts and renewable-energy operations. Ethnic landscapes have been molded by the nation's shifting immigration policies, and interstate commerce connections and regulations helped integrate the national economy. Cities, suburbs, and amenity-rich playgrounds also have been fundamentally transformed by federally funded mass transportation and highway systems, commercial and residential investments influenced by federal tax policies, and easy access to federally insured mortgages.

Second, western political space has evolved under the umbrella of the federal government. Its creation began with international borders that were imposed across the West between the Louisiana Purchase in 1803 and the Gadsden Purchase in 1854, a process that involved subjugating Native peoples, waging a war with Mexico, and negotiating international agreements. From Oregon's Fort Clatsop to Montana's Big Hole National Battlefield, you can see the legacies of that formative era in the presence of military outposts and battlefields across the West. Once the United States gained effective political control, it had to foster development, provide political institutions, and guide settlement. Shifting territorial boundaries gave way to a more stable geography of state lines (fig. 6.3). The township-and-range system made it possible to provide efficient land surveys of the public domain, and you can still spot the familiar grid of cardinally oriented section and lot lines across the West. Federally sponsored legislation also established land disposal programs, including the Homestead Act (1862) and the Desert Land Act (1877); organized the political landscape; and produced a federalized geography of political space that still shapes everyday life in the West.

Third, particular federal programs have had a striking impact on the west-

FIG. 6.3 Four Corners monument. Federally imposed state lines created enduring, often Euclidean regional landscape features, nowhere more dramatically encountered than at Four Corners, the only North American locality where a person can be in four states (Arizona, Utah, Colorado, and New Mexico) at once.

FIG. 6.4 Navajo Indian Irrigation Project south of Farmington, New Mexico. Begun by the federal Bureau of Reclamation in the 1960s, the hundred-thousand-acre project supports irrigated farming on the Navajo Indian Reservation. Southwestern Colorado's La Plata Mountains are visible in the distance.

ern landscape in a variety of ways. Federal dam and water projects were signature elements of Uncle Sam's management of public resources, the vast majority of them in the West (fig. 6.4). Also, during the Great Depression, while assistance went to all parts of the country, residents of the western states received the largest per capita federal expenditures. The expressions of this New Deal legacy remain a part of the region's everyday landscape, visible in county courthouses, new roads and schools, and a myriad of other investments in regional infrastructure (figs. I.2 and 6.1). Later on, the rush to respond to the Japanese attack on Pearl Harbor was felt most strongly in the West, where huge military bases and factories were built, some almost overnight. The breathtaking proliferation of modern military spaces continued after the war in a region whose open space, isolation, varied terrain, and dry climates are preferred for military bases. An educated workforce and access to urban infrastructure create other opportunities for defense-related manufacturing. Most recently developed was the atomic West, with its own historical events, institutional settings, and landscape changes focused on the nation's nuclear imperative.

The last broad imprint of Uncle Sam's hand across the West comes from the government's role in managing federal lands. Through much of the nineteenth century, federal land policies in the West were designed to *dispose of* the public domain, to get it into private hands. The proceeds from land sales went into the federal treasury to further the nation's economic development. But large portions of the region remained unsold, a common situation across sizable chunks of the arid and mountainous West, and settlers thought little about using portions of this unsold public domain for their own purposes. Cattle and sheep grazed the hills, miners panned color from the streams (though the General Mining Law of 1872 imposed some basic guidelines), and lumberjacks hauled timber off unclaimed slopes.

A cadre of politically powerful conservationists—including John Muir, Gifford Pinchot, and Theodore Roosevelt—and other concerned people were determined to protect the West's resources. Between 1880 and 1940,

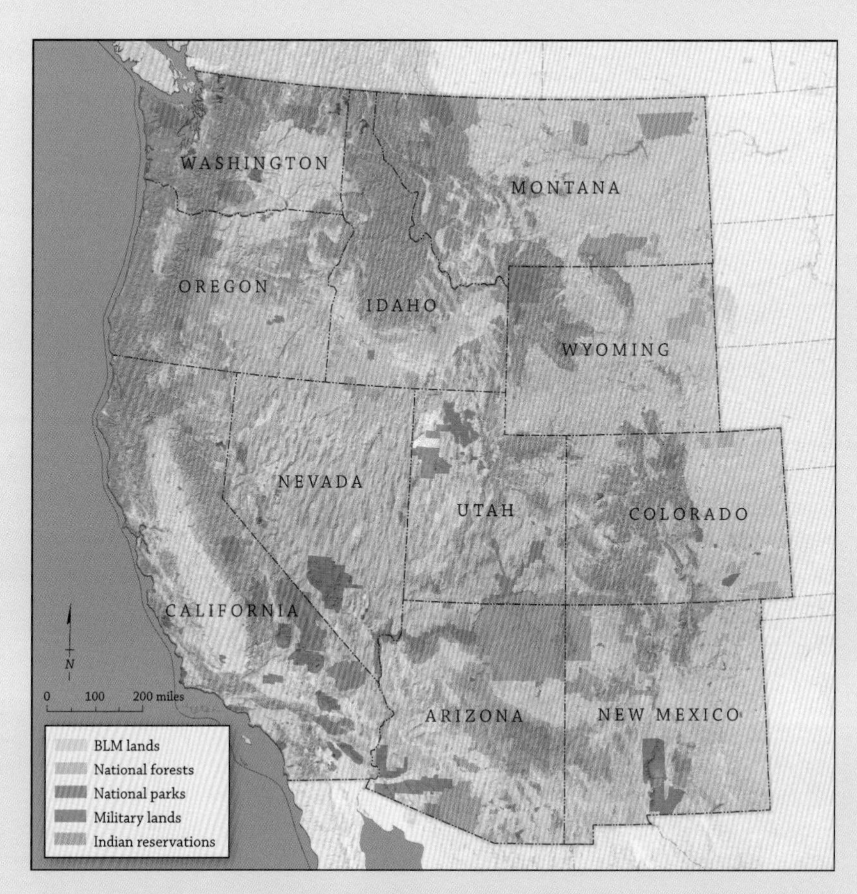

FIG. 6.5 Federal lands and Indian reservations. Federal lands within the region retain the legacies of shifting policies of land disposal and management. The United States Forest Service and Bureau of Land Management are the largest stewards of western acreage, with each agency controlling more than 150 million acres.

the so-called gospel of efficiency evolved as federal policy and land ethic, proclaiming that the efficient, scientific management of public lands would produce the greatest economic gain for the nation as well as protect the resources from wanton despoliation. A growing legacy of protective legislation shaped the region and its landscapes (fig. 6.5). Yellowstone National Park was established in 1872, the forest reserves (which later became the national forests) in 1891, and the National Park Service in 1916; the designation of wilderness areas began (with the Gila Wilderness) in 1924, and the Taylor Grazing Act was passed in 1934. In addition, federal wildlife refuges, administered since 1940 by the U.S. Fish and Wildlife Service, protected wetlands and other habitats and helped maintain bird, fish, and other animal populations (fig. 6.6). More recently, federal regulation of the public domain and the growing role of ecosystem management and wilderness preservation can be traced in the Multiple Use–Sustained Yield Act of 1960, the Wilderness Act of 1964, the National Environmental Policy Act of 1970, the Federal Land Policy and Management Act of 1976, and the Omnibus Public Land Management Act of 2009.

Not surprisingly, the political process has been leavened with conflict. Preservationists such as John Muir recoiled from too much development on public lands, while local and corporate resource users resented federal interference. That early twentieth-century narrative continues more than a cen-

FIG. 6.6 Viewpoint, Malheur National Wildlife Refuge, southeastern Oregon. Hundreds of federal wildlife refuges dot the West. These managed landscapes are designed to maintain bird, fish, and other wildlife populations as well as offer recreational opportunities.

tury later. The mosaic of public lands can be both fascinating and bewildering, but you can start to make sense of it in the contact zones between public and private lands and among the local, state, and federal agencies that manage public land. Local land-use issues often simmer along these boundaries.

Separately, the nation's largest private land trusts (also called land conservancies) are found in the West, where they manage land to limit development and promote sustainable uses. Since 1980, the Nature Conservancy and other organizations have increased land acquisitions in the West, with more than seven million acres of private lands conserved in land trusts as of 2013. Essentially, these private organizations have acted in the public interest, paralleling and reinforcing similar federal efforts. These organizations also form partnerships with federal agencies to create larger blocks of protected lands. The privately funded American Prairie Reserve, for example, acquired (through ownership and lease agreements) more than 270,000 acres of northeastern Montana prairie and is working together with the federally managed Charles M. Russell National Wildlife Refuge to reintroduce North American bison to one of the continent's largest restored grassland ecosystems. ❋

58

TOWNSHIP-AND-RANGE SURVEY SYSTEM

FIG. 6.7 Rectangularity on the landscape, Seattle suburbs, Washington. This aerial view shows how a cardinally oriented rectangular grid shapes the suburban landscape of streets north of Seattle. Note how the angled layout of the cemetery (see the irregular clumps of evergreens, center) interrupts the pattern.

MANY PEOPLE FLYING CROSS-COUNTRY HAVE LOOKED OUT THE WINdow and seen evidence of the cardinally oriented, rectilinear, federally imposed township-and-range survey system, which organizes much of the American landscape (fig. 6.7). As one of the inventors of the system, Thomas Jefferson would be delighted with the modern western landscape, with its simple squared-off geometry of roads, property, and fields.

Much of the West, with exceptions in Spanish-settled New Mexico and California (**39**), bears the imprint of the Public Lands Survey System (PLSS). As mandated in the Land Ordinance of 1785, the survey imposed six-mile-square townships, each containing thirty-six sections of 640 acres, numbered so that local officials, surveyors, and settlers could locate and identify parcels (fig. 6.8). Intersections of east-west baselines and north-south principal meridians organize successive range and township lines. First used in eastern Ohio's Seven Ranges in the late eighteenth century, the system expanded westward, sometimes leading and sometimes following settlement. It imprinted the continent with rectangular logic, a visual ode to Age of Enlightenment thinking.

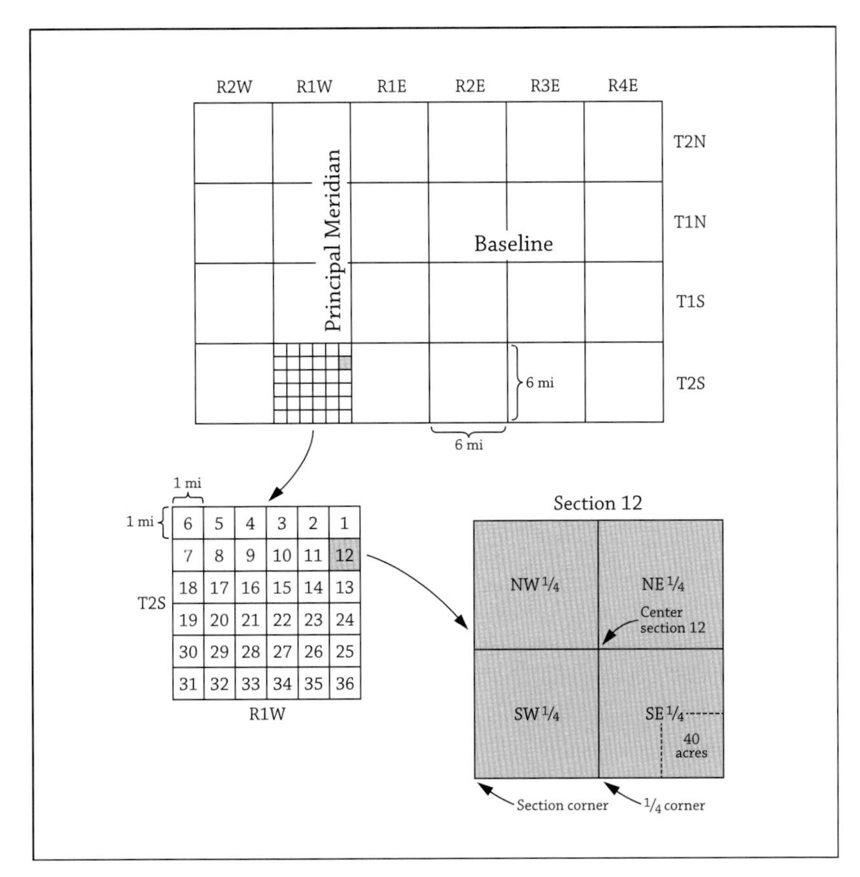

FIG. 6.8 Township-and-range survey system. Within the Public Land Survey System (PLSS), land parcels can be located by section number in relation to a larger cardinally oriented rectangular grid of townships that are numbered in relation to regional baselines (east-to-west) and meridians (north-to-south).

The system reinforced American notions about land. The public domain—the vast landholdings claimed by the federal government after the Revolutionary War—was a commodity, a real estate venture of epic scale that enabled an orderly disposal of acreage into private hands. While the West still has the nation's largest share of public lands (**62**, **66**, **67**, and **68**), millions of acres were "alienated"—that is, put into private ownership—through federal land laws, all organized around the PLSS. Thus, both public and private lands in the West bear the rectilinear stamp of the first federal surveys.

Western lands were alienated in several ways. The Homestead Act of 1862 offered settlers a quarter-section of free land (160 acres) if they paid filing fees and occupied and improved the parcel for five years. Other laws allowed settlers to purchase acreage, often at $1.25 per acre. States that received grants of federal lands (for schools, etc.) also sold acreage to raise money. Several western railroad companies were granted alternate sections of land near the tracks—more than a hundred million acres (**50**)—in exchange for building lines across the West's inhospitable terrain. The resulting checkerboard pattern of land ownership has endured, complicating land-use patterns and management strategies.

Several land laws acknowledged that 160 acres were often insufficient for viable agricultural operations in the rugged, arid West. For example, the Timber Culture Act (1873) offered an additional 160 acres to settlers who planted

FIG. 6.9 Grid on the ground: east of Twin Falls, Idaho. More subtle than in an aerial view, this simple intersection of cardinally oriented section-line roads is a quiet reminder of the federal presence. Utility lines, fences, property lines, and field boundaries add linearity.

40 acres in trees; the Desert Land Act (1877) provided quarter-sections to farmers willing to irrigate (**23**); the Enlarged Homestead Act (1909) opened parcels of 320 acres to dryland farmers (**21**); and the Stock-Raising Homestead Act (1916) provided 640 acres for ranching operations (**20**).

The cumulative visible and legal manifestation of the PLSS created a ubiquitous human imprint on the western landscape, yet its presence is subtle, suffusing the ordinary scene with straight lines and right angles that we navigate without question (fig. 6.9). What to look for? Trained eyes can spot survey corner markers—small metallic or concrete benchmarks—at strategic points, but it is the predictable rectangularity of the section-line roadscape that most clearly reveals the survey system in the rural West, with similar cardinally oriented streets and subdivision boundaries in metropolitan settings (**70**). Many roads simply follow section lines. Watch for road grids, often with intersections spaced one mile or six miles apart (fig. 6.9), and straight-line highways (fig. 6.10). Utility lines, fences, property lines, and field boundaries reinforce the linearity. Even within fields, orchards and crops are often cardinally oriented (fig. 6.11). Interruptions also attract attention, such as quick jogs in north-south roads and correction lines where surveyors adjusted their course to account for converging meridians (the challenge of fitting a grid onto a spherical planet). Antecedent lines that predate the survey—railroad alignments and early, noncardinally oriented urban plats (**50** and **70**)—can also disrupt the grid's simplicity.

FIG. 6.10 Section-line road east of Tucson, Arizona. Long, straight road segments that run across varying terrain often indicate the underlying grid. North Soldier Trail follows the grid through this exurban neighborhood. Note in the distance (upper left) how the road ends at a dry wash with another segment of section-line road picking up beyond.

FIG. 6.11 Cardinally oriented onions southeast of Bakersfield, California. Elements of the grid are translated throughout the landscape. These onion sacks are lined up for transport along lines perpendicular to the nearby east-west highway.

Advocates of the grid argue that it enabled the rapid and efficient disposal of land, fostered development, and created an elegant linear landscape that is easy to read and navigate. Critics contend that loosely monitored federal land policies invited abuse, fraud, and speculation and that the inflexible rectangularity of the system often made a poor fit with topographic and ecological diversity. Whether bane or blessing, the grid survives, shaping western patterns of property ownership, flows of daily travel, and the appearance of the ordinary landscape. ❋

59

INTERNATIONAL BORDERS

FIG. 6.12 International border, Montana, USA (left), and Alberta, Canada (right). This view looks westward along the forty-ninth parallel near a quiet neighborhood shared by Sweetgrass, Montana, and Coutts, Alberta. Small pylons are gentle reminders of the international boundary line. Nearby ports of entry house more formal customs and immigration facilities.

THE WEST SHARES INTERNATIONAL BORDERS WITH CANADA (1,017 miles) and Mexico (almost 700 miles). On those borders, environment, history, and politics have produced disparate landscapes (figs. 6.12 and 6.13). Visit these settings and think about how each boundary is really three borders in one.

Above all, borders define discrete political and economic spaces with different laws, regulations, and practices. The historical geography creating these boundaries is straightforward. To the north, British-American agreements hashed out the forty-ninth parallel boundary in 1818 (from the plains to the Rockies), 1846 (the Oregon Treaty extended the line westward), and 1872 (when the border in the Strait of Georgia was adjusted between the Gulf Islands of British Columbia and Washington's San Juan Islands). To the south, the Treaty of Guadalupe Hidalgo (1848) acknowledged U.S. expansion across California and the Southwest, and the Gadsden Purchase (1854) secured an added strip of southern Arizona and New Mexico. Distinct human geographies have evolved in these three nations, the southern boundaries of California, Arizona, and New Mexico marking one of the sharpest economic divides on the planet.

Looked at differently, borders also define zones of contact and mixing, reflecting cultural and economic connections between countries. Listening to the Canadian Broadcasting Corporation in Montana or visiting one of the world's busiest borders in the Southwest provides reminders that borders create mutual interdependence and shared human experiences (**41**). The 1994 North American Free Trade Agreement codified some large-scale economic dimensions of these interactions, but there are many local expressions. In Nogales, Mexico, for example, *farmacias* dispense low-cost prescriptions to ailing Americans. In nearby portions of Arizona, many residents face increased dangers from drug-related violence that has spilled northward from Mexico. Especially along the Mexican border, you can also see the diversity of these contact zones—Douglas, Arizona, for example, is very different from San Diego, California.

Finally, borders are material artifacts. Elaborate customs and immigration facilities characterize larger ports of entry in Nogales, Arizona, and Blaine, Washington, while smaller ports at places like Naco, Arizona, or Scobey, Montana, need only a small staff. The boundaries themselves vary greatly (contrast figs. 6.12 and 6.13). On the Canada-U.S. border, the International Boundary Commission (1925) produced a mandate to keep clear a twenty-foot-wide corridor that is sometimes paralleled by access routes and closely monitored by the Border Patrol (white vehicles with green stripes) but generally not fenced. On the Mexico-U.S. border, Operation Gatekeeper (1994) and the Secure Fence Act (2006) created pedestrian fencing (steel walls, chain link, wire mesh) and vehicle fencing (post-and rail, concrete barriers) along much of the California, Arizona, and New Mexico line. There is also virtual fencing—that is, cameras, towers, road sensors, and unmanned aircraft—and Border Patrol agents conduct line watches (surveillance) and staff traffic checkpoints close to the border.

Enforcement successes in California have propelled drug runners and people who smuggle undocumented immigrants, called coyotes, into isolated parts of Arizona. About half of all illegal drugs and undocumented immigrants coming into the United States are captured there. Be watchful. Some rugged desert and mountain zones in the southern part of the state remain beyond the effective control of the United States government. ❋

FIG. 6.13 International border, Arizona, USA (left), and Sonora, Mexico (right). This view looks eastward along the dividing line between Nogales, Arizona, and Nogales, Mexico. The high wall and nearby lights and surveillance equipment discourage casual border crossings.

60

STATE LINES

FIG. 6.14 State line above the Columbia River near The Dalles, Oregon. The Washington-Oregon state line follows the Columbia River for miles in the Pacific Northwest, one of the few settings in the region that departs from the simplicity of straight-line boundaries.

MUCH OF THE FAMILIAR JIGSAW PUZZLE OF WESTERN STATE LINES, largely an exercise in political horse-trading and Euclidean serendipity, was shaped between 1850 and 1880 (fig. 6.3). The Northwest Ordinance of 1787 created a two-step process for establishing new territories and states and provided a model of federal expansion. Except for California, which was admitted directly as a state in 1850, western states followed the template, completing the process with Arizona and New Mexico statehood in 1912. Precise territorial and state borders were the product of negotiations and often involved several stages. Washington, Nevada, and Utah territories, for example, changed their boundaries multiple times before settling into their recognizable forms by 1870. Most western state boundaries are straight lines, although a drainage divide (between Idaho and Montana) and substantial river courses (separating California, Arizona, and Nevada and Washington, Oregon, and Idaho) created exceptions (fig. 6.14).

Once established, state lines have enduring impacts. Distinctive state laws, regulations, and traditions have shaped land uses and cultural landscapes. From the wording of state slogans and license plates—Colorful Colorado, the Cowboy State (Wyoming), the Grand Canyon State (Arizona)—to the clustering of casinos around the edges of wager-friendly Nevada (fig. 6.15), state

boundaries and state-based place identity have produced distinctive regional signatures. For travelers, signs and welcome centers offer cues and symbols, along with a flurry of changed speed limits, billboard designs, and driving regulations, especially for truckers. You can see the changes from state to state in laws that generate state-line liquor stores, bars, fireworks stands, discount cigarette stores, and strip clubs (**94**). Pavement condition, roadside mileage markers, and the number and condition of interstate rest areas and state parks also vary with state budgets (**55**). Long ago, California's strong agricultural lobby (**24** and **25**) persuaded the state to build inspection stations along major routes to query travelers about whether they have fresh fruit that may carry diseases or pests (fig. 6.16). Such gestures of state-based protectionism and pseudo-sovereignty no doubt reached a low point in the 1930s when Colorado Governor Edwin C. Johnson sent the Colorado National Guard to the border with New Mexico to halt the northward movement of "Mexicans" during the Depression, an initiative subsequently determined to be illegal. ❋

FIG. 6.15 Whiskey Pete's Hotel and Casino, Nevada state line near Las Vegas. Casinos crowd the Nevada state line at several strategic points along borders with Idaho, Utah, Arizona, and California. For impatient Southern Californians who can't wait to reach the Vegas Strip, Whiskey Pete's has long been a state-line tradition on the highway from Los Angeles.

FIG. 6.16 Agricultural inspection station, California border north of Reno, Nevada. The Golden State's multibillion-dollar agricultural kingdom is vulnerable to exotic diseases and pests. All major highways into the state have inspection stations similar to this one along U.S. 395 north of Reno. Any fresh fruits or vegetables to declare?

61

HISTORIC MILITARY LANDSCAPES

FIG. 6.17 Fort Laramie National Historic Site, Wyoming. Much of this trading and military fort (1849–1890) has been preserved and reconstructed.

HISTORIC MILITARY LANDSCAPES—ESPECIALLY FORTS AND BATTLE-grounds—represent another manifestation of federal largesse in the region (fig. 6.17). Western expansion, especially during the 1840s (**59**), produced an increased military presence. Forts also acted as early nuclei of settlement, principally to secure strategic trails and waterways, mountain passes, and zones of development. After 1865, new areas were settled, railroads were built (**50**), and the military focused on defeating Indian resistance (**38**). Army forts garrisoned troops and organized attacks against Native people who refused to live on reservations. Between 1865 and 1891, more than a thousand military engagements in the West led to 2,500 white and 5,500 Indian deaths, with many more caused by disease and starvation (fig. 6.18). By the 1890s, the conflicts had ended and many forts had been abandoned.

The distribution of forts and battlegrounds signifies areas of persistent conflict and shifting federal strategies of containment and control. Where Native peoples were subjugated early, such as coastal California, federal military power was less evident. Elsewhere—in southeastern Arizona, the

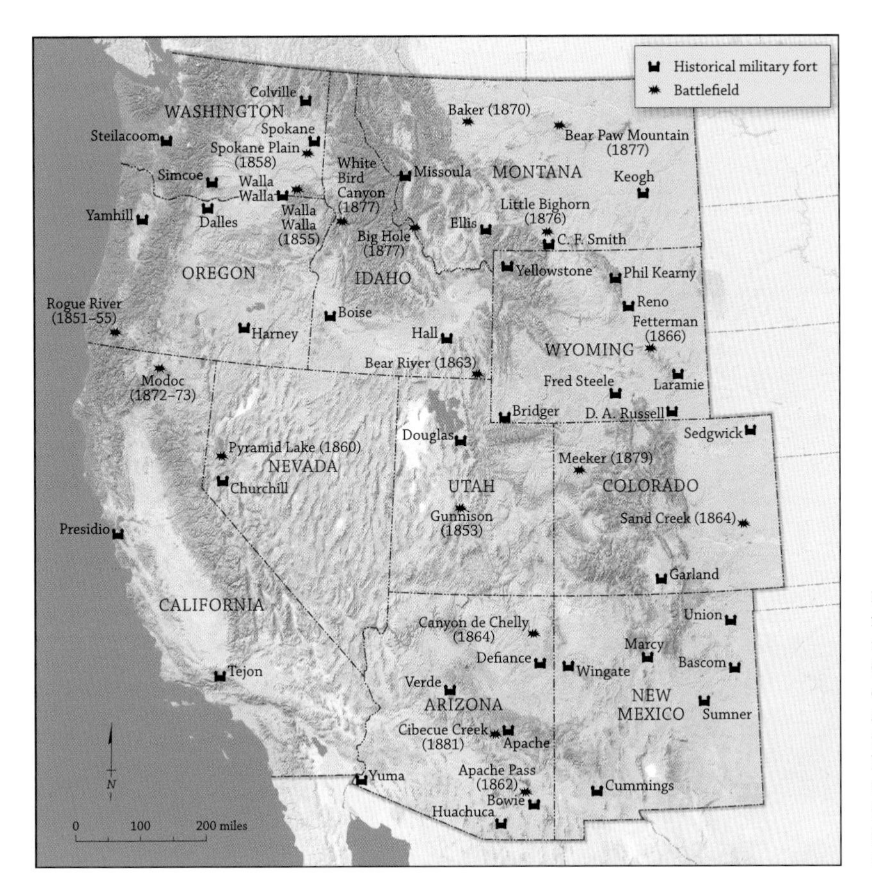

FIG. 6.18 Selected military forts and battlefields, 1850–1890. While coastal California was free of major conflicts between Indians and Euro-Americans, areas that experienced more hostile relations in Arizona, the western plains, and the interior Northwest became focal points for military forts and battlefields.

eastern plains, and the interior Northwest—tensions between Indians and federal authorities led to more numerous conflicts and a greater number of long-term military outposts.

A few historic military landscapes have been converted to modern military spaces (**62**), and some forts (typically built of logs or earthworks such as adobe) that were abandoned have been partially restored or rebuilt (fig. 6.17). Many battle sites were looted and became overgrown and were later reconstructed or reinterpreted as historic sites (fig. 6.19). Many are on the National Register of Historic Places and are managed by the National Park Service and the states as historic parks. At both forts and battlegrounds, you can see archaeological remains and find ways to understand the broader geographical setting and explore the modern cultural meanings of these sites for Native Americans and Euro-Americans.

Federally managed sites that are especially rich in cultural significance are Fort Laramie, Wyoming; Fort Bowie, Arizona; Little Bighorn Battlefield and Big Hole National Battlefield, Montana (and other units of the Nez Perce National Historic Park); the Presidio of San Francisco, California; and the site of the Sand Creek Massacre in Colorado. Excellent state facilities include Fort Tejon, California; Fort Churchill, Nevada; Fort Garland, Colorado; Fort Verde, Arizona; Fort Phil Kearney, Wyoming; and Fort Bridger, Wyoming. ❋

FIG. 6.19 Sand Creek, Colorado. Interpretive panels help visitors reconstruct the Sand Creek Massacre in 1864, where dozens of mostly unarmed Native American women and children were killed by the Colorado militia. Today the isolated spot is designated as the Sand Creek Massacre National Historic Site, administered by the National Park Service.

62 MODERN MILITARY SPACES

FIG. 6.20 Former coast guard station, Presidio, San Francisco, California. One of the West's oldest Euro-American forts, the Presidio was founded under Spanish control (1776) and later served as an outpost for both Mexico and the United States. In 1994, the U.S. military transferred the Presidio to the National Park Service, and it became part of the Golden Gate National Recreation Area.

WESTERNERS HAVE BECOME ACCUSTOMED TO THE PRESENCE OF low-flying jets, military convoys, and the sights and sounds of modern military spaces. The West, which is home to some of the planet's most militarized real estate, contains the vast majority of the Defense Department's 30 million acres (as well as related lands managed by the Department of Energy). The power of those landscapes—expressed through technological know-how, weapons production, and a coordinated assembly of personnel and equipment—is projected daily around the globe. This close, sometimes uneasy relationship between the military and the West was sealed by the Japanese attack on Pearl Harbor in 1941 (**45**). Since then, the region has played a peculiarly strategic role in the nation's defense, including the development of atomic weapons (**65**). Regional landscapes still reflect that World War II–era scaffolding, with successive layers added as the Korean War, Cold War, Vietnam War, and post-1990 conflicts ensued.

How did this geography evolve? An older scattering of military bases survived the frontier era (**61**). San Francisco's Presidio, which earlier had been under Spanish and Mexican control, became a launching pad of oceanic power as the nation expanded into the Pacific starting about 1900 (fig. 6.20). Wyoming's Fort D. A. Russell, near Cheyenne, was built to battle Indians; it

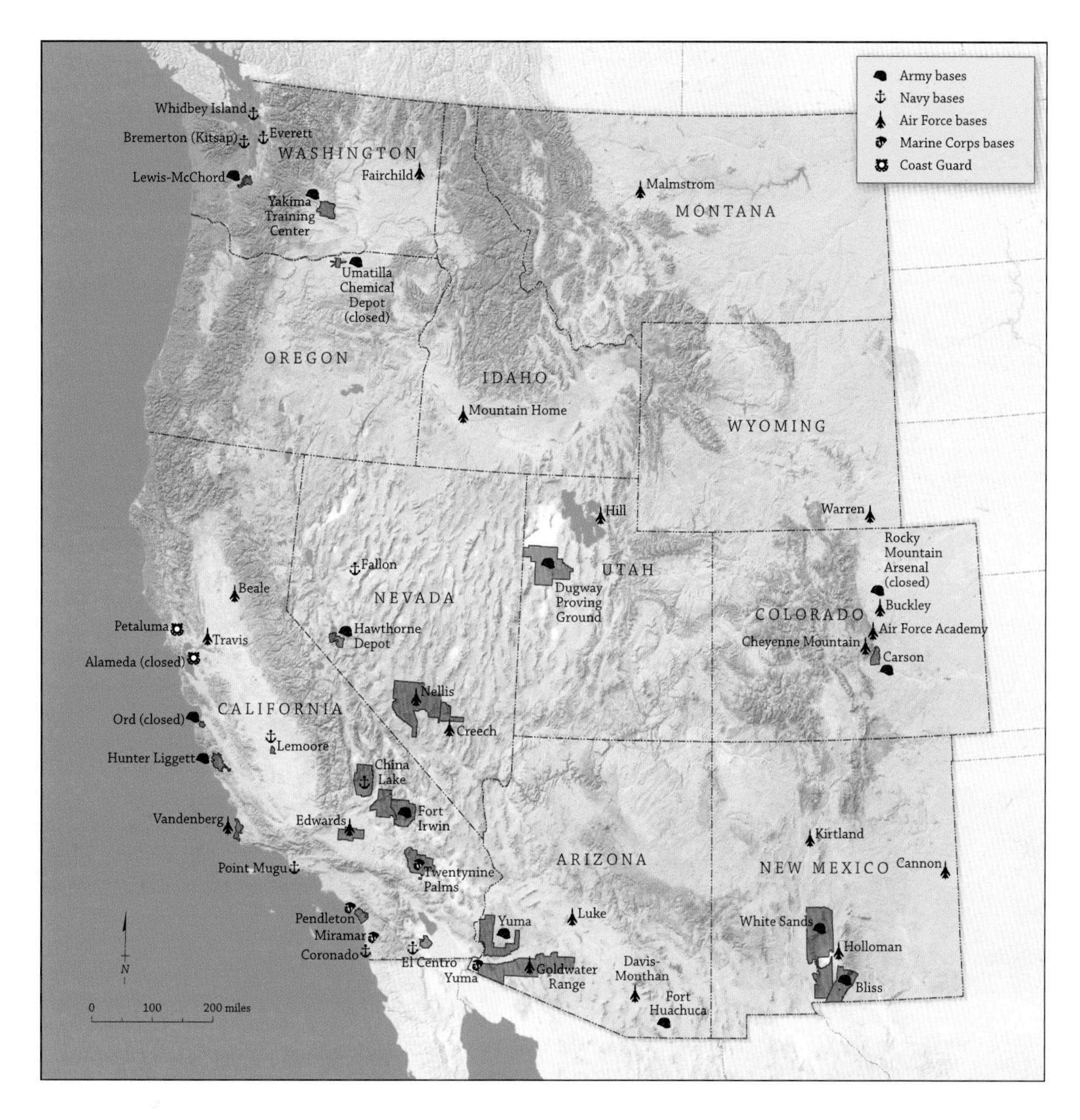

FIG. 6.21 Selected military bases. Since World War II, California has been an important location for western military bases. Other facilities are scattered throughout the region and include major army, air force, navy, coast guard, and marine bases. Nearby towns are also affected by the military presence.

has evolved to fit modern needs, and today, as Fort Warren (it was renamed in 1930), it is one of the nation's largest strategic missile facilities. But in most cases, modern military spaces had their origins in World War II and the Cold War (fig. 6.21).

The war created an extraordinary need for new bases, defense industries, and urban infrastructure, including thousands of wartime houses and apartments. Places like Southern California—called the Blue Sky Metropolis by aerospace historians—benefited most, but huge investments also poured into the Bay Area, Seattle, Denver, Portland, Salt Lake City, and a constellation of smaller places. Postwar years brought reinforcements to the region as more conflicts in the Pacific and the Cold War required new investments in what had become a military-industrial complex of defense-related facilities and industries. In the West, strategic coastal sites and a protected interior; a

FIG. 6.22 White Sands Missile Range near Alamogordo, New Mexico. The vast reaches of the White Sands Missile Range include both rocky hills and desert playas. This army facility, the largest in the western United States, played a central role in the development of the atomic bomb during World War II.

largely dry climate; open spaces of isolated, federally controlled land; scientific expertise at universities such as Stanford and the University of California at Berkeley; and enough political clout to obtain bases and defense contracts all combined to bring the region more than its share of military spending.

There are three distinctive expressions of military influence you can look for: major military bases and their adjacent towns; other support facilities controlled by the military, some inaccessible to the public; and affiliated landscapes more broadly connected to military programs and presence in the region. More than sixty major military bases are still active in the West, half of them in California (fig. 6.21). Landscapes in these areas are carefully controlled, managed by their own rules and regulations, and shaped by their own brand of architecture and design. Access varies, often limited by fences and gates. Air force installations expanded in the region after World War II, oriented to ongoing deployment (Luke, Nellis, Holloman, Hill) as well as to strategic air and missile defense (Malmstrom, Warren). In the vicinity of these bases, you can see isolated, fenced missile silo sites. Both the Air Force Academy (1954) and the underground North American Aerospace Defense Command facility at Cheyenne Mountain (1966) were built near Colorado Springs. Key navy (San Diego, Puget Sound), marine corps (Southern California), and coast guard (Bay Area) facilities are concentrated along the Pacific coast. Large U.S. army bases are widely dispersed, including the huge White Sands Missile Range in New Mexico, with 3,150 square miles (fig. 6.22), and the Yuma Proving Ground in Arizona, which covers 1,300 square miles of desert complete with simulated Middle Eastern settlements. In towns located adjacent to military bases, you will find off-base family housing and other services that single military personnel demand (**94**).

One variant includes bases that were closed after the late 1980s, most of them in California. These former military spaces are being reinvented as university campuses, golf resorts, nature preserves (Fort Ord), wildlife refuges (Rocky Mountain Arsenal), and national park and recreation areas (the Presidio; figs. 6.20 and 6.21). Ecologists have found that large, relatively

undisturbed portions of these lands—those areas not laced with nuclear waste or unexploded ordnance—contain some of the West's most remote, pristine environments, places that have been beyond the grasp of weekend dune buggies and condo developments.

Even less accessible and decipherable are military support facilities, which are strictly controlled by the military or the Department of Energy. Often you can visit military hospitals, cemeteries, and energy reserves (petroleum and oil shale), but visitors are prohibited at testing laboratories, bombing ranges, weapons storage depots, repair yards, and intelligence complexes (fig. 6.23).

Beyond direct government control is a myriad of broader military landscapes. You can see the influences of World War II, for example, at federally subsidized aluminum plants in the Pacific Northwest or in regional ports and shipyards from San Diego to Puget Sound. Other derivative landscapes include sprawling defense-related manufacturing plants and service operations, including those of Boeing, Lockheed Martin, Northrop Grumman, Honeywell, Raytheon, and General Electric. Defense-related research is being done at western universities and think tanks. It is also important to recognize the local Veterans of Foreign Wars post, the American Legion hall, and other expressions of patriotism as part of the military landscape in the West (fig. 6.24). ❋

FIG. 6.23 Hawthorne Army Depot, western Nevada. Since 1930, this depot has functioned as a major weapons and ammunition storage facility. Aging chemical weapons have also been decommissioned here. Beyond the barbed-wire fences, dozens of storage sheds and bunkers dot the high-desert landscape.

FIG. 6.24 Veteran's Memorial Park, Winnemucca, Nevada. This small city park on Winnemucca's northern outskirts celebrates the military's enduring importance for local residents. Nearby structures (background) supported an air force radar station atop Winnemucca Mountain (in the distance) during the Cold War.

65

NEW DEAL

FIG. 6.25 CCC project: Painted Desert Inn, Petrified Forest National Park, Arizona. Dozens of New Deal workers refashioned the inn between 1935 and 1940, giving it the pueblo-style look still enjoyed by visitors today.

THE GREAT DEPRESSION LEFT NO CORNER OF THE WEST UNTOUCHED. Especially hard hit were farm and ranch communities as prices for grains and livestock plunged. Making matters worse were drought conditions that affected large portions of the region—the Dust Bowl era on the plains—between 1931 and 1937 (**12**). Mining, logging, fishing, and manufacturing economies also suffered as demand for commodities fell, and unemployment and foreclosure rates soared. Banks failed and cities had Hoovervilles, squatter towns where the unemployed struggled to survive.

Beginning in 1933, Roosevelt's New Deal radically refashioned the region's economy, transformed its cultural landscape, and brought the federal government into the daily lives of many—indeed, most—westerners. The federal initiative jump-started the economy and stabilized the circumstances of workers, families, and communities. The West far outpaced the rest of the country in terms of per capita New Deal aid, with Nevada, Montana, Wyoming, and Arizona benefiting the most. The Civilian Conservation Corps (CCC), the Works Progress Administration (WPA), and the Public Works Administration (PWA) created jobs and invested in infrastructure projects, while the Emergency Banking Relief Act (1933), the Federal Emergency Relief Administration, and the Social Security Act (1935) stabilized the nation's financial system and put money into the pockets of millions of Americans (fig. 6.25).

In rural areas, the New Deal made utility lines possible, sparked by the Rural Electrification Administration. Shelter belts were created to limit erosion, the Soil and Conservation Service promoted contour plowing in hilly zones, and new irrigation districts (**23** and **64**) and hydroelectric projects (**56**) were created as part of dam-building and reclamation initiatives. Migrant worker housing (**27**) and new agricultural communities also were funded under farm resettlement programs, including Bosque Farms, New Mexico; Kinsey Farms and Fairfield Bench Farms, Montana; Casa Grande Valley Farms, Arizona; and San Luis Valley Farms, Colorado. On reservation lands, programs that were part of what was called the Indian New Deal (under the Indian Reorganization Act of 1934) stabilized the economy and increased tribal authority (**38**).

In cities and towns, in addition to sidewalks, bridges, and street improvements, thousands of local and state buildings and hundreds of public structures were built (figs. I.2 and 6.26), including courthouses, schools, community centers, hospitals, libraries, and post offices (fig. 6.27). Other elements of community infrastructure benefited from federal expenditures, including swimming pools, city parks (fig. 6.1), high school stadiums, power grids, public housing, and municipal water plants. The New Deal also created cultural capital through the Federal Arts Project and Federal Writers' Project (under the WPA), which produced paintings, city and state guidebooks, and public art (fig. 6.28).

On western public lands, the New Deal brought wide-ranging policy changes and investments. Consider the impact of timber management and livestock grazing programs on land that is now managed by the Forest Service

FIG. 6.26 Natrona County administrative center, Casper, Wyoming. The handsome Natrona County administrative center (formerly the courthouse) was built by the Public Works Administration in 1940 and has many Art Deco flourishes and bas-relief stone insets (top left) that chronicle the state's history.

FIG. 6.27 School and gymnasium, Kim, Colorado. Constructed during the 1930s under the New Deal's Civil Works Administration, these fine stone buildings are still used in this tiny southeastern Colorado town. Many similar schools, post offices, and other public buildings from the era dot the region.

FIG. 6.28 Mural, Coit Tower, San Francisco. This San Francisco scene (one of a series at Coit Tower) was commissioned as part of the New Deal's Public Works of Art Project, which employed thousands of artists during the Great Depression.

(**67**) and Bureau of Land Management (**68**). These policies emphasized sustainable grazing and timber harvesting practices and a shift toward scientific management strategies. In addition, under the Taylor Grazing Act of 1934, President Roosevelt initiated the withdrawal of more than 160 million acres from the public domain in 1935, further solidifying Uncle Sam's role as western land manager (most of these lands became BLM acreage). More broadly, there was an emphasis on nature preservation over resource extraction and privatization, with lasting consequences for western national parks (**66**) and wilderness areas (**69**).

Finally, infrastructure investments refashioned state and national parks, forests, and other public lands in the West (fig. 6.29). You can see that investment in roads, trails, utility lines, flood-control projects, tourist lodges, picnic shelters, campgrounds, interpretive centers, scenic viewpoints, fire

FIG. 6.29 The Castle, Guernsey State Park, Wyoming. Fine craftsmanship is evident in this elaborate stone picnic shelter built by the CCC as part of a project designed to upgrade park infrastructure. Several CCC structures within the park have received national recognition for their simple, elegant, rustic-style architecture.

lookout towers, fish hatcheries, and tree-planting initiatives. Many of these improvements were in the form of CCC projects. Hundreds of CCC work camps housed thousands of young men working in isolated rural, mountain, and coastal settings. Later, many of the camps were made into public campgrounds and resorts.

While the literary, artistic, and architectural contributions of the New Deal are increasingly being preserved, the broader cultural and political values that were part of those programs have weathered with time. Despite the desperate economic challenges of the Depression, New Deal programs were an expression of a society willing to invest in a shared notion of the public good. The New Deal kept the nation's economic system intact, quickened processes of national integration, and created a civic landscape that reflects the enduring value of rewards that were gained in a more progressive past. ❋

64

FEDERAL DAMS AND WATER PROJECTS

FIG. 6.30 Low water on Shasta Lake, California. A brown ring of exposed soil signals another drought year in Northern California. Federally built Shasta Dam, completed in 1945, remains a key part of California's Central Valley Project, which provides irrigation, municipal water, and electricity for state residents.

THE CREATION OF FEDERAL DAMS AND WATER PROJECTS IN THE WEST began with a trickle (fig. 6.30). By 1875, irrigation's potential had been demonstrated (**23**) and boosters were calling for federal involvement. The Desert Land Act of 1877 rewarded individual farmers with land if they irrigated; at the state level, California's Wright Act (1887) encouraged farmers to form cooperative irrigation districts; and the Carey Act (1894) made land available to western states that were willing to develop irrigation projects. But the crucial legislation was the Newlands Reclamation Act (1902), which created a federal system for financing and building dam and irrigation projects to support small farms. While the bill created a flurry of early and successful irrigation initiatives—for example, the Salt River, Arizona; Minidoka, Idaho; and Uncompahgre, Colorado, projects—the aggregate impact of the act on western irrigation was initially modest (fig. 6.31).

The tide turned with the Boulder Canyon Project Act, approved in 1928 (fig. 6.32). The act launched a new era of dam-building focused on massive, multipurpose projects that integrated flood control, hydroelectric power gen-

eration (**56**), irrigation (**23**), municipal water use, and recreation. The timing of the Boulder Dam (later renamed Hoover Dam) initiative was critical. Completed in 1935, the project was quickly folded into New Deal initiatives (**63**), providing a template for other federal projects, including the Bonneville (1938) and Grand Coulee (1942) dams on the Columbia, the Fort Peck Dam (1940) on the Missouri, and the Shasta Dam (1945) on the Sacramento, part of California's Central Valley Project (figs. 5.36, 6.30–6.32).

FIG. 6.31 Selected federal dams, 1902–1975. While some early (pre-1930) reclamation projects remain important today, it was the West's megaprojects (1930–1975) that most dramatically refashioned the regional scene.

After World War II, the pace of dam building quickened. In the Rocky Mountains, the Colorado–Big Thompson project (the C-BT, 1954) included transbasin movements of water from the Colorado to the Platte valleys through a system of 10 reservoirs, 25 tunnels, 3 pump plants, and 6 power plants. Both the Bureau of Reclamation and the U.S. Army Corps of Engineers were key players during those dam-building years between 1945 and

FIG. 6.32 Hoover Dam. Completed in the mid 1930s, Nevada's Hoover Dam marked an important shift in the scale of federally supported water projects in the West. The project served as a model for later initiatives around the region. Note the intake towers (above top of dam), which guide water to power turbines below (lower right). *Courtesy of Arizona Historical Society/Tucson, image no. PC196 F182–04-A.*

1975, when virtually every major western watershed was altered (figs. 2.2 and 6.31).

There have been countercurrents. Environmental lobbyists blocked some dam-building initiatives, including the proposed Echo Park Dam in western Colorado, which was defeated in the early 1950s. In Arizona, the controversial Glen Canyon Dam was eventually completed in 1966; but the popularity of Edward Abbey's novel *The Monkey Wrench Gang* (1975), which described the virtues of blowing up the dam, suggested that times were changing. In the late 1970s, the Carter Administration halted several federal dam projects, and dam building has slowed since 1980 because of rising costs, continuing protests, and environmental regulations. Increasingly, rhetoric has shifted toward dam removal on such rivers as the Klamath and the Elwha, particularly to restore native fish habitat.

There are at least three ways of looking at a western dam (figs. 5.36 and 6.32). First, dams tell us about cultural attitudes regarding nature. For many people, dams symbolize the successful management of the natural world, an ability to harness nature for beneficial use. Rivers are a western resource (**7**), a commodity that demonstrates how science has harnessed the power of the natural world for the public good. The spectacular placement of Hoover Dam in Boulder Canyon and Grand Coulee Dam on the Columbia are latter-day paeans to Manifest Destiny. According to this view, every dam is an altar to what historian Samuel Hays calls the "gospel of efficiency," a reminder of the benefits gained when progressive peoples unite the twin assets of capitalism and enlightened government. Opponents of dam construction and supporters of dam removals must still battle these beliefs, which run deep in the nation's utilitarian psyche. Many Americans love both large concrete structures and the engineering genius that built them.

Second, every federal dam has a local political story. Dam creation in the West never resulted from a carefully planned document that outlined the goals of integrated regional and national development. Efforts at regional planning—through the Missouri Valley Authority, the Bonneville Power Administration, and the Colorado River Compact—often yielded more legal conflicts than clarity. Americans both flourish and flounder within a diverse, pluralistic system of interest-group politics, bureaucratic competition, horse trading, and lawsuits that outlive their litigants. Every federal dam has a long history of planning, construction, and use: Where should it be located? Which federal agency should manage the project? Which engineering companies should be awarded the contracts? Who should reap the benefits and bear the costs? How should economic and environmental consequences be addressed?

Third, every dam is an intricate cultural landscape. If you look at a dam by following the water, you begin above the dam, at the reservoir where water is stored (fig. 6.30). Taking a broad view, you can see how the reservoir is a part of a larger watershed. Water levels can give you information: a bare shoreline may suggest drought, drowned trees and shrubs tell you the water is abundant. Consider the shoreline (**88**). It is a 1,520-mile hike around Montana's Fort Peck Lake. What land uses define this shoreline environment—resort homes, public beaches, fishing boats, or sailing craft? Finally, imagine the landscape without a dam.

Next, study the dam itself. The height of the dam can tell you about its hydroelectric potential. Arch-gravity dams, such as Hoover Dam, are convex structures typically wedged into narrow canyon settings (fig. 6.32), while gravity dams, such as Grand Coulee, hold back water through sheer size (fig. 5.36). Intake towers and penstocks (sluices, gates, and pipes) take water from above the dam to power turbines, which are turned by falling water to generate electric current. The turbines are typically located at the bottom of the dam to take maximum advantage of the "hydrostatic head," that is, the distance water falls, to generate power (fig. 6.32, lower right). Depending on reservoir levels and dam design, you may also see the spillway, where excess water is released. Some dams also have ship-navigation locks and fish ladders.

Finally, explore downstream. Dams dramatically change rivers, their sediment loads, currents, color, nutrients, plant life, and fish populations. The routes of high-voltage power lines can dominate the landscape around a dam, and water can also be diverted beyond the river (fig. 6.4) through pumping stations that lift water from reservoir levels. Major aqueducts and feeder lines can be open and concrete-lined or funneled through aboveground or buried pipelines. Downstream impacts can be varied and extensive. In elaborate irrigation works, such as California's Central Valley Project or the Columbia Basin Irrigation Area, and metropolitan systems such as the C-BT or the Central Arizona Project, water is transported many miles from the dam to faraway farms and thirsty suburban lawns. ❋

65

THE ATOMIC WEST

FIG. 6.33 Nevada Test Site (now the Nevada National Security Site) northwest of Las Vegas. This 1,300-square-mile federal facility hosted hundreds of nuclear tests between 1951 and 1992. Portions of the site remain enduringly (sometimes invisibly) poisoned, while other areas are breeding grounds for rare desert wildlife.

PART OF THE LEGACY OF THE ATOMIC WEST IS ITS ALMOST MIRACULOUS consummation along with its twinned qualities of potency and invisibility, all born from the trajectories of physics, war, and serendipity. Its regional landscapes both reflected and produced a disturbingly modern world (**62**). The infrastructure that sustained it—largely the result of federal imperatives—is now part of a global nuclear geography.

Dating the origin of the atomic West is problematic, but a reasonable beginning is June 28, 1941, when President Roosevelt signed Executive Order 8807, ramping up what became the Manhattan Project, a top-secret initiative designed to produce a nuclear bomb. What unfolded was a national program with profound regional consequences (fig. 6.34). Those consequences can be seen in a variety of settings across the West: think tanks, nuclear processing and waste storage sites, blast zones, yellowcake mining areas, nuclear military arsenal sites, and nuclear power plants (fig. 6.34).

A useful starting point is to consider those places where the atomic West was invented, beginning with the Los Alamos National Laboratory (fig. 6.34). This government settlement, which was built on an isolated northern New Mexico mesa in 1943, was the home for scientists such as J. Robert Oppenheimer. The planned community remains a vibrant research center, an

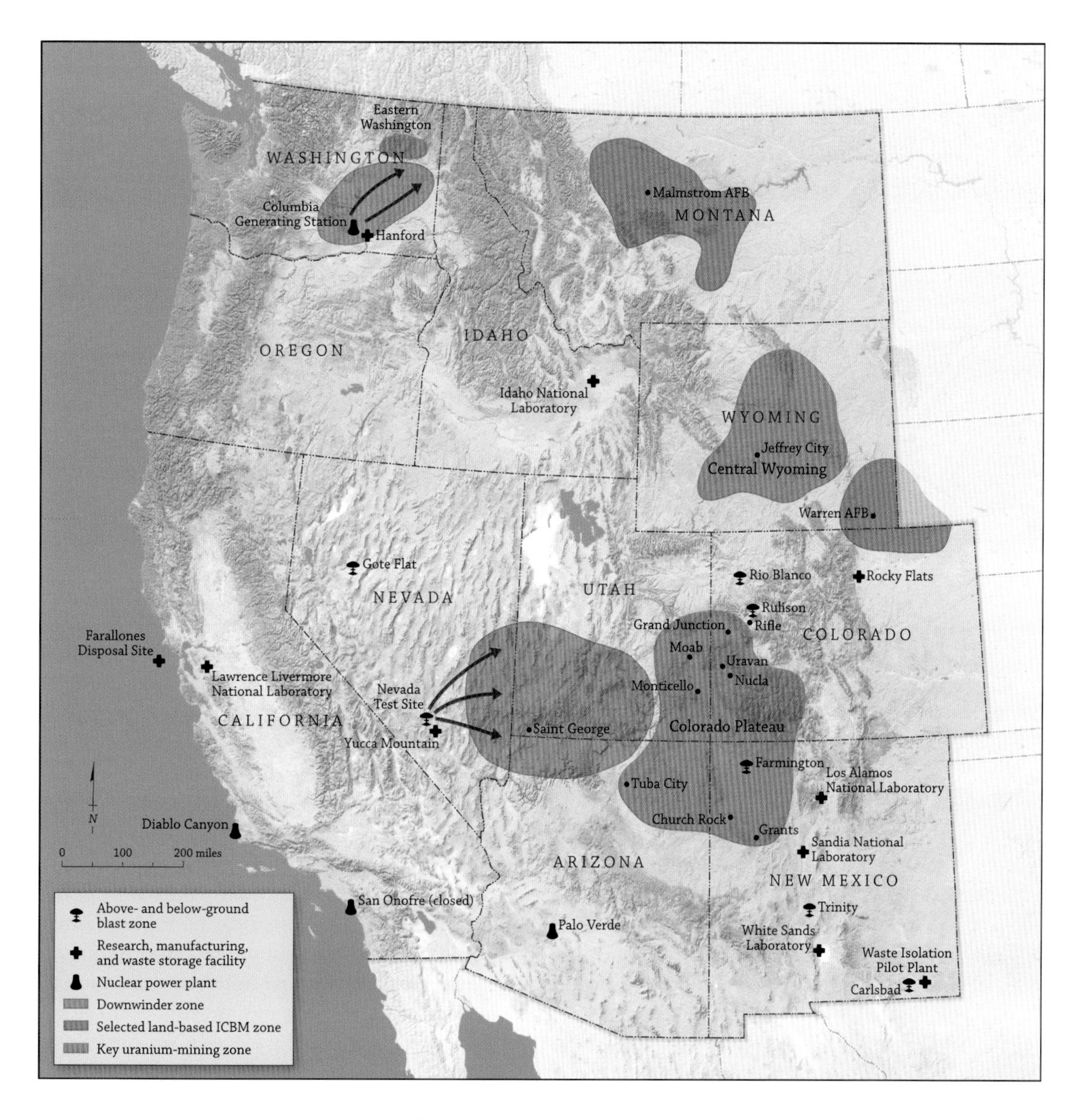

FIG. 6.34 The atomic West, 1941 to the present. The locations shown on this map are a sampling of features from the nuclear age that can be seen in the West.

exotic settlement far removed culturally from nearby Hispano villages (**39**). The Lawrence Livermore National Laboratory (1952) in California and the Idaho National Laboratory (1949) near Idaho Falls are other sites of nuclear research. Many universities—especially the University of California at Berkeley—and industrial labs (**82**) also helped shape the atomic West.

Research led rapidly to development; begin at Hanford (established in 1943), part of the Manhattan Project (fig. 6.34). This sprawling 586-square-mile slice of southeastern Washington was home to reactors that produced the enriched plutonium for nuclear weapons, including the first atomic bomb. Nearby Richland housed thousands of government workers. Radioactive waste (much remains here) eventually leaked into the groundwater and the air, making Hanford infamous as the nation's largest environmental cleanup project.

FIG. 6.35 The Waste Isolation Pilot Plant (WIPP) near Carlsbad, New Mexico. With little fanfare and predictable security, the federal WIPP facility is surrounded by some of the state's least populated land. Much of the nation's nuclear waste travels here to be buried quietly far beneath the desert landscape.

Other key nuclear processing and waste disposal landscapes include Rocky Flats near Denver—the onetime producer of plutonium bomb triggers is now a wildlife refuge—and the Waste Isolation Pilot Plant, a giant underground salt deposit near Carlsbad, New Mexico, which is a key destination for nuclear waste (fig. 6.35). In California, the Farallones, off the coast of San Francisco, are a group of islands whose surrounding waters for decades served as the nation's primary dumping ground for nuclear material. Despite local opposition, Yucca Mountain, Nevada, northwest of Las Vegas, is a candidate for the long-term disposal of discarded nuclear material.

The world's first atomic blast (July 16, 1945) took place at the Trinity Site in southern New Mexico's desolate Jornada del Muerto Desert (the name means "route of the dead man"; fig. 6.34). Now a national historical landmark, Trinity allows limited annual visits. During the Cold War, about one hundred aboveground (between 1951 and 1962) and more than nine hundred belowground (after 1962) blasts took place in remote western settings, most of them at the 860,000-acre Nevada Test Site—today known as the Nevada National Security Site (figs. 6.33 and 6.34). Public access is restricted for impact zones, subsidence craters, and test sites where damaged buildings remain, though tours are occasionally offered. Contact with released radioactive fallout at both the Nevada Test Site and Hanford caused serious health problems, particularly cancers, for a generation of "downwinders" in nearby Nevada, southwestern Utah, and Washington. You can explore more at the National Atomic Testing Museum, an affiliate of the Smithsonian, in Las Vegas.

What about the raw materials used to fuel the atomic West? Follow the yellowcake trail. The Navajo word for uranium is *leetso,* "the yellow dirt," leading to the term *yellowcake,* a mineral that made the Four Corners area a central landscape in the atomic West. Yellowcake—a radioactive mineral in carnotite, a yellowish sandstone found on the Colorado Plateau—was an essential raw material: nuclear reactions depended on radioactive uranium, an atomic building block for highly refined plutonium. Although initial supplies of uranium came from outside the United States, federal officials pressed for the development of domestic sources. The post–World War II uranium booms across the West created yellowcake towns (**31**) in southeastern Utah (Moab and Monticello), western Colorado (Grand Junction, Rifle, Nucla, and Uravan), northeastern Arizona (Tuba City), western New Mexico (Grants and Church Rock), and central Wyoming (Jeffrey City; figs. 3.25 and 6.34). The towns and associated processing facilities (**30**) produced milling operations and tailings piles.

Sadly, while the yellowcake trail gave life to the nuclear age, it also took life away from many who traveled its length: thousands of workers who worked in or near the mines and mills, including Navajos, were exposed to radioactive dust. For them, the trail ended within their own bodies, and elevated cancer rates remain a regional fact of life. In addition, federal and industrial demands for uranium have swung from boom to bust, creating twentieth-century ghost towns (**32**). If you explore these landscapes, be careful near yellowcake zones, which are full of elaborate environmental cleanup operations.

The West is also home to products of the nuclear age. In modern military spaces (**62**), you can see all parts of the nuclear triad, a three-pronged defense strategy to protect the nation. There are land-based ICBM (intercontinental ballistic missile) silos, most common across remote areas of Montana, Wyoming, and the Dakotas; Trident submarines near major bases on the coast; and long-range strategic bombers and their associated air force installations. The Titan Missile Museum, south of Tucson, Arizona, offers the Cold War context.

Nuclear power also satisfies some of the West's hunger for electricity (**56**; fig. 6.34). Reactor vessels and cooling towers are seen in California at Diablo Canyon near San Luis Obispo and at San Onofre in San Clemente (now closed); in Arizona at Palo Verde, a huge complex west of Phoenix; and in Washington at the Columbia Generating Station at Hanford. One interesting tidbit of isotopic intermingling: after 1995, highly concentrated uranium from thousands of discarded nuclear weapons from the former Soviet Union was diluted to produce lower-grade fuel that was then sold to the American power industry. Today, thanks to the Megatons to Megawatts Program, North American skies are lit up in ways never contemplated by Cold War strategists.

To move from physical to metaphorical landscapes, the recognizable symbol of a mushroom cloud, signs warning of radioactivity (from hospital labs to test sites), Cold War homilies to "duck and cover," the Miss Atomic Bomb beauty contests of the 1950s, and political conversations surrounding the antinuclear movement are all part of the cultural fallout of our atomic history. There are also atomic-themed literary works, including Leslie Silko's *Ceremony;* films such as *Dr. Strangelove, The China Syndrome,* and *Them!;* and video games such as *Missile Command, DEFCON,* and *Call of Duty: Modern Warfare 2.* In addition, Richard Misrach, Peter Goin, and Carole Gallagher are among the photographers who have powerfully documented discarded nuclear localities and victims. ❋

66 NATIONAL PARKS

FIG. 6.36 Old Faithful Inn, Yellowstone National Park. Designed by Robert Reamer and completed in 1904, the Old Faithful Inn was an early and important example of rustic national park architecture (or "parkitecture") that made extensive use of naturalistic design elements and local materials. It is the largest log hotel on earth. *Source: author's collection.*

THE WEST'S MOST SPECTACULAR, ICONIC NATURAL LANDSCAPES—Yosemite Valley, Grand Canyon, the redwoods, Old Faithful—are in national parks (fig. 6.36). The National Park Service (NPS) manages about twenty million acres of western federal lands in more than 130 separate units (fig. 6.37). National parks are defining elements of regional character and representation, and they function as national symbols suggesting how nature is entwined with American identity.

National parks are also distinctive political spaces, shaping land use, landscape features, and recreational and preservation practices. Yellowstone was the nation's first national park, created in 1872 (portions of Yosemite were set aside in 1864 as a state park, but not federally controlled until 1890), and for almost twenty years it was literally *the* national park (fig. 6.36). As interest grew in western conservation, the government established other parks between 1890 and 1916 (fig. 6.37), and the American Antiquities Act of 1906 gave the president the power to declare national monuments, which helped create more park units, including many that preserved sites of historical interest.

The Organic Act of 1916 formally established the NPS and initiated an integrated approach to park development and promotion. But the act also

set ambitious, seemingly contradictory goals. The parks were to preserve resources while also providing for their enjoyment by the public. Early NPS directors Stephen Mather and Horace Albright believed it was vitally important to promote the parks to the public—an effort echoed by railroad interests (**50**)—by opening them to automobile travel; providing campgrounds, lodges, and hotels (developed as private concessions); and educating visitors with campfire talks, ranger walks, and other interpretive programs (fig. 6.38). But they also strove to maintain a rustic atmosphere in which visitors could enjoy a park's unique character without damaging it. More large western parks were created under their influence between World War I and World War II (fig. 6.37). New Deal programs, especially the CCC and WPA (**63**), also added park infrastructure in the 1930s.

FIG. 6.37 Selected national parks. Major units created prior to 1940 remain among the most popular parks in the West. While the NPS has expanded its definition of park-worthy characteristics, preferences for high mountain environments and deep, spectacular canyon settings still shape the system.

After a period of little funding for national parks from 1942 to 1955, the

FIG. 6.38 Ranger Walk, North Rim, Grand Canyon National Park. Particularly after 1930, guided ranger walks, evening campfire presentations, and other interpretive programs became an integral part of the national park experience in the West.

NPS was reenergized by Mission 66, an ambitious ten-year plan (1956–1966) to modernize the parks. Between 1960 and 1990, more western parks were created, but the philosophy shaping them shifted to an ecosystem management approach that focused on preserving their wilderness character (**69**). The NPS also added national recreation areas, national historic parks and sites, battlefields (**61**), and historic trails (**49**).

What should you look for when entering a park? National parks—especially the large, older units—have gateway communities and formal entrance stations. Uniformed rangers will collect fees and offer you park maps and circulars explaining rules and regulations. Most visitors arrive by automobile, and national parks include scenic drives and viewpoints, many constructed with Mather's inspired idea that "wilderness" can be experienced through a car windshield. Famous auto excursions include the Grand Loop Road in Yel-

FIG. 6.39 Visitor center, Zion National Park. Completed in 2000, the visitor center is a mix of old and new, combining traditional rustic-style building elements with energy-efficient construction technologies. It is promoted as one of the hallmark "green buildings" of the National Park Service.

FIG. 6.40 Visitor boardwalks, Yellowstone National Park. This winding trail carefully guides visitors through the hazards of Yellowstone's Upper Geyser Basin (home to Old Faithful). The National Park Service has produced its own characteristic federal signature across the West.

lowstone, Going-to-the-Sun Road in Glacier, Mount Carmel Highway in Zion, the Trail Ridge Road in Rocky Mountain, the Paradise Valley Road in Mount Rainier, and the Rim Drive in Crater Lake.

Early visitor centers and lodging facilities reflect the popularity of rustic design elements—the use of local materials that harmonize with the natural setting, sometimes called *parkitecture* (fig. 6.36). Mission 66–era facilities built from the 1950s through the 1970s emphasize modernity and convenience and have a postwar suburban feel. More recent structures (post-1980) display a neotraditional fusion of modern and earlier architectural styles (fig. 6.39). Park infrastructure, interpretive programs—ranger talks, roadside exhibits, trail signs, and guided tours—and natural and cultural features combine in national parks to create a carefully crafted experience for visitors (figs. 6.38–6.40).

National parks remain contentious places, and you can see the evidence of how local or national management issues are playing out. External threats to a park—nearby population centers, resource pressures, global-scale changes—and internal threats—inadequate infrastructure and potentially incompatible activities—complicate management. Imperfect fragments of nature and history, national parks may be accidents of haphazard political bargaining and changing national priorities, but as geographer Thomas Vale puts it, they are "the most successful landscape protection project ever attempted." ❋

67

NATIONAL FORESTS

FIG. 6.41 Campsite, Beaverhead-Deerlodge National Forest, Montana. Thousands of Forest Service campgrounds have been constructed since 1920. A typical campsite includes a picnic table, fire ring, and tent site. Sites are normally spacious, cleared of undergrowth, and strung out along loop roads. In this setting, the fence separates the camp from a nearby cattle-grazing allotment.

AS THE U.S. FOREST SERVICE SLOGAN CLAIMS, NATIONAL FORESTS ARE managed as "lands of many uses." A long-standing commitment to multiple use is a defining mantra of the Forest Service, which manages these parts of the public lands. Beginning in the 1780s, the federal government focused on the disposal of lands in the public domain, selling to settlers, farmers, and ranchers who populated the West (**58**). That process worked for land with agricultural possibilities, but by the late nineteenth century most of the region's forested mountains remained unclaimed. A coalition of preservationists, who favored preserving forests, and conservationists, who supported the responsible use of forests, worked to set aside large areas of forest land. Today, there are more than seventy-five national forests and several national grasslands totaling more than 165 million acres across the West. National forests make up 39 percent of all land in Idaho, 25 percent in Oregon, 22 percent in Colorado, 21 percent in Washington, and 20 percent in California (fig. 6.5).

The historic shift from land disposal to public stewardship unfolded between 1890 and 1910, when much of the extent, administrative structure, and management philosophy of the national forest system fell into place. The Forest Reserve Act (1891) created the executive authority, beginning with President

FIG. 6.42 Logging landscape, Lolo National Forest, Montana. Timber sales on public lands within the northern Rockies remain an important source of Forest Service income but often generate local conflicts between the logging industry and environmental groups.

Benjamin Harrison, to reserve western public lands and protect their timber and watershed resources. The blueprint for using the new forest reserves was laid out in the Forest Management Act (1897), which created a detailed template for managing the lands for the benefit of multiple users. The system's vitality was secured during the administration of Theodore Roosevelt (1901–1909), when national forest acreage increased from 46 million to 151 million acres. Chief Forester Gifford Pinchot, who was responsible in 1905 for having the national forests placed under his control in the Department of Agriculture, created the Forest Service to develop regulations and guidelines designed to manage the lands sustainably and scientifically for multiple users.

The tradition of multiple forest uses has continued since then. Farmers needed dependable watersheds, for example, and disastrous wildfires in 1910 led to an emphasis on fire prevention. The growing popularity of outdoor recreation sparked road building and campground construction, which accelerated in the 1920s and benefited from New Deal infrastructure spending in the 1930s (**63**; fig. 6.41). Livestock grazing on leased allotments (**20**) expanded, but drought and overuse required tighter controls after 1934. At the urging of preservationist advocates such as Aldo Leopold, the Forest Service also began designating wilderness or primitive areas, beginning in 1924 with the Gila Wilderness (**69**).

Following World War II, there were new demands for recreation and increases in commercial timber sales (**33**) to meet the needs of the growing home-building industry (fig. 6.42; **76**). The Multiple Use–Sustained Yield Act of 1960 reaffirmed the agency's commitment to managing the forests for multiple uses while mandating that it pay attention to sustained yield. The National Environmental Policy Act (1970) and the National Forest Management Act (1976) mandated ecology-based management, greater regulation of forest resources, and an increased emphasis on recreation and preservation.

The boundaries of national forests are usually marked on the landscape with familiar brown and yellow signs, but there are also complicating geographies that reflect the agency's bureaucratic evolution. First, each forest lies within one of six larger western regions—Rocky Mountain, Southwestern, Intermountain, Northern, Pacific Northwest, Pacific Southwest—and every forest has its own administrative geography of ranger districts (fig. 6.43) and a coordinating supervisor's office. Second, you may notice a checkerboard pattern of alternating Forest Service and privately owned sections, evidence of the enduring power of nineteenth-century land grants to railroads (**50** and **58**). There are also small, oddly shaped parcels of private inholdings, evidence of lands alienated before the national forest was designated or of settlers who took advantage of the Forest Homestead Act of 1906. Special-use areas are recognizable through the presence of grazing allotment fences, boundary signs for wilderness areas, or landscapes of mining and energy development as well as logging.

Earth tones dominate the national forest landscape: brown and tan signage, green patrol vehicles, khaki-green ranger uniforms, and pine-tree-studded badges. Fire danger signs, under Smokey Bear's watchful eye since 1944, remind visitors of that seasonal hazard (fig. 6.43). An institutionalized material landscape includes ranger stations and cabins, campgrounds,

FIG. 6.43 Rocky Mountain Ranger District Office, Lewis and Clark National Forest, Choteau, Montana. Many western towns are home to local district offices, which manage nearby national forests. This view includes a familiar institutional landscape of signs, structures, and vehicles.

FIG. 6.44 Bridge over the Lochsa River, Nez Perce–Clearwater National Forests, Idaho. This sturdy bridge offers easier access to the nearby Selway-Bitterroot Wilderness and is one of many examples of federal infrastructure found on western public lands.

forest highways, bridges, and trails that follow standardized practices, but also includes variations that reflect regional tastes and the use of local materials (figs. 6.41 and 6.44).

The national forests also reflect the changing policies of public-lands stewardship. Consider the cumulative breadth of Forest Service manipulation. Examples include millions of logged acres reseeded in varied species; the impacts of livestock grazing; wildlife management such as historical predator controls, reintroduced species (such as wolves), recreational hunting regulations, and planted fish populations; weed eradication and insect pest-control programs; environmental impacts of regulated mining and energy developments (**28**, **29**, and **36**); and the steadily growing role played by recreational activities and resort development.

Will the national forests survive the twenty-first century? Will a warmer, drier West and an increase in regional populations threaten forest health? And most important, how will tomorrow's generations interpret the notion of multiple uses to create future national forest landscapes? ❋

68

BLM LANDS

FIG. 6.45 BLM grazing allotment west of Quemado, New Mexico. In western New Mexico, large tracts of BLM land are integrated with private ranch acreage to provide seasonal cattle pastures. Fenced pastures help organize the ranching landscape, and it is not easy to separate private and public acreage just by looking.

TAKE A SLICE OF THE WEST. REMOVE FARMS, TOWNS, AND CITIES. NEXT, select monumental landscapes for national parks and monuments. Claim crucial forested watersheds for national forests. Eliminate marginal lands populated by Native peoples for Indian reservations. Finally, excise acres for military bases and atomic blast zones. What's left over? Welcome to the West's BLM (Bureau of Land Management) lands.

Despite the inviting river, mountains, and trees depicted on the BLM's official logo, most of its 150 million acres in the West have a tawnier, arid look. About one-third of BLM lands are in Nevada (68 percent of that state's area). Similar tracts of desert and rangeland are in southeastern California, southeastern Oregon, southern Idaho, Wyoming (29 percent of that state's area), Utah, and the Southwest (fig. 6.5).

Unlike the National Park Service with its scenic jewels (**66**) or the United States Forest Service with its compelling mission articulated by Gifford Pinchot (**67**), the BLM has always suffered from an identity crisis. It has no riveting focus or charismatic founder. Nor do iconic landscapes define it. Since becoming part of the public domain, BLM lands have passed through three broad phases of federal tenure.

Before the passage in 1934 of the Taylor Grazing Act, which initiated reg-

FIG. 6.46 Jawbone Off-Highway Vehicle Open Area north of Mojave, California. While critics point to the ecological damage evident at these BLM sites, off-road enthusiasts limit their larger regional footprint by concentrating use on lands reserved for recreation.

ulated grazing districts and closed the public domain to settlement, most of these lands were available for homesteading or purchase. But farmers typically looked elsewhere, and most users were cattle and sheep ranchers, who grazed their stock on what was identified as the "open range" (**20**). There were also mining and oil and natural gas interests, which operated under the General Mining Law of 1872 and the Mineral Leasing Act of 1920 (**28**, **29**, **35**, and **36**).

The second phase began when the BLM was created by the 1946 merger of the Grazing Service and the Government Land Office, but its mission within the Interior Department remained focused on the extractive economy (fig. 6.45). For thirty years, mining, grazing, and energy development dominated the BLM landscape.

The most current phase began in 1976 when the Federal Land Policy and Management Act mandated that BLM lands should be managed for multiple uses under a sustained-yield philosophy that also emphasized recreational activities, wilderness preservation, and ecosystem restoration. Since then, despite internal resistance, the BLM has developed more recreational landscapes and moved to preserve its western wildlands—1,500 miles of rivers, 8.7 million acres of wilderness (**69**), and more than 12 million acres designated areas of "critical environmental concern."

The BLM is the least visible public land agency, but there are clues you can follow to find its landscapes. Small agency signs direct you to recreational sites where off-road vehicles are allowed (fig. 6.46). Also look for "watchable wildlife" sites and "back-country byways." Some national monuments and wilderness areas—key elements of the agency's sprawling 27-million-acre National Landscape Conservation System, christened in 2000—represent BLM's expanded role. Traditional BLM activities endure in grazing allotments and mining and timber sales, while energy-related activities (through leases) are increasing, particularly in shale-rich zones being developed for their fossil-fuel potential (**36**). ❋

69 FEDERAL WILDERNESS

FIG. 6.47 Primitive campground, Lee Metcalf Wilderness, Montana. This Montana wilderness area includes units managed both by the Forest Service and by the BLM. This simple campsite along the Madison River is within the BLM's Bear Trap Canyon Unit, the agency's first tract of designated wilderness.

THREE STRANDS OF HISTORY, INEXORABLY INTERTWINED, PRODUCED the map of federal wilderness areas that we see today (figs. 6.47 and 6.48). First, trace the story of settlement: Between 1850 and 1920, huge portions of the West were opened to development, regional populations grew, and resources were exploited. Within two generations, wildlands had been vastly diminished at the same time that the nation's population was becoming increasingly urban. Second, recognize how cultural shifts over the course of American history shaped American attitudes toward wilderness: Writers such as Henry David Thoreau, John Muir, Aldo Leopold, and Edward Abbey articulated these shifting attitudes. The cultural values they advocated resonated with many urbanized Americans who saw the nation's remaining wildlands threatened. Finally, you can see the creation of wilderness as the result of political action: As demands for wilderness grew, a contentious legislative process began with late nineteenth-century urban elites (the Sierra Club originated in 1892) and gathered inspiration from preservationists such as John Muir. Early chapters in that story are embedded in the creation of national parks (after 1872) and primitive areas (the Gila Wilderness in 1924).

The statutory creation of federal wilderness crystallized in the Wilderness Act of 1964 (fig. 6.49). Passed amid growing post–World War II concerns over

FIG. 6.48 Federal wilderness areas. California contains the largest wilderness acreage in the West (a mix of mountain and desert lands). Other large blocks of wilderness include tracts in the northern Rockies, along the northern Cascades, and across the Colorado high country.

the survivability of wildlands, the act provided the legal framework for the fifty million acres of wilderness lands in the West today (about 43 percent of the nation's total). About 60 percent of designated wilderness acreage in the region is within national forests, the largest tracts in the California and Idaho mountains (**67**), with roughly another 20 percent each inside national parks (**66**) and on BLM lands, primarily in the deserts of California, Nevada, and Arizona (**68**; fig. 6.47). More than two million acres were added to the system with the Omnibus Public Land Management Act of 2009.

Explore the West's wilderness areas. Recreation is nonmotorized, campsites are primitive, and logging is prohibited. While you may encounter seasonal hunting and fishing and even selective livestock grazing and mining, most wilderness areas are roadless. Wilderness areas offer solitude and silence, precious amenities to many Americans who spend most of their lives in busy cities and suburbs. ❋

FIG. 6.49 Wilderness boundary sign, Zion National Park, Utah. Landscape expressions of federal wilderness in national parks and forests include simple trailside signs marking the boundaries of these specially protected areas.

7

CITIES AND SUBURBS

FIG. 7.1 Suburbs under construction, Sonoma Ranch, New Mexico. This master-planned community near Las Cruces takes shape in several phases. Completed neighborhoods and a golf course offer immediate amenities while new streets and lots are still being carved from the desert landscape.

A CITY'S COLLECTIVE MEMORY IS ITS CULTURAL LANDSCAPE, BUT IN the urban West historic landscapes have often been bulldozed, burned down, flooded, rebuilt, redesigned, and reimagined. The pace of urban change has been remarkable in the region, as landscapes have been transformed by vigorous growth, much of it since World War II (fig. 7.1). Some elements of the urban West reflect larger American preferences, while others reveal regional peculiarities. In a western American city with a population of about half a million people (fig. 7.2), you can consider yourself in a bona fide metropolitan area, defined by the U.S. Census Bureau as a central city of at least fifty thousand people plus adjoining counties.

If you begin at the city's edge and navigate a course from the far suburbs to the urban center, miles on the ground translate into decades on the landscape, from new developments on the suburban periphery to the older downtown. The first challenge is to identify the city's edge. Every city has what is known as a tributary area (or functional region) within which its economy and influence dominate. But this can often extend far beyond its adjacent suburbs. Newspapers and billboards connected to the city are found sometimes hundreds of miles away, as are the water sources, dairy products, and crops that sustain people who live there. Recreational zones and amenity retreats—coastal resorts, ski hills, campgrounds, cabin communities—that cater to the city's residents are also a part of this larger zone. Sometimes these hinterland relationships are articulated in benign ways, such as the his-

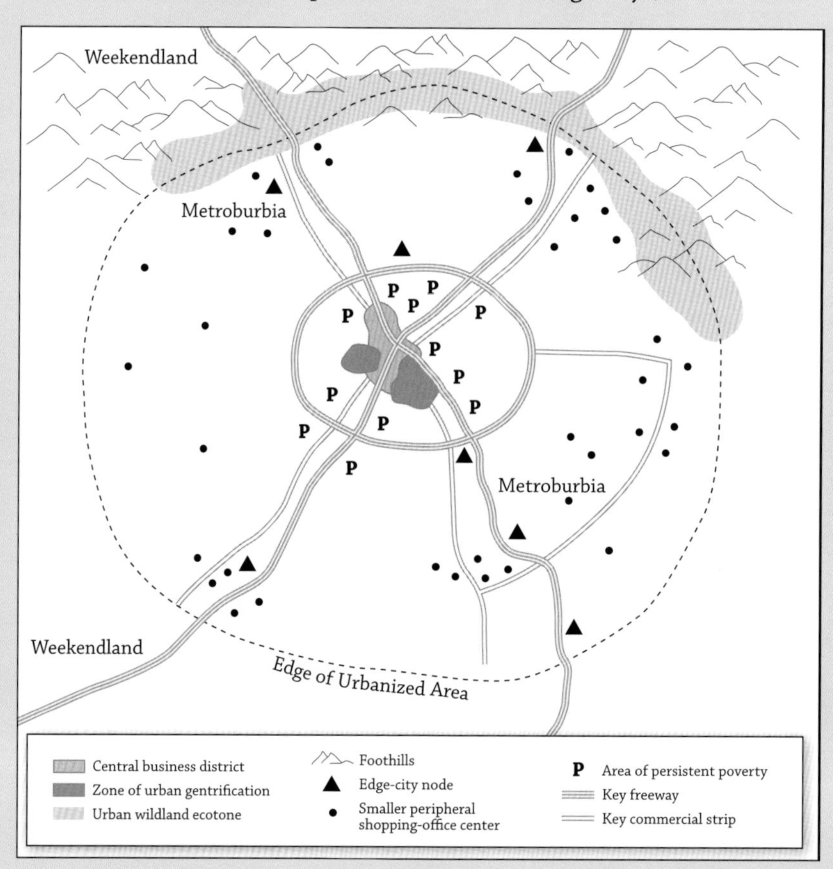

FIG. 7.2 The western American city. The typical western metropolis retains a vibrant downtown, but dynamic urban growth is on the expanding periphery.

FIG. 7.3 Commercial strip, East Colfax Avenue, Denver, Colorado. Daily necessities beckon suburban commuters on this busy stretch of Colfax that offers convenient short-term loans, piercings, fast food, liquor, and gasoline.

torical embrace of the Inland Empire in Spokane, Washington. Elsewhere, people who live on the periphery have a less harmonious relationship with their urban neighbors; consider the story of how Owens Valley water was captured by thirsty Los Angeles. However imagined, these larger urban footprints are worth tracking far beyond the first suburban interchange.

Finally, there is the urbanized area, the actual limit of denser urban settlement where new subdivisions meet open land (figs. 1.6 and 7.1). In the West, topography, mountains, and coastlines often define these urban margins, shaping patterns of expansion and sprawl. For many western localities, this link to the natural world is a defining element of urban character. In cities such as Tucson next to the Santa Catalina Mountains, Seattle on Puget Sound, Denver facing the Front Range, and Phoenix in the Lower Sonoran Desert, physical setting helps produce both land-use patterns and place identity (fig. 7.2).

Western cities are christened and sustained by construction, fueled by cycles of speculative building that over decades add layers to the urban fabric (fig. 7.1). On the periphery of a city, you'll likely see new homes, condominiums, and apartment buildings. Nearby are shopping centers, strip malls, perhaps a cluster of offices and retailing (edge cities), and an assortment of restaurants. This mix of residential, commercial, and industrial land use is not restricted to the West, but fast-growing western cities offer superb examples of these multiple peripheral nuclei of concentrated activities (sometimes called *metroburbia*), especially near interstate highways and metropolitan freeways.

In established suburbs, commercial strips may lead you through older suburban districts (figs. 7.2 and 7.3), with shopping centers and strip malls that sometimes cater to travelers on nearby interstates. As the landscape ages toward the city's center, distinctive regional cultural influences may be more visible. With the possible exception of Texas and Alaska, the characteristic regional mix of Anglo, Latino, Asian, Native American, and African American residents appears in no other North American setting.

FIG. 7.4 Southside bungalows, Denver, Colorado. Often built between 1905 and 1930, the California-style bungalow can be found in dozens of older residential neighborhoods across the West. Brick versions are common in the Rockies.

FIG. 7.5 Victorian landscape, San Francisco, California. While older housing stock has been demolished in many western cities, San Francisco's famed "painted ladies" (often adorned with bright, varied colors) survive in many parts of the city.

More generally, houses get older as you move toward the city center. You can find more peripheral neighborhoods dominated by ranch houses from the 1950s and 1960s, a popular architectural style with strong roots in the West. The regional love affair with the California bungalow, many built between 1905 and 1930, is evident nearer to the city center (fig. 7.4). Especially in older cities such as San Francisco, you might even see surviving residential signatures from the Victorian era, often close to downtown (fig. 7.5).

Apartment houses are a nationwide phenomenon, but many apartment complexes in the West have design elements and landscape features that give them a distinctive regional look. The relatively high cost of land in western cities has made apartment houses and higher-density suburban living common. Surprisingly, this means that Los Angeles, San Diego, Phoenix, Sacramento, and Las Vegas are among the most densely settled large urban areas in the country.

At the city's core, you will find a great deal of block-to-block diversity: old warehouse districts, new sports stadiums, neighborhoods of persisting poverty, and upscale brewpubs, all within a block or two of one another. The central business district remains a dense collection of tall buildings housing a mix of office-related, retailing, entertainment, and civic functions (figs. 1.25 and 7.6). The scale of the downtown landscape is notable, containing examples of mega civic landscapes, where

public activities are centered, and mega consumer landscapes, where shopping and consumption are emphasized. These landscapes carry symbolic value and are important elements in urban place identity. Inner-city landscapes also reflect recent investments—sometimes called gentrification—in shopping venues, sports and entertainment complexes, civic centers, upscale central-city housing, and open space (fig. 7.7).

In addition to assessing this geographical variability around the city, you can also appreciate how historical processes have shaped the urban landscape. Despite its reputation for wide-open spaces, the West is one of the most urbanized parts of the nation in the early twenty-first century. But the region's human geography has changed a great deal over time. In 1830, the map of the urban West would reveal only small settlements at Santa Fe (dating from 1610), Los Angeles (1781), and Monterey (1770), the British trading post at Fort Vancouver (near present-day Portland, Oregon), along with small urban settlements in San Antonio (then in Mexico) and Sitka (part of Russian Alaska).

Between 1840 and 1920, the look and distribution of western American cities were transformed as the region became fully incorporated within the United States. Dozens of towns, typically laid out in an expansive grid pattern with oversize streets and an overabundance of lots, guided regional settlement as migrants moved west. The raw numbers are breathtaking: in 1860, San Francisco

FIG. 7.6 Central business district, Denver, Colorado. Western American downtowns have well-defined clusters of high-rise office buildings, hotels, and shopping districts. Sports facilities and recreational activities (Pepsi Center and Elitch Gardens, left) also offer centralized attractions.

FIG. 7.7 Urban renewal, Tucson, Arizona. This pleasant open space in El Presidio Park near the Pima County Courthouse is part of a larger effort to renovate downtown that began in the early 1970s.

(57,000 residents) overwhelmingly dominated the West's urban system, with Sacramento (14,000), Salt Lake City (8,000), and Los Angeles (4,000) mere towns in comparison. By 1920, a largely urban West boasted two sprawling California metropolises, Los Angeles (577,000) and San Francisco (507,000); a Pacific Northwest dominated by Seattle (315,000) and Portland (258,000); and an interior with several major urban centers, including Denver (256,000) and Salt Lake City (118,000). Only the Southwest had few urban residents, awaiting later Sun Belt growth. In 1920, Phoenix was still a modest town of 29,000 and Albuquerque had a population of 15,000.

Each city possessed its own flavor. All had well-defined downtowns surrounded by neighborhoods differentiated by class and ethnicity. In larger centers, streetcar lines promoted suburbanization, especially after 1890. Denver, Los Angeles, and San Francisco saw suburbs blossom along the new transportation lines.

New expressions of ideal urban life also emerged selectively between 1900 and 1930, as planners promoted downtown renewal and lobbied for tree-lined boulevards and grand buildings. Visionaries imagined an emerging middle class positioned in the suburbs between a world of urban conveniences and the amenities of open space, lower population density, and social exclusivity. Architecturally, the affordable, efficient modern bungalow was considered a perfect home for the suburban family.

Between 1920 and 1960, the automobile transformed American life, spurring the expansion and improvement of streets and highways and creating new demands for automobile-related services. As environmental historian Christopher Wells describes the process, the United States became the world's quintessential "car country," forever connected to its Fords and Buicks. In western cities, that meant urban sprawl and changing the landscape to create spaces for parking, billboards, homes with garages, commercial strips, freeways, suburban shopping centers, and drive-in theaters. Los Angeles became a prototype for this auto-oriented urban life, and Southern California's suburban lifestyle assumed its role as a new ideal in the mid-twentieth century. The proximity of other western cities to open land invited extensive suburbanization and an economic and cultural reorientation that celebrated the conveniences of mobility.

Federal expenditures during the New Deal and World War II favored western cities. By 1940, six out of ten westerners lived in cities, and over the next few years huge investments in wartime infrastructure, especially along the Pacific Coast, encouraged in-migration from the rural West and beyond, especially the South. After the war, Cold War military funding poured into Los Angeles, Seattle, Denver, Colorado Springs, and other western cities. Changing federal regulations in the 1950s also offered huge tax incentives for commercial real estate speculators, who invested in shopping centers and malls, and the Interstate Highway Act (the Federal-Aid Highway Act of 1956) created thousands of miles of urban-oriented arterial roads.

Postwar demands for affordable homes produced a new scale of investment in housing and planned residential subdivisions. Among the earliest, Tucson's Pueblo Gardens, a seven-hundred-home community built by Del Webb, drew national attention when it opened in 1948 (fig. 7.8). Southern California's Lakewood community, near Long Beach, opened in 1950 and was

FIG. 7.8 Pueblo Gardens, Tucson, Arizona. Del Webb's suburban vision is now a low-income Latino community on the city's south side. The original one-story, wood-frame "patio homes" (based on six floor plans) remain, now incorporated into a latinized landscape that includes many fenced front yards (41).

one of the West's largest subdivisions, with 17,500 homes and eighty thousand residents. Many ranch-style houses date from this postwar construction boom (fig. 1.50), when the peacetime economy expanded and cities attracted new people. Impacts of the housing boom also reverberated to other parts of the region, for example in the increased demand for lumber from key timber-producing areas in the Pacific Northwest.

The 1950s and 1960s were also an era of clearing what was known as "urban blight," a process that dislocated thousands of the urban poor and aimed to produce central-city settings conducive to economic growth. Skyscrapers, modern parking garages, and refashioned civic centers were marks of success. Downtown skylines changed and growth-oriented coalitions of public officials and private investors promoted the modernity, conveniences, and quality of life of their cities.

By the late 1960s, that modernist vision of the western city was being challenged. Unfettered growth, fostered through the powerful forces of industrial capitalism and the federal government, brought with it an assortment of urban ills—air pollution, traffic congestion, and social instability, as race riots occurred in both Los Angeles and the Bay Area during the 1960s. Some critics blamed growth, and Portland, Oregon; Seattle, Washington; Boulder, Colorado; and Petaluma and Berkeley, California, created land-use policies and planning initiatives designed to slow growth and guide it in more purposeful ways. Others leveled critiques against the very suburbs that were once celebrated as postwar paradises: less comely images described them as the "valium belt" or the "Plains of Id," a sprawl of ticky-tacky houses where people lived lives of quiet desperation.

For some, challenges to the growth mantra created new priorities: urban landscapes with healthy cultural diversity, a clean environment, an appreciation for neighborhood character and place-based amenities, and a willingness to sacrifice growth for livability. At the experimental margins of these inclinations, architect Paolo Soleri designed Arcosanti, a settlement in the hills north of Phoenix (fig. 7.9). Dating from 1970, Arcosanti was a tightly clustered

FIG. 7.9 Arcosanti, north of Phoenix, Arizona. The utopian urban forms are the inspiration of visionary Paolo Soleri, whose ideas for a cohesive urban community stand in contrast to the sprawl of nearby Phoenix. Slender Mediterranean cypresses were imported from Soleri's native Italy.

settlement that was sustainably placed in the landscape. Today, Arcosanti residents maintain gardens, manufacture metal and ceramic bells for sale, and foster an urban lifestyle based on shared community ideals.

Elsewhere in the West, profound changes in the ethnic population yielded new urban communities. Cities also changed and developed to accommodate a diversifying array of high-tech manufacturing, bioengineering, financial services, recreation- and leisure-based activities, natural-resource industries (especially energy), and federal military installations. Regionally, much of this growth occurred in the desert Southwest (Phoenix and Las Vegas), in older metropolitan areas (on Puget Sound, the Wasatch Front, the Front Range, and interior Southern California), and in smaller amenity-rich urban settings (Boise, Idaho; Las Cruces, New Mexico; Bend, Oregon; Missoula, Montana; and Saint George, Utah).

A new wave of downtown renovation and redevelopment has further transformed western cities since 1980, often with attention to distinctive local histories, neighborhoods, and natural environments. More generally, many neighborhoods have been shaped by legal mandates imposed by historic preservation interests. Private capital has joined with public money to make such places possible.

Suburban settings reveal increasingly corporate signatures of investment and place branding. Post-1980 residential and commercial construction came with easy money and powerful development incentives, and national home

FIG. 7.10 Life on the waterfront, Scottsdale, Arizona. The pleasures of restaurants and shops as well as of condominium living (background) are part of the Waterfront development (along the Arizona Canal), a large mixed-use project east of Phoenix.

builders and real estate franchises peddled new subdivisions. Construction initiatives were often coordinated with highway improvement efforts and improved rapid-transit systems. Countless corporate-managed malls, shopping centers, office parks, and leisure-time venues now punctuate the metroburban periphery (fig. 7.10). This corporatized vision of the good life—reproduced and shared in advertising, popular culture, and social media—seems likely to replicate itself beyond the present metropolitan fringe, the pace of urban expansion varying with the realities of the postindustrial economy, demands for housing, and the enduring marriage between the region's political and economic institutions. ❋

70

URBAN GRID

FIG. 7.11 The urban grid, San Francisco, California. This view looks westward from Coit Tower across the city's original gridded landscape, which did its best to ignore the hills. Columbus Avenue cuts diagonally across the rectangular pattern (center) and leads past Washington Square (left).

Euclidean order infuses American culture and vernacular language: we have gridlock in traffic, we think outside the box, we resist being too square but enjoy three square meals a day, and when we've had our fill of the conventional routine we live off the grid. The urban grid, with its repeating imperatives of parallel streets and blocks, defines urban landscapes across the country. Deeply impressed across the continent as a design template for metropolitan space, various versions of the grid—similar to the federal township-and-range survey system (**58**)—have influenced patterns of land use and ownership, molded and organized features on the urban landscape, and determined the movement of people through cities.

This enduring connection between American life and the urban grid had its beginnings most prominently with William Penn's platting of Philadelphia in 1682. Simple and efficient, his layout of streets, lots, and squares became a model for frontier towns. Grid surveys were quick and cheap, and they offered an easy way to assign land, which appealed to both buyers and sellers of real estate. Grids facilitated commercial activity, offered flexible movement through the city, and could be expanded as a city grew. While critics found grids dull, predictable, and insensitive to variations on the land, boosters extolled the democratic virtues of the grid, a landscape open to all and an

FIG. 7.12 Alley, Denver, Colorado. Combine the regularity of the suburban grid with the standardized features of the residential subdivision and even back alleys reveal a landscape with predictable arrangements of garages and weekly garbage collection.

urban expression of the Jeffersonian agrarian ideal in which tailors, tavern-keepers, merchants, and laborers could dwell together and mingle with one another. While reality did not always match this ideal, especially as urban social space grew more complex, the grid endured.

Expressions of the grid are visible in most western cities. The layouts of both Santa Fe and Salt Lake City are reminders of the grid's appeal to people as different as Spanish conquistadors and Mormons. Its use by Jasper O'Farrell in the late 1840s gave San Francisco a unique landscape of steep hills overlaid with a rectangular street pattern better suited to the Nebraska plains (fig. 7.11). Railroad towns also had a penchant for rectangular order, though their orientation often deviated from cardinal alignments to parallel the tracks (**50**). Other complexities arise when multiple grids intersect (watch for sharp breaks in the grid). Navigate Denver's Five Points neighborhood or the angled, often anguished flows of traffic north and south of San Francisco's Market Street, where O'Farrell added a second grid at a forty-five-degree angle to the first.

From Boise and Burbank to Salem and San Diego, however, rectilinearity reigns. Beneath the obvious geometry of intersections and evenly spaced blocks, combined with mass-produced housing, the grid also offers a larger, more encompassing visual order that suffuses alleyways, the alignment of garage roofs, and even the placement of garbage cans (fig. 7.12). ❋

71 CITY BEAUTIFUL

FIG. 7.13 City Beautiful: Civic Center, Denver, Colorado. This 1940s-era view of Denver's Neoclassical Civic Center shows the City and County Building (top), Greek Theater (left), and Voorhies Memorial (right), all facing the symmetrical open space of Civic Center Park. Much of the landscape remains intact today. *Courtesy of Denver Public Library, Western History Collection, image no. 11002089.*

THE CITY BEAUTIFUL MOVEMENT SHAPED URBAN LANDSCAPES ACROSS the country in the late nineteenth and early twentieth centuries. As cities grew larger and more congested after the Civil War, urban planners, landscape architects, municipal reformers, and progressive politicians searched for ways to improve metropolitan life. Between 1900 and 1920, many found common ground in the notion that creating a beautiful, functional, and efficient urban landscape would improve the lives, morals, and civic virtues of residents. In 1893, Chicago's World's Columbian Exposition, with its elegantly planned White City, crystallized and popularized many of these impulses, laying the foundation for urban planning and design imperatives that would shape dozens of American cities.

Broadly conceived, the material expressions of these initiatives included city parks, tree-lined boulevards, civic buildings, and open space as well as playgrounds, water supplies, and municipal sewage systems (**72**). While the larger goals of the movement eventually fell short of expectations, fragments of this urban beautification impulse survive in the modern West.

The most unadulterated City Beautiful landscapes are in downtown Denver (fig. 7.13) and San Francisco (fig. 7.14). Denver's elaborate Civic Center complex, designed between 1904 and 1933, was influenced by Frederick Law

Olmstead, who favored Neoclassical architecture and an emphasis on symmetry and open space. Nearby city parks, monuments, and tree-lined boulevards display similar impulses. Even the mountain parks (**6**) and associated boulevards leading into the Front Range west of the city are surviving fragments of these early urban improvement efforts. San Francisco's grandiose City Hall, which opened in 1915, is an elegant example of Beaux-Arts architecture. Once one of the city's great public spaces, the lawn spread out before it was seen as an integral part of the design, all shaped by the vision of chief architect Arthur Brown Jr. In more recent years, the symbolic significance of the site has been further reinforced as many of the city's parades and protest marches have started or ended there.

Elsewhere in the West, urban improvement campaigns, attempts at comprehensive planning, and more mundane expressions of the City Beautiful movement are visible in older inner-city parks, open spaces, historical monuments, and ornamented boulevards. In the Northwest, examples include urban parks in Portland and Spokane, lakeside boulevards in Seattle, and the Washington State Capitol complex in Olympia. You can also see the movement at work at Julia Davis Park in Boise, at Griffith Park and the Greek Theater in Los Angeles, and in the Hispano-inspired reinvention of Santa Fe's central plaza, which was codified in the city's Plan of 1912 (**39**). While not a cure-all for urban ills, City Beautiful retains its visibility, a reminder of both the benefits and the limitations of the utopian impulse in modern city planning. ❋

FIG. 7.14 City Hall, San Francisco, California. Opened in 1915, this Beaux-Arts building celebrates the classic style of the City Beautiful movement. The fact that many poor and homeless people live nearby complicates the locality's contemporary social significance.

72

MEGA CIVIC LANDSCAPES

FIG. 7.15 Los Angeles Music Center. Modernist architecture and attractive open space shape this downtown Los Angeles mega civic landscape. The Mark Taper Forum (center), one of several performance venues, offers an intimate setting for live theater. Jacques Lipchitz created *Peace on Earth* (left), one of the city's most photographed pieces of public art.

MEGA CIVIC LANDSCAPES ARE THOSE LARGE AND VISIBLE PUBLIC places that serve important urban functions and shape a city's collective sense of place (fig. 7.13). These landscapes inevitably intermingle with nearby mega consumer areas (**73**) as downtowns increasingly orient themselves toward amenities, tourism, and consumers with leisure time.

Most mega civic landscapes are twentieth-century innovations. Between 1900 and 1920, some downtowns created new public space, influenced by the Progressive Era's embrace of social reforms, municipal improvements, and urban planning (**71**). Slum clearance and downtown modernization initiatives between 1955 and 1970 offered more opportunities for change, displacing thousands of inner-city poor people in the process. You can see examples of such landscapes at Seattle's Space Needle, which was part of the 1962 World's Fair, and at the Los Angeles Music Center, completed in 1967 (fig. 7.15).

Since 1975, mega civic landscapes have become increasingly visible regional signatures as metropolitan governments, often with the help of private capital, have attempted to create place-defined urban identities in their downtown areas. Many projects reflect a city's environmental features. Waterfront settings, often with rundown port and warehouse districts, have been trans-

FIG. 7.16 Historic Arkansas Riverwalk of Pueblo, Colorado. Many older rundown districts in western cities have witnessed dramatic face-lifts. In this linear public space, visitors stroll along the water, relax at neighborhood bars and restaurants, and enjoy public events at a nearby pavilion.

formed into parks, walkways, open spaces, and entertainment venues from San Diego and San Francisco to Puget Sound. Portland's Governor Tom McCall Waterfront Park along the Willamette River, which opened in 1978, and Pueblo's Riverwalk development along the Arkansas River (the Historic Arkansas Riverwalk of Pueblo), begun in 1995, have dramatically redefined these urban settings (fig. 7.16).

In many other places, public space has been built around historical and cultural resources. Portland's Pioneer Courthouse Square, opened in 1984, integrates urban open space, historical themes, and community activities amid an older cityscape of office buildings and retailers (fig. 7.17). Public promenades, museums, parks, and open space associated with Old Sacramento or Old Town Fort Collins reflect similar impulses.

Some large-scale initiatives have produced entirely new symbolic venues, cultivating a sense of place where none existed. In California, Orange County's Segerstrom Center for the Arts, opened in 1986, is centered on its spacious Arts Plaza and includes a concert hall and theater (fig. 7.18). It is a bold, visually eclectic space that provides a sense of place for nearby suburban residents.

Government buildings are frequent features of mega civic spaces. Some older courthouses, state capitols, and municipal office buildings have Neoclassical design elements dating to the City Beautiful movement (**71**). The Los Angeles City Hall, opened in 1928, remains an iconic Southern California landmark. Newer structures are often more architecturally diverse.

Many cities have rushed to build large convention centers, most of which are cavernous and visually uninteresting. Memorable examples include San Francisco's Moscone Center (1981), which anchored redevelopment efforts in the city's SoMa (South of Market Street) district. You can also get lost in the Las Vegas Convention Center (**100**), one of the world's largest.

Museums and performing arts complexes, funded partly or completely with public money, have also become integral elements of urban identity. The Seattle Art Museum, for example, holds collections of Northwest and Asian

STARBUCKS COFFEE

art that resonates with the region, and Denver's Performing Arts Complex has dramatically transformed a four-block area of the city into a premier cultural center.

Finally, professional sports stadiums and arenas and larger university facilities, often constructed with a mix of public and private funding, are popular components of urban and regional identity. Denver, for example, has venues for the Rockies (baseball), the Broncos (football), the Nuggets (basketball), and the Avalanche (ice hockey; note how the team names aim at reinforcing place identity). These venues also anchor nearby economic activities (fig. 7.19) and have utterly transformed surrounding neighborhoods and business districts. Nearby communities have been changed in different ways by these mega civic landscapes. In Los Angeles, for example, the controversial construction of Dodger Stadium in Chavez Ravine (opened in 1962) annihilated a low-income Latino community to make way for the glitzy new ballpark. On the other hand, sports stadiums in Seattle (Seahawks and Mariners), San Diego (Padres), Phoenix (Diamondbacks), and San Francisco (Giants) have injected new investment and economic activity into surrounding areas. Critics sometimes deride such structures as "ball pork," opportunities for public tax dollars to create private-sector profits.

In some cities, mega civic landscapes are more modest. San Jose's revitalized downtown is focused on a convention center, museums, and the Plaza de César Chávez. Smaller towns such as Sedona, Arizona; Bend, Oregon; and Kalispell, Montana, have built new community centers, pedestrian paths, and parks. They have also set aside places for public markets, downtown strolls and art walks (**97**), seasonal festivals and parades, and local sports and hobby activities such as cycling events and car rallies. ❋

FIG. 7.17 Pioneer Courthouse Square, downtown Portland, Oregon. This central public space, surrounded by older office buildings and large retailers, celebrates the city's history and offers a popular daytime spot to enjoy downtown.

FIG. 7.18 Segerstrom Center for the Arts, Orange County, California. These imaginative buildings are home to a large concert hall and smaller theater and serve as a new cultural center in suburban Orange County. Nearby office buildings are a part of the South Coast Metro edge city (81).

FIG. 7.19 Coors Field, LoDo (lower downtown) neighborhood, Denver, Colorado. Home of the Colorado Rockies baseball team, Coors Field is an economic and cultural anchor of urban revitalization efforts in Denver's LoDo district.

75

MEGA CONSUMER LANDSCAPES

FIG. 7.20 Shops in Old Town, Albuquerque, New Mexico. Since 1980, city leaders and merchants have renovated Albuquerque's downtown (west of today's central business district). Brick walkways, redesigned building fronts, and regionally oriented shops and restaurants combine with the area's historical fabric to make it a pleasant environment for locals and tourists.

MEGA CONSUMER LANDSCAPES ARE CENTRAL-CITY SETTINGS WHERE shopping, eating, and drinking shape the visual scene and help define urban identity. While closely connected to downtown public space (**72**), these landscapes are private places open to the public and focused on consumption. As geographer Michael Conzen notes, these developments demand corporate investments, but the boundaries of postindustrial capitalism are rarely crisp. Many mega consumer landscapes benefit from public financing, such as loans to renovate a blighted area, and are interlaced with public parks, museums, and open space.

While suburban malls and big-box stores emphasize uniformity, downtown retail space is more diversely conceived, playing to local setting and history. These settings are still designed to relieve people of their money, but they do so in ways that capitalize on the place-centered nature of the shopping experience. Running into a WalMart to pick up lightbulbs is a very different experience from strolling the brick walkways of a place like Old Town Albuquerque (fig. 7.20).

Department stores dominated downtown retailing between 1900 and 1950. As people and buying power moved to the suburbs, however, commercial strips, edge-city nodes, and strip malls threatened downtown retailers (**80** and **81**), and many downtown stores closed or moved. The decline sparked changes that were part of larger plans for central-city revitalization. The design of San Francisco's Ghirardelli Square (1964), a collection of shops and restaurants housed in an old industrial complex, proved prophetic. After 1970, most larger American downtowns began to focus on creating opportunities for people to shop and consume in a place that was also entertaining.

Mega consumer landscapes use place in a variety of ways. Some sites emphasize open-air venues, while others offer safe, enclosed environments. At a finer grain, notice the ways retail space is segmented—the size, density, composition, and arrangement of stores—and how retailing is interwoven with other urban activities. Restaurants and minimarts are located in office buildings, souvenir shops are next to arenas, and enclosed malls are on the ground floors of condominium towers. Some frequently seen features include:

Signature department stores. These dinosaurs from the urban Jurassic era still thrive in some settings, often integrated with other retailing opportunities. Chains such as Macy's and Nordstrom have stores in cities such as Seattle, Portland, San Francisco, and Los Angeles. Elsewhere, Salt Lake City's ZCMI—the nineteenth-century Zion's Cooperative Mercantile Institution—survives in a larger complex of downtown stores.

Enclosed urban malls. In these downtown facilities, retailers lease space, adjacent parking garages offer convenience, and shoppers are protected from the elements and the uncertainties of the street. Popularized in the 1980s, enclosed urban malls are typically multistory and less horizontally extensive

FIG. 7.21 Sixteenth Street Mall, Denver. In the early 1980s, portions of Sixteenth Street were redesigned as a pedestrian mall. A mix of modern office buildings, entertainment centers, shopping opportunities, and historic buildings has created an attractive downtown setting. Walkways, benches, public art, and free bus service offer additional amenities.

than malls are in the suburbs, and enclosed walkways and skywalks connect to adjacent department stores and hotels. No umbrellas are required in Portland's Pioneer Place Mall, and it never snows inside Denver's Tabor Center.

Downtown pedestrian malls. To reinvigorate western downtowns, some cities have closed their streets to automobile traffic, creating pedestrian malls. These corridors are leavened with new open space and contain architectural signatures, statues, and other landscaping elements with place- or history-oriented themes. Older examples (from the 1960s) include malls in Burbank and Riverside (later reopened to automobiles) in Southern California, as well as Boulder's Pearl Street Mall (1977), Denver's Sixteenth Street Mall (1982), and Fort Collins's Old Town Square (1984) in Colorado (fig. 7.21). One variant on this kind of space is the street fair, where downtown streets are closed for a day or weekend of special shopping and entertainment activities.

Lifestyle shopping centers. These versions of the downtown mall frequently include both indoor and outdoor settings. Shopping is supplemented with entertainment, eating, and drinking options, all within a carefully landscaped, themed environment. Sometimes called festival markets, these shopping centers aim for a village atmosphere. San Diego's Horton Plaza shopping center, with its vaguely Mediterranean touches, is an early example (1985). The Grove in Los Angeles (2002) and the Ferry Building Marketplace on San Francisco's Embarcadero (2003) have become shopping resorts and major tourist destinations.

Mixed-use development. In some downtowns, places for shopping and entertainment have been integrated into office and residential space. Promoters boast about such places as examples of smart growth, developments that allow people to walk from home to work and shops. In Salt Lake City,

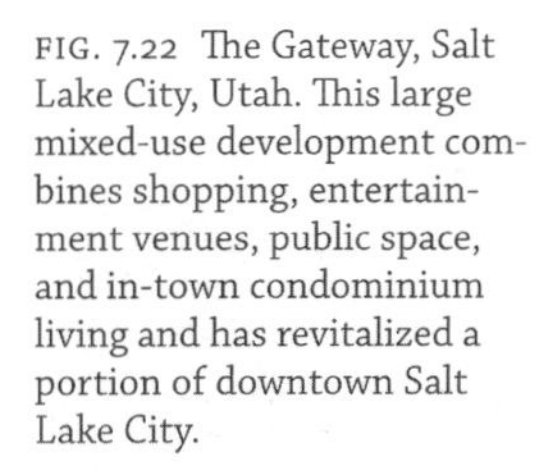

FIG. 7.22 The Gateway, Salt Lake City, Utah. This large mixed-use development combines shopping, entertainment venues, public space, and in-town condominium living and has revitalized a portion of downtown Salt Lake City.

FIG. 7.23 The Burbank Collection mixed-use development, Burbank, California. Movie theaters (left), pleasant outdoor walkways (center), restaurants (lower right), and upscale urban lofts (upper right) combine to create an appealing downtown environment for residents. Nearby picnic areas, a swimming pool, and a fitness center offer more attractions.

the Gateway, completed to coincide with the 2002 Winter Olympics, offers a cluster of shops, restaurants, entertainment venues, public space, and condominiums (fig. 7.22). You can also find this mixed-use ambiance at Los Angeles–area developments Paseo Colorado in Pasadena and the Burbank Collection; at the Arizona Center in Phoenix; at the Marina District in San Diego; and at Riverplace Marina in Portland (fig. 7.23).

Adaptive reuse. Consumption in these gentrified inner city landscapes is packaged in renovated neighborhoods and structures (fig. 7.2), powerfully shaped by the post-1975 appeal of historic urban landmarks and neighborhood preservation. Albuquerque's Old Town (founded in 1706) includes buildings like the San Felipe de Neri Catholic Church (1793) along with structures that were part of ambitious redevelopment initiatives after 1980 (fig. 7.20). Denver's Lower Downtown—known as LoDo—has gentrified inner-city landscapes and mixed-use neighborhoods that repurpose older structures such as warehouses and factories to create a zone of upscale residences, brewpubs, entertainment venues, and specialty retailers. Seattle's Pike Place Market preserves the traditional public market but also has retailing and entertainment options on nearby streets. Other examples are the Gaslamp Quarter in San Diego, the Pearl district in Portland, and Pioneer Square in Seattle. ❋

71

CITY INVISIBLE

FIG. 7.24 City invisible, downtown Los Angeles, California. Life on the street for thousands of homeless residents in the urban West creates its own landscape of enduring urban poverty.

EVERY WESTERN URBAN LANDSCAPE HAS A VAST COLLECTION OF ORDINARY places, nameless streets and alleyways, low-rent districts, and parking lots. In these landscapes, the boundaries between City Beautiful (**71**) and city invisible can be sharp. In Tucson, Pueblo, Los Angeles, and Spokane, for example, you can pass beneath the freeway, across the tracks, down an alley, or around a corner and find yourself in a very different part of town, where gentrified shops give way to homeless residents, shuttered storefronts, and vacant lots (fig. 7.24). Planners have invented clever terms to describe these places: *lulus* (locally unwanted land uses) and *toads* (temporary obsolete abandoned derelict sites), or, more bluntly, *sinks* and *dumps*.

Reading these landscapes is an important part of getting to know any urban setting in the West. You might begin by looking at the larger unsung elements of urban infrastructure. Warehouse districts have wide streets, loading docks, fenced property, limited access, and sprawling, neutral-toned buildings. Such uncelebrated landscapes intermingle with railyards, ports, air freight facilities, postal annexes, and express shipping centers, along with strip clubs and massage parlors (**94**). Even less visible may be the accumulations of industrial waste generated by manufacturing and industrial operations. In a more mundane way, consider how much urban space is occupied by parking lots and park-

ing garages (fig. 7.25). From block to block, also keep an eye on hourly and daily rates for parking. There are few better gauges of urban centrality.

There are also the spaces in between, especially alleys (fig. 7.26). Long a part of urban landscapes, especially in older parts of a city, alleys have a slightly seedy reputation. Characters congregate there, children play unsupervised, and utility lines dangle overhead. While some alleys have been gentrified or have small shops, most remain places that receive little upkeep and attention. Vacant lots are also in-between landscapes characterized by feral vegetation and neighborhood detritus, including aluminum cans, drug paraphernalia, plastic cups, and random seed pollen.

FIG. 7.25 Lots of parking, Denver, Colorado. Both on-street parking lots and parking garages take up sizable acreage in and near western downtowns. Keep an eye on posted hourly rates (visible center left) as a quick measure of urban centrality.

In their own way, places of urban poverty are in-between landscapes (figs. 7.2 and 7.24) that often reveal the gap between the ideal and the real. Enduringly, central cities are home to the majority of the urban poor. Whether it's Tucson's south side (and nearby South Tucson), Portland's outer eastside neighborhoods, Pershing Square in Los Angeles, or West Oakland, these settings are discouragingly similar. Latino and African American neighborhoods (**41** and **44**) often carry the weight of these urban challenges. These settings typically are home to run-down housing, evidence of gang territories (often visible in graffiti), and drug dealing (with a daily street-corner rhythm all its own). These neighborhoods also have anchors of local community, including small grocery stores, fast-food restaurants, churches, and gathering points for day laborers (fig. 7.27).

FIG. 7.26 Alley off East Colfax Avenue, Denver, Colorado. These "in-between" spaces play important roles in local social life. Colorful graffiti and tagged buildings suggest the importance of urban territoriality and links between place identity and the urban landscape.

Skid rows—the term was first used on Seattle's Skid Road, where workers skidded logs downhill—feature downtown neighborhoods with dilapidated buildings and older, often homeless populations. Street life reveals its own tangible, yet ephemeral world of resting places—cardboard boxes, Dumpsters, cheap hotels—prized corners for panhandling, and a scattering of support services such as rescue missions, food banks, soup kitchens, and warming shelters (fig. 7.28). These landscapes are painfully revealing, reminders that a city's character is made up of both its residents' successes and their suffering. ❋

FIG. 7.27 Play Fair Market, Bakersfield, California. Nondescript localities such as neighborhood markets are important landmarks in the vernacular landscape and serve the needs of nearby residents.

FIG. 7.28 Bread of Life Mission, Seattle, Washington. Many of Seattle's homeless live near Pioneer Square. Their daily routines incorporate an assortment of stopping places such as rescue missions, soup kitchens, and food banks, a landscape that remains invisible to many of the city's middle-class and wealthy residents.

75

CALIFORNIA BUNGALOWS

FIG. 7.29 California bungalow, Merced, California. This classic wood-frame bungalow on Merced's near north side has a porch and decorative brick piers that support the structure's attractive overhanging eaves at the front of the house.

THE CALIFORNIA BUNGALOW IS CONNECTED WITH A PARTICULAR period of urban expansion in the West, between 1905 and 1930, when the region experienced a building boom that produced thousands of new homes. These horizontally designed homes were dramatic departures from the earlier Victorian-era houses found in many western cities and towns. Bungalows carried powerful cultural messages about the people that built them and called them home. With informal, open, and flexible interiors, bungalows had domestic space that was focused on modern conveniences, connections with the outdoors, and healthy, efficient lifestyles popularized during the Progressive Era. The style was so popular that it became part of the national landscape.

The bungalow, adopted by the British in colonial India (the name is derived from Bangalore), is a rectangular, low-profile, one- or one-and-a-half-story house with overhanging eaves, rooflines parallel to the street, and a veranda or porch, particularly at the front of the home (figs. 7.4 and 7.29). Flourishes include porch piers, Craftsman-style shingles, and decorative dormer windows. Bungalows blossomed early in Southern California, where large-scale housing developments after 1905 shaped suburbs in Los Angeles, Glendale, and Pasadena. Bungalow house plans and building kits were sold through Sears, Roebuck and Co. catalogs, and popular magazines touted their virtues (fig. 7.30).

FIG. 7.30 The Vallonia Bungalow, 1925. Retailing for $1,982, the Vallonia model from Sears, Roebuck and Co. displays many of the style's characteristics and was advertised by the company as a "prize bungalow home." *Source: Sears, Roebuck and Co., Honor Bilt Modern Homes (Chicago: Sears, Roebuck and Co., 1925), 26.*

FIG. 7.31 Modern-day bungalow, Bozeman, Montana. Echoing the classic style, this twenty-first-century bungalow (completed in 2002) suggests the enduring popularity of the style across the West.

Regional variants of the bungalow can be found across the West. In California and the Southwest, bungalows were influenced by Spanish Colonial architecture, with tiling, archways, stucco walls, and Mediterranean-style gardens (**40**). In colder Denver, Boise, and Salt Lake City, they were constructed of brick and had sunlit basements (fig. 7.4). In Northern California, hillside bungalows with sleeping porches were built on small lots in the Berkeley Hills. Bungalows in Portland and Seattle emphasized wide shingled roofs, dark-stained siding, rock porch piers, and exposed log beams.

In addition to the enduring popularity and upgrading of original bungalows in older parts of cities, houses in newer western subdivisions and planned communities (**79**) have incorporated many elements of the design, ensuring that bungalows will remain a part of the region's residential landscape (fig. 7.31). ❋

76

RANCH HOUSES

FIG. 7.32 Ranch-style house, Sun Valley, California. This classic Southern California rancher is home to a two-car garage, single-story floor plan, and backyard patio and pool reserved for private family life.

LOW-SLUNG AND FAR-FLUNG, CALIFORNIA-STYLE RANCH HOUSES MULtiplied after World War II, becoming the house of choice for millions of Americans. The single-story, rambling, horizontal elements of the ranch house, often organized around an interior-facing patio, have their roots in vernacular western and Hispanic traditions (**39**). Between 1945 and 1975, frame- and stucco-built ranch-style houses were mass-produced and became the model home for middle-class homeowners across the country (fig. 7.32). Architect and San Diego native Cliff May, who fell in love with the "hacienda style" during the 1930s, wrote *Sunset Western Ranch Houses* for *Sunset* magazine's publisher in 1946. The lavishly illustrated guidebook was a best seller and helped propel the ranch house from its Southern California heartland into virtually every American suburb.

For postwar Americans, the ranch house had several attractive qualities. Basic thousand-square-foot ranch houses were affordable, especially with federal housing incentives; they were modern; and they had an appealing informality that centered living on a backyard patio, a brick barbecue, and often a swimming pool. Healthy, outdoor-oriented western living became a national obsession. More subtly, ranch houses symbolized the virtues of privacy (backyard living, fenced yards), social exclusivity (most ranch houses

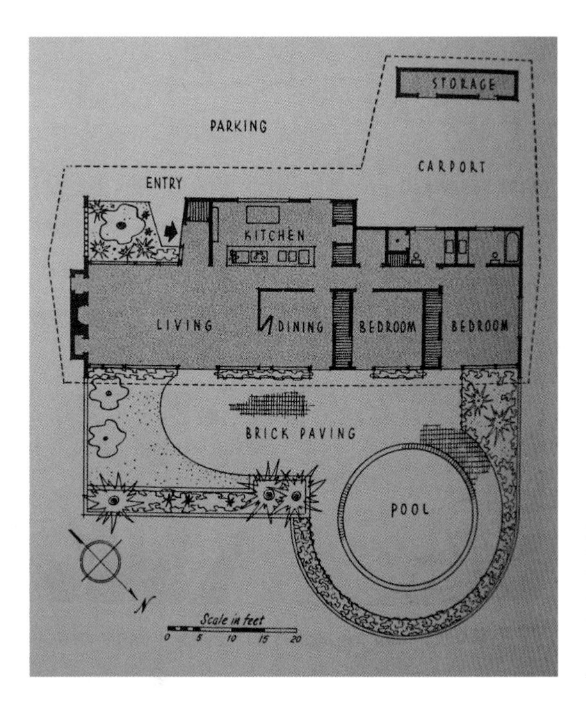

FIG. 7.33 Desert rancher, California. This simple floor plan integrates indoor and outdoor living space, a perfect fit for modern life in the desert. *Source:* Sunset: Ideas for Planning Your New Home *(Menlo Park, CA: Lane Books, 1968), 88.*

were in white-dominated subdivisions), and the independence and autonomy of the nuclear family.

Desert versions of the ranch house—built in the Desert Modernism style in Palm Springs, California, for example—had brick and stone exteriors, breezeways, covered carports and patios, pools, and native vegetation (**77**; fig. 7.33). Split-level variants were built in hillside locations. After 1965, as suburban land values escalated and buyers demanded more interior space, multistory versions appeared, and larger garages were built to accommodate cars, lawnmowers, and recreational toys such as boats and pop-up trailers. As neighborhoods evolved, Latino variations also appeared (**41**), sometimes with a greater emphasis on front-yard and streetside living (fig. 7.34).

Overall, the basic ranch-house design has been remarkably flexible, and the elements of hacienda living that Cliff May so admired have been thoroughly absorbed into western American neighborhoods and into broader national preferences for informal suburban living. ❋

FIG. 7.34 Latinized rancher, Orosi, California. Many older ranch houses in this Central Valley community are now home to Latino families who make fuller use of space in the front yards. In this setting, tables, chairs, and the family barbecue invite more interaction with neighbors and the street.

77

FRONT YARDS

FIG. 7.35 Front yard, Tucson, Arizona. This older home on Tucson's north side has a xeriscaped landscape of gravel, cacti, yucca, and palms, complementing the home's Mediterranean styling.

NO ONE DENIES THE DOMINANCE OF THE AMERICAN LAWN. LAWNS still define the domestic landscape of millions of front yards. About two-thirds of American households possess lawns, and they cover a vast national acreage roughly equivalent to the area of New England. The lawn was a national passion inherited from European (especially French and English) landscape tastes that, once germinated in the New World, grew during the late nineteenth century to become a near ubiquity on the residential scene. As urban parks, outdoor sporting events, and planned suburbs flourished, so did the lawn. In many western American suburbs, carefully manicured turf is an expected and sometimes mandated part of the landscape (fig. 7.29). Still, the West offers more than its share of challenges to the grassy norm. Across the suburban West there is evidence of the growing vitality of distinctly regional front yards.

The roots of this departure are found, most obviously, in the surrounding physical environment. In a xeriscaped yard in Tucson (**13**), for example, yucca, palm trees, barrel cactus, and ocotillo (set in gravel) can thrive where grass cannot (fig. 7.35). In other locations, elevation, moisture regimes, temperature, topography, and drainage dramatically shape front yards. Spruce in Seattle and cottonwoods in Denver simply make environmental sense.

FIG. 7.36 Front yard, Carson City, Nevada. Native sagebrush, cottonwoods, and pines dot this suburban property just east of the Sierra Nevada (visible in the distance, left). Many westerners select native plant varieties to integrate their yards with their surroundings.

FIG. 7.37 Front yard, Glendale, California. Exotic and lush, this front yard mingles the traditional lawn with palms, cypresses, and pepper trees harvested from afar. Planters brim with succulents and ivy, framing the Mediterranean doors, windows, and tiled roof.

Cost imposes its own logic. As water costs rise in semiarid western suburbs, large lawns lose their appeal and residents look for cheaper, low-maintenance alternatives, choosing native shrubs and grasses over pricey turf. Arthur Carhart, a native-plant enthusiast in the 1920s and 1930s, popularized the notion that it was best to plant local vegetation in yards. Planting native plants literally brought nature home and helped integrate a house into its surroundings. Yards reflected regional place identity (fig. 7.36).

Exotic cultural preferences also successfully took root in western soil. Californians' love affair with the Mediterranean ideal (**40**), for example, encouraged the wide use of exotic junipers, palms, ivy, and succulents in front yards (fig. 7.37). And Asian landscape aesthetics (**46**) grew in popularity, sometimes expressed in extensive rock work, bridges, water elements (ponds, streams, and waterfalls), carefully coiffed conifers, intricately shaped shrubs, and shrines. ❋

78

MODERNIST APARTMENT BOXES

FIG. 7.38 Culver City dingbat, Los Angeles, California. The simple dingbat was made popular in Southern California in the 1950s and 1960s. The classic stucco box offered sheltered parking (front or rear) and basic one- or two-bedroom apartments within.

THE RECIPE FOR THE MODERNIST APARTMENT BOX IS SIMPLE: PLACE A cube on the urban grid (**70**) and adjust height, width, and length to lot limits, zoning regulations, and pocketbook. Two- or three-story wood-frame structures are typical. Dice up the interior into four to twelve apartments. Baste with stucco and emphasize simple, rectangular forms in windows and exterior lines. Decorate with stylistic flourishes and landscaping—names, jungle vegetation, Asian motifs, and tikis—and add a courtyard and pool. Surround the building with a parking lot or create a covered carport or underground garage. Set timer between 1955 and 1975, and build quickly and cheaply. Bake until fully occupied.

While urban tenement houses and luxury high rises have a long national pedigree, the rapid postwar growth of western American cities set the stage for a formative era in apartment-house construction and design. Southern California led the way. As urban land values rose (**75** and **76**), developers built thousands of inexpensive, high-density apartment complexes (those with open-air parking beneath are called dingbats: fig. 7.38), both in established

neighborhoods, where they replaced older single-family homes, and in newer suburbs in places like the San Fernando Valley. These simple, modern apartment buildings emphasized convenience, versatility, and fun, and they were immensely popular among single people and young couples (fig. 7.39).

By 1990, many older apartment boxes had acquired new occupants, as African Americans, Latinos, and Asian Americans moved into suburban communities (**41**, **44**, and **46**). In areas of new construction, the modernist cubes featured more postmodern variations, larger and more elaborate layouts, and a more complicated infrastructure that often saw them as part of mixed-use developments (**73**) and planned communities (**79**; fig. 7.40). Even with these changes, elements of the simple modernist apartment box survive, along with the lifestyles that generated an appetite for them in the first place. The long-term upward direction of land prices and the growing number of single-member households in western American cities suggest that these options for higher-density living will continue to shape future metropolitan landscapes. ❋

FIG. 7.39 Desert Modernist apartments, Palm Springs, California. Bordered by signature palms, this apartment box emphasizes simple horizontal lines that reflect a distinctive local style popular in both single homes and multifamily units.

FIG. 7.40 Apartment homes, Orange County, California. While no humble dingbat, this large landscaped complex has handy access to South Coast Metro offices and shopping and retains the convenience and adaptability of the modernist apartment box.

79

SUBURBAN MASTER-PLANNED COMMUNITIES

FIG. 7.41 Entrance to Foothill Ranch, Orange County, California. The archetypal nuclear family stands in bold white relief at the entrance to one of Orange County's many master-planned communities. Residents enjoy pleasantly landscaped surroundings, convenient shopping, and relatively low crime rates.

AT FOOTHILL RANCH, ONE OF SOUTHERN CALIFORNIA'S MANY SUBURban master-planned communities (fig. 7.41), a statue of an imagined nuclear family stands in bold white relief against a background of carefully landscaped grounds and homes. The advertised strengths of master-planned living are its insularity, low crime rates, affluent neighbors, convenient shopping, entertainment, outdoor activities, and well-tended streets, homes, and front yards (often enforced through covenants).

Planned communities blossomed in the West in the 1960s and 1970s. Proponents of comprehensive urban planning point to their economic, social, and environmental virtues, but critics worry about their social exclusivity and question whether they provide enough urban vitality and diversity. However they strike you, master-planned communities may signal what many of tomorrow's suburbs will look like.

Master-planned communities have deep American roots. The creation of exclusive nineteenth-century suburbs allowed urban elites to find refuge from the city's environmental and social challenges. Planned communities such as Llewellyn Park, New Jersey; Riverside, Illinois; and Tuxedo Park, New York, were laid out by some of the nation's leading landscape architects, including the Olmsteads. These developments had gated entrances, gently curving streets, open space, and fashionably landscaped residences. In the West, examples of early elite suburbs include Bay Area developments at Burlingame and Southern California's Palos Verdes Estates, designed in the 1920s by Frederick Law Olmstead Jr.

Most master-planned suburbs were developed after World War II, and four variables set the stage for their success. First, the demand for suburban housing in the West was sparked by postwar prosperity and sustained population growth. Second, the federal government offered powerful financial incentives that benefited homebuyers, the real estate and home building industries, and commercial developers. Taken together, these incentives were designed to mass-produce an affordable middle-class dream of private home-ownership. Third, in the 1960s and 1970s real estate development corporations assembling planned communities acquired a sophisticated set of economic and political skills that joined private-sector thirst for profit with popular demands for thoughtful, balanced suburban growth (versus sprawling grids and unplanned commercial strips; **80**). Crucial to this process was the willingness of municipal governments to create land-use and zoning regulations codifying these planned unit developments. And after 1965, upper-middle-class whites in western cities and retirees moving to the West saw these communities as a refuge, a safe haven from the hostile, heterogeneous world they were fleeing.

Southern California was an incubator of master-planned suburbs, and the combined entrepreneurial and marketing skills, planning templates, and cul-

FIG. 7.42 Downtown Verrado, Arizona. This planned community west of Phoenix has a traditional compact downtown with convenient pedestrian access to residential areas. Markets, drugstores, banks, and everyday services are within an easy stroll of one another.

FIG. 7.43 Residential area, Highlands Ranch, Colorado. Carefully designed open spaces, parks, and walkways combine with upscale housing options to offer a pleasant residential landscape in this master-planned community.

tural landscapes created there diffused to other parts of the West. Lakewood near Los Angeles is considered an early (1950) planned suburban development, complete with parks, schools, and shopping center. Still, communities such as Westlake Village, Valencia, Irvine Ranch, and Mission Viejo, which appeared between 1965 and 1980, were more formative in shaping the master-planned communities we see today. Westlake Village, northwest of Los Angeles near Thousand Oaks and begun in 1963 (population eight thousand), was a "city in the country"; the upscale community remains more than 80 percent white. Mediterranean-style homes and pleasant *paseos* (walkways) characterize Valencia, located in the Santa Clarita Valley northwest of San Fernando. To the south, Orange's County's Irvine Ranch, with a population of two hundred thousand people, was designed in the 1960s around a new University of California campus (**82**), two commercial districts, and a mosaic of subtly different residential "villages." Nearby, Mission Viejo, with a population of ninety-five thousand people, integrated profitable office and retailing developments (**81**) and embraced Spanish place-names and Spanish Colonial Revival architecture (**40**).

Southern Arizona also proved pivotal. In Tucson, the sprawling Pueblo Gardens (fig. 7.8), now largely a Latino community, was developed in 1948 by developer Del Webb. McCormick Ranch in Scottsdale, Arizona, was begun in the 1960s with the cooperation of local governments and the inspiration of California planner Simon Eisner. McCormick Ranch devel-

oped more than four thousand acres of desert land into an oasis of parks, lakes, and residences. To the south, the Lakes in Tempe drew directly on capital and expertise from the Mission Viejo Company, as did Highlands Ranch south of Denver. Thirty miles north of Phoenix, the Anthem development offered varied residential choices, including a gated community of homes, luxury estates, affordable real estate, and eco-friendly properties. To the west, near Buckeye, Verrado became a community strongly centered around the principles of New Urbanism, a design movement that emphasizes pedestrian-scale, high-density landscapes and the virtues of small-town communities. Residents walk from their tree-lined neighborhoods of California bungalows (**75**) and Spanish Colonial Revival homes (see fig. 4.20) to a compact downtown with brick streets, palm trees, and Mediterranean-style buildings (fig. 7.42).

FIG. 7.44 City hall, Rio Rancho, New Mexico. Rio Rancho is the self-proclaimed "City of Vision." Building a city hall amid undeveloped acreage is designed to spur growth in this high-desert landscape northwest of Albuquerque.

These communities are characterized by integrated and themed architecture and signage, large-scale construction zones (fig. 7.1), real estate flags, curving streets, and amenities such as bike paths, golf courses, and designed open space (fig. 7.43). They include Snoqualmie Ridge in Seattle, Hidden Springs in Boise, Stapleton and Highlands Ranch in Denver, Summerlin and Inspirada in Las Vegas, Mesa del Sol in Albuquerque, and Otay Ranch in San Diego. Sometimes evidence for planning can truly be breathtaking, including the construction of a city hall (at Rio Rancho near Albuquerque) amid still-undeveloped land (fig. 7.44). ❋

80

COMMERCIAL STRIPS AND STRIP MALLS

FIG. 7.45 Traveler's strip, Holbrook, Arizona. Appearing before an era of sign ordinances and strict zoning regulations, this eclectic commercial strip developed along U.S. 66 and today serves traffic from nearby Interstate 40. Local businesses (right) combine with a scattering of larger fast-food and gasoline retailers (distant left).

JUST AS THE GRID (**70**) SHAPES CITIES, THE LINEAR IMPERATIVES OF commercial strips and strip malls shape traffic flows, connect central city and suburb, and produce built environments that celebrate automobiles and mass consumption (figs. 1.9, 7.2, and 7.3). Landscapes in motion, strips are designed to be seen from behind the steering wheel of a car. The strip's visual shorthand is apparent in the procession of large signs, corporate symbols, turning lanes, and parking spaces. Above all, strips reflect the plasticity and segmentation of urban social space. Canaries in the coal mine of consumer whim, they provide evidence of material abundance and the love of individualism, speed, and affordability. While many highway strips of the classic era (1930–1960) were created for travelers (fig. 7.45), strips today are more oriented to suburban populations (fig. 7.3). The common denominator is convenience, all negotiated from the cockpit of a moving vehicle.

The antecedents of today's strips could be seen along suburban streetcar lines, which encouraged retailers to locate at stops that served neighborhoods and towns at the periphery of cities. But the transformative technology was the automobile. By the 1930s, garages, gas stations, tire stores, cabin camps, cottage courts, and diners lined main highways. As cities expanded, suburban residents also needed shopping centers, beauty salons, gas stations, and fast-food restaurants. Automobile and oil companies, real estate developers, and federal tax breaks for businesses after 1954 greased the wheels, but this was no shotgun marriage: Americans willingly wedded themselves to what strips offered.

After 1970, mass advertising and economic restructuring produced simplified, franchised strip landscapes—with McDonald's stores, Best Western motels, and Home Depots—that are less flamboyant than their predecessors (figs. I.9 and 7.46). Newer strips also have larger edge-city properties (**81**), where supersize big-box stores congregate in retailing power centers.

For smaller businesses that can't afford street-front properties, strip malls are a godsend, offering parking, affordable leases, and smaller storefronts (figs. 4.46 and 7.47). The result is specialized retailing, with everything from tattoo and nail parlors to churches and health clinics.

Walk the strip to unravel the details. Look at the height and placement of signs, calculate the ratio of parking spaces to businesses, and count the frequency of "For Lease" signs. Think about how strips have changed over time, as motels converted to apartments or insurance offices. Read the semiotics of business names, and notice the variety. Octogenarians at Sun City (fig. 7.48; **98**) produce different social geographies than multicultural suburbanites do in south San Jose (fig. 7.49). There are also single-purpose strips, where all the lampposts and palm trees seem to match (fig. 7.50).

Well-established strips include the Strip in Las Vegas (**100**). You can also cruise Colfax Avenue in Denver; Sunset, Foothill, and Sepulveda boulevards

FIG. 7.46 Corporatized commercial strip, Chubbuck, Idaho. Many businesses serving this strip in the Pocatello suburb of Chubbuck display the standardized signatures of corporate America. Both travelers (from nearby Interstate 86) and local residents patronize the strip.

in Los Angeles; State Street in Salt Lake City; El Cajon Boulevard in San Diego; Van Buren Street in Phoenix; Central Avenue in Albuquerque; and Virginia and Fourth streets in Reno.

Strips are landscapes that everyone uses but no one quite claims. Although many urban planners criticize their appearance and uniformity, strips reflect with alacrity an honest, eclectic self-portrait of who westerners are and what they seem determined to become. ❋

FIG. 7.47 Strip mall, San Fernando Valley, California. Many commercial strips across the West offer open-air strip malls with convenient off-street parking, varied leasing space for retailers, and specialized shopping opportunities and services for local residents.

FIG. 7.48 Strip mall, Sun City, Arizona. Reading strip-mall business signs is a sure way to capture a quick sense of local cultural geographies.

FIG. 7.49 Strip mall, San Jose, California. This south San Jose strip mall caters to the area's diverse local population, including a sizable Vietnamese American contingent.

FIG. 7.50 Automobile strip, Tucson, Arizona. Single-purpose strips make it easier for consumers to do comparison shopping. Tucson's Auto Mall Drive offers an opportunity to kick the tires amid a landscape designed for convenience and accessibility.

81 EDGE CITIES

FIG. 7.51 Edge city: South Coast Metro, Orange County, California. Near the junction of several busy freeways, South Coast Metro offers convenient office and retailing space and is one of the largest edge cities between Los Angeles and San Diego.

YOU KNOW IT WHEN YOU SEE IT: THE EDGE CITY JUST OFF THE FREEway, a generic cluster of suburban office buildings surrounded by swathes of green, encircled by boulevards and parking lots (fig. 7.51). In 1991, journalist Joel Garreau wrote that a true edge city has at least five million square feet of office space, more than six hundred thousand square feet of retailing space (a good-size shopping mall), and has emerged as a well-defined locality for urban residents (fig. 7.52).

One of the West's earliest edge cities is the Denver Tech Center (DTC), developed in the 1970s on the city's southeastern margins near two interstates (fig. 7.53). Originally on the periphery, the DTC is now surrounded by suburban infilling and retailing. In California, Orange County's South Coast Metro, which blossomed in the 1980s (fig. 7.51), is close to interstate highways, an airport, and South Coast Plaza, one of the largest upscale shopping centers (begun in 1967) in the country. There is also a performing arts center (fig. 7.18) and multiple apartment and condominium complexes (fig. 7.40).

Edge cities are noteworthy for their density, size, and legibility. Often financed with speculative capital, they are easy to spot on the periphery of a larger city—examples include developments in Bellevue and Redmond (Seattle), Beaverton (Portland), Palo Alto (San Francisco), Scottsdale and Tempe

FIG. 7.52 Optima Camelview Village, Scottsdale, Arizona. Promoters of this mixed-use edge-city development east of Phoenix proclaim that it "leads the way in urban chic." It includes office and retail space, a nearby mall, and convenient condominium living.

FIG. 7.53 Denver Tech Center, Colorado. A mixed collection of office buildings, open space, hotels, parking garages, and nearby shopping, the Denver Tech Center is one of Colorado's largest and oldest edge cities, located southeast of downtown along Interstate 25.

(Phoenix), and Rancho Cordova (Sacramento). Variants, called edgeless cities, are smaller clusters of one- and two-story office parks and suites, and these localities are places of business for everything from insurance agents and financial planners to orthodontists and dermatologists.

These landscapes have a cropped look where sameness prevails. They are attractive, functional, and predictable. In street and business signs and in the placement and design of structures, rectangularity dominates—although angular flair and a creative use of glass can make them interesting architectural spaces. Look at the density and height of offices, hotels, shopping areas, and parking lots. Examine the carefully tended shrubs and trees that ornament walkways and spaces between buildings. Note the similarities between these carefully manicured landscapes and the master-planned residential communities (**79**) many workers call home. ❋

82

SUBURBAN RESEARCH PARKS

FIG. 7.54 Cisco Systems, San Jose, California. Cisco's main San Jose campus contains Neoclassical office buildings and research facilities. Convenient parking and light-rail service provide handy access for employees. Open space and employee health and fitness centers are on-campus amenities.

LANDS OF LAWNS AND LABORATORIES, THE WEST'S SUBURBAN research parks have produced landscapes of national and global significance. Where the best and brightest minds come to invent and innovate, they are centers of research and development that occupy an ambiguous conjoined space in the interdependent worlds of corporate capitalism, university-based and federally sponsored research, and private think tanks.

Many features of suburban research parks resemble a modern university campus. These landscapes are generally clusters of low-rise buildings, landscaped grounds, and convenient parking (fig. 7.54). Signage is muted and standardized, and open space—often with basketball courts, soccer fields, and fitness centers—is carefully managed, much like a master-planned community (**79**; fig. 7.55) or edge city (**81**). Workers dress casually, resonating with the choreographed informality of the grounds and work spaces. Surrounding communities provide amenities such as bike paths, restaurants, and entertainment. Above all, these are places where capital creates knowledge and knowledge creates capital.

FIG. 7.55 Tektronix campus, Beaverton, Oregon. One of Oregon's older examples of a suburban research and manufacturing facility, the Tektronix Industrial Park opened in 1959 and offers employees landscaped grounds and recreational opportunities.

FIG. 7.56 University of California, Irvine. Opened in 1965, this campus resembles a large suburban research park. Famed for its concentric, decentralized layout, the campus is also known for open space and eclectic Modernist and Brutalist (use of repetitive angular forms, often in concrete and brick) architecture.

The brainchild of Provost Frederick Terman, Stanford University's industrial park in Palo Alto (begun in the early 1950s) set in place the design elements of today's research parks. Terman emphasized a suburban look, where light manufacturing and scientific research took place in modern buildings set in parklike grounds. He advocated for close relationships among researchers, the corporate world, and flows of federal funding. A broader template emerged for research-oriented corporate facilities (often called campuses), other research parks (the term appeared after 1960), and new universities such as the University of California, Irvine (fig. 7.56).

Investigate California research parks from Palo Alto, San Jose, and Livermore in the north to Pomona and San Diego (where the University of California, San Diego, Science Research Park was built near Torrey Pines) in the south. Large research parks have also been built in Portland (between Beaverton and Hillsboro), Seattle (home to Microsoft and more), and Phoenix (Arizona State University Research Park). Smaller cities such as Provo, Boise, Albuquerque, and Colorado Springs have modest versions of the Stanford miracle. Indeed, the innovation has gone global: research parks in Singapore and India produce ideas trumping earlier American innovations, all in settings directly transplanted from California. ❋

85

URBAN-WILDLAND ECOTONE

FIG. 7.57 Leapfrog development, Fountain Hills, Arizona. This fashionable suburban community northeast of Phoenix was founded in 1970. Fairways and upscale homes stand in sharp contrast to open desert land within the nearby Salt River Pima–Maricopa Indian Reservation immediately to the south (in the distance).

As WESTERN CITIES EXPAND INTO SURROUNDING FOOTHILLS, MOUNTAINS, deserts, and prairies, they carve out an urban-wildland ecotone, a penumbra of peripheral landscapes that are no longer wild but are not yet fully urban (figs. 7.2 and 7.57; an ecotone is a transition zone between two disparate ecological communities). These front lines of suburban expansion are revealing indicators of regional change. They are hybridized zones of varied ecological niches, political battles, and complex land uses. Expect rapid change. These boundaries are not fixed Maginot Lines of trenches and barbed-wire fences, but oscillating, fast-changing frontiers shaped by blitzkrieg advances and shifting local alliances.

Surrounding Los Angeles, for example, is an urban-wildland perimeter more than seven hundred miles long, much of it abutting rugged mountain slopes. You experience that landscape on a drive through Sylmar, Altadena, Upland, Rialto, and Yucaipa, an all-day trek. Other southwestern cities with impressive, complex examples include Phoenix, Tucson, Las Vegas, and Albuquerque. Colorado Front Range cities (west and northeast) and those along

Utah's Wasatch Front (east and southwest) also feature these landscapes (fig. 7.58), as do smaller cities from Boise, Idaho, and Bozeman, Montana, to Las Cruces, New Mexico, and Grants Pass, Oregon.

You can explore these settings from three different perspectives. First, urban-wildland ecotones are, by definition, an ecological mixing ground. Newly disturbed soils and altered topography are invitations for fresh juxtapositions of plants and animals (fig. 7.1). The wild settings that are being modified may be forested mountain wilderness, open foothill slopes, desert valleys, or prairie swales. The density and spatial extent of the human advance into wild areas might be an onslaught of suburbanites (fig. 7.59) or a less-dense scattering of pioneers (fig. 7.58). Ecological change in these dynamic settings can be marked by dramatic events such as wildfires (**16**) and floods. More subtly, lawns usurp prairie, fairways trump saguaros, and a marijuana patch blends into a slope. Juniper and cinquefoil become decorative shrubs around condos, and imported aspens shade annuals from a nursery. Household pets mingle and often tangle with wild birds, deer, voles, cougars, and field mice (**18**). Raccoons scavenge for garbage, and rabbits discover flowerbeds. The easy adaptation of wild creatures to these domesticated settings produces what journalist Jim Sterba calls "nature wars" (fig. 1.72).

Second, you can see these transitional areas as zones of competing land uses and political interests. There is no regional master plan for the West's urban-wildland ecotone, nor is there any consensus about how these settings should evolve. Conflicting visions, along with a mosaic of landowners and a fragmented geography of local jurisdictions—created when western suburbs readily annex nearby private rural lands to expand—have produced a fascinating and fractured political landscape. In one setting, developers may partner with a land trust to make plans, while elsewhere, a joint city and multicounty planning district may cobble together long-term goals only to be challenged by competing voices and the heady sway of money. Always ask who owns the land and what regulations shape its use. Even on private lands, land-use and development restrictions govern uses. There are also private

FIG. 7.58 Urban-wildland ecotone west of Denver, Colorado. This view, looking west from Interstate 70 in the Front Range foothills west of Denver, reveals a mixed landscape of Colorado conifers, commuter residences, and weekender homes. Wildfires and bears raiding the garbage are periodic concerns.

FIG. 7.59 Suburbs meet the Sandia Range east of Albuquerque, New Mexico. A sharp boundary exists between open land and suburban streets. National Forest lands and municipally owned open-space corridors draw the line on development. Note how large circular water tanks are discreetly placed within neighborhoods.

organizations committed to purchasing open space and wildlife habitat and other landowners who are committed to preserving farms and ranches. If the land is public, is it a mountain park, a municipal open space, a wildlife preserve, or BLM or Forest Service acreage (**67** and **68**)?

Third, look at the basic spatial patterns and land uses that characterize these places. As you navigate the transition zone, notice its breadth and the broader patterns of interplay between suburbs and wildlands. You can usually find intriguing examples of four different patterns: sharp-edged boundaries, in which ownership patterns or use restrictions prohibit development (figs. 7.57 and 7.59); so-called interdigitated advances, in which topography or planning regulations produce a mix of open and developed tracts (figs. 1.6 and 7.60); leapfrog settlements, in which isolated developments jump ahead of existing suburbs (fig. 7.57); and frontier isolates, in which small numbers of homes colonize fresh ground (fig. 7.61).

Tomorrow's urban-wildland ecotone will look different from today's. Its new geography will be based on local terrain and land ownership, shifting appetites for real estate speculation, and constituencies that are committed to keeping land wild. In this era of dam removals and restoration ecology, perhaps wilder lands may someday advance again. Where will we see the first abandoned suburb that is bulldozed back to sage or chaparral? ❋

FIG. 7.60 Suburbs in the hills east of Concord, California. Grassy hills are interlaced with suburban streets, creating a hybrid landscape in which house cats and trees from the nursery meet coyotes and native oaks.

FIG. 7.61 Pioneer on the fringe, Carefree, Arizona. In some cases, isolated homes on the urban fringe remain surrounded by expanses of open land. This house commands a view in the north Phoenix suburb of Carefree.

FIG. 7.62 Hillside letter, Butte, Montana. Visible from much of the city and dating from 1910, Butte's hillside letter sits above the campus of Montana Tech of the University of Montana. The letter is illuminated at night.

84

HILLSIDE LETTERS

THE MOST VISIBLE LANDMARK OF SOME WESTERN CITIES AND TOWNS is on a hillside. The widespread regional practice of constructing hillside letters is more than a century old and may be another take on an even older western tradition of connecting urban identity to surrounding high points such as Pikes Peak (Colorado Springs), Mount Rainier (Seattle), or Mount Timpanogos (Provo; **6**). Today, hundreds of these letters decorate slopes above western towns, almost always sponsored by a local college or high school (fig. 3.17 and 7.62). The letters are most commonly built from rocks, concrete, wood, and metal. In some desert settings, they are simply painted on rock faces and cliffs. In a few cases—such as above Redlands, California—they are cut out from surrounding vegetation.

The tradition of constructing hillside letters is strongly linked to portions of the West where accessible steep slopes, preferably treeless, make possible such monuments to place. The letters become highly visible symbols of urban and community identity. Once letters appear, traditions reinforce their visibility: hiking trails sometimes link the site to the town, and annual treks to maintain them or to illuminate them for events like holidays, homecoming, and graduation give them a continuing role in maintaining civic pride.

Hillside letters (also termed mountain monograms) blossomed in the early

twentieth century, paralleling the popularity of letterman sweaters between 1890 and 1920. This connection between the alphabet and place identity apparently began in 1905 in Berkeley, when students at the University of California erected a large masonry *C* on Charter Hill. Within a few years, schools from Brigham Young University (1906) and the University of Utah (1907) to the Colorado School of Mines (1908) and the University of Montana (1909) followed suit. Today more than five hundred hillside letters exist in the country, the vast majority of them in the West. While less common in forested portions of Oregon and Washington, hillside letters are thick across Montana, Utah, Nevada, Arizona, New Mexico, and California.

Variations on the tradition include the famed Hollywood sign (originally "Hollywoodland") on Southern California's Mount Lee, put in place in 1923 (renovated in 1978) as an advertisement for a real estate venture (fig. 7.63). In many settings (including Hollywood), pranksters periodically remove, change, and add letters, mostly in good fun. Globe, Arizona, has a copper-colored letter that celebrates its history as a mining town. After the September 11, 2001, terrorist attacks, dozens of other letters were modified in red, white, and blue. Simply put, hillside letters have become geoglyphs, iconic signatures into which place-based cultural identity is etched. ❋

FIG. 7.63 Hollywood sign, Los Angeles. A California variant on the hillside letter tradition, the Hollywood sign began as a real estate promotion in 1923 but later became more identified as a regional landmark that remains visible from many portions of the city.

8

PLAYGROUNDS

FIG. 8.1 Life in the exurbs, Bitterroot Valley, Montana. Large areas of the rural West, including sizable portions of western Montana's Bitterroot River Valley, have been refashioned as playgrounds for amenity-seeking residents and visitors.

IN 1904, THE THIRD REVISED EDITION OF KARL BAEDEKER'S *UNITED States* was one of the mostly widely consulted volumes on American travel. A good deal of the space in the guidebook is devoted to the West's natural cathedrals—Yellowstone National Park, Yosemite Valley, and the Grand Canyon. There is also information about Colorado, its hot springs resorts, mountains, mining towns, and narrow-gauge railroads. In the Pacific Northwest, Mount Rainier is featured, along with the Columbia Gorge. For California, there are San Francisco and Yosemite, of course, but also Lake Tahoe, the Big Trees (redwood forests) above Santa Cruz, golf at Monterey's Del Monte Hotel (opened in 1887), the Mount Lowe Railway above Los Angeles, Spanish missions, and the beaches at the Hotel del Coronado near San Diego (opened in 1888).

More than a century later, much of the West still focuses on its playgrounds and our penchant for play. In many settings, peculiarly western environments and western history are integrated into that experience and in the process produce places oriented toward leisure. Indeed, it is crucial to see "leisure" not merely as some discretionary activity on the region's scenic margins, but rather as a key component of the regional economy, a shaper of landscapes and environments, a powerful influence in defining regional character, and a creator of cultural capital (fig. 8.1). Summer sojourns to dude ranches and lakeside resorts, restorative pilgrimages to spas, seasonal hunting and fishing trips, and ski, golf, and beach vacations are all rooted in an abiding connection between the West and the human need to escape, explore, and enjoy. We can assess these settings in a variety of ways, looking at their history, their economic and environmental impacts, and their role in shaping place identity.

Every playground has a history. Sometimes the past is a very visible, integral part of a playground's setting, while other localities have evolved over time. The character of the ski town of Aspen, Colorado, for example, was forged in mining, and that history remains a central part of the resort's place identity (fig. 8.2). Even a brand-new western resort may tap into well-established

FIG. 8.2 Downtown, Aspen, Colorado. The town's preserved architectural fabric, forged in silver-mining days, retains street appeal for present-day pleasure seekers. Ski slopes await autumn snow.

playground traditions, many of which may have roots in places such as the Adirondacks, the New England coast, or European hot-springs retreats.

Every generation also reinvents what it means to be a tourist, and every generation creates its own playground landscapes. Yesterday's search for the "sublime" by a wealthy traveler to the Catskills parallels today's quest for being "amped" by a twenty-something snowboarder getting an aerial rush on the half pipe. Both are cultural conventions shaped by language and practices that define our playgrounds and the expectations we bring to them.

Every playground creates economic consequences. Regardless of their history, playgrounds are situated in the present-day economy. Think about them as focal points of capital and workers, organizing a variety of natural and cultural resources in ways that encourage consumers to pay to play. Exploring the West's playgrounds helps you appreciate the ways capitalism commodifies place, creating social capital—experiences and memories, for example—for consumers and a return on capital for producers.

Real estate is often a source of speculative wealth in the creation of a playground, but the boom-and-bust character of land speculation can wreak havoc on an economy based on leisure activities. Other forms of consumption also drive the playground economy, such as clothing, gear, food, lodging, and entertainment. As historian Hal Rothman points out, however, one irony of these investments, which he terms a "devil's bargain," is that landscapes are often irrevocably changed when they are developed for recreation. The economic gains achieved may benefit only a favored few, while many long-time residents lament the loss of community and new residents struggle with low-wage service jobs and expensive housing.

Every playground transforms its environment. Consider the difference between the environmental consequences of an open-pit copper mine (fig. 3.8) and a desert golf resort (fig. 8.3). The mine is dramatic and spatially focused, while the resort sprawls across a large area and includes nearby amenities and sub-

FIG. 8.3 Golf course, The Phoenician, Scottsdale, Arizona. Palm trees, water hazards, and green fairways belie southern Arizona's arid environment but offer a pleasant escape for sun-loving duffers.

divisions. One is a big hole in the ground and the other is an expanse of green. But each setting is the site of profound environmental change.

Perform an informal ecological assessment on the playgrounds you visit. How have golf courses and ski hills altered topography and vegetation? How do wood-burning stoves in a resort community affect air quality? How much water is consumed in a desert retirement community, and how does a beach town manage wastewater and sewage? Playgrounds have reshaped settlement geographies, population densities, and viewsheds. Exurban mobile homes and ranchettes are now part of the landscape along scenic back roads in the West, and resort communities have larger environmental footprints than their population totals suggest. One can argue that playgrounds in the West are transforming regional ecologies in ways that are at least as profound as the changes generated by mines or lumber camps.

Every playground produces place identity. As symbolic landscapes, playgrounds transform places and the people who experience them. Playgrounds become part of who we are and how we think about the West. Many of these landscapes are rooted in regional myths, cultural traditions, and historical practices. Dude ranches, rodeo grounds, and fishing streams all offer iconic connections with cowboy culture and frontier traditions. The shared cultural meanings of the frontier past make every western tourist a participant in that national story.

Many playgrounds are also centered on the region's varied physical settings. The West's public lands are magnets for outdoors enthusiasts. Hot springs resorts, coastal and mountain playgrounds, and lakes reveal the persistent attractions of water in the West (fig. 8.4). Look in a westerner's garage: you will likely see inner tubes, boogie boards, swim fins, boats, beach umbrellas, canoes, kayaks, water skis, and more—all designed to bring people into closer recreational contact with their playgrounds of choice.

Western playgrounds also reflect lifestyle and social preferences. Westerners embrace amenity-oriented exurbs, attracted by open space and low-

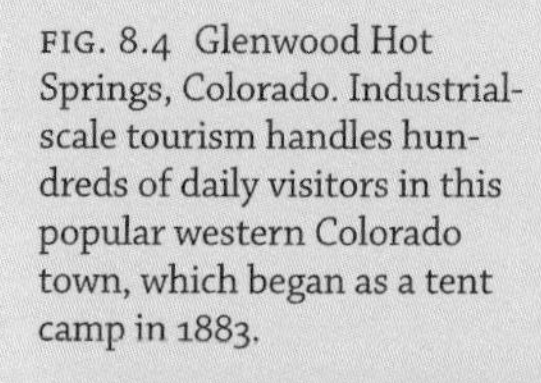

FIG. 8.4 Glenwood Hot Springs, Colorado. Industrial-scale tourism handles hundreds of daily visitors in this popular western Colorado town, which began as a tent camp in 1883.

FIG. 8.5 Entrance to Leisure World, Seal Beach, California. Residents of this gated retirement community (opened in 1962), one of the West's oldest, enjoy security and various recreational activities in their golden years.

density living. Retirement communities and gated subdivisions offer special insularity and privacy (fig. 8.5). Over time, many western playgrounds have also emphasized their cultural and class-based homogeneity through legal restrictions (all-white resorts, restricted beaches), cultural practices (private club memberships), and restrictive covenants (minimum lot and building sizes).

Historically, many early western playgrounds were created for people who had the time and money to travel for health and leisure. By 1890, Southern California's "sanatorium belt" near Pasadena and the hot-springs resorts around Colorado Springs attracted many easterners in search of a salubrious western setting where they could regain their health and treat their disabilities. The wealthy were also attracted to ostentatious destination resorts that resonated with Gilded Age aesthetics. After 1900, the Southwest emerged as another distinctive western playground, with its exotic vegetation, Indian culture, and a climate that purportedly could restore the vigor of those who suffered from asthma and other respiratory diseases.

Popular writers used magazine articles, novels, and guidebooks to describe the region's mythical past, scenic wonders, and cultural charms. Charles Nordhoff (*California: For Health, Pleasure, and Residence,* 1873), Helen Hunt Jackson (*Ramona,* 1884), Charles Lummis (*The Land of Poco Tiempo,* 1893), and George Wharton James (*In and around the Grand Canyon,* 1900; *California, Romantic and Beautiful,* 1914) created idealized images of California and the Southwest, stamping these locales with a place identity that many found irresistible. Parallels to exotic locales were often stressed. In promotional campaigns initiated by state and local tourist bureaus, Colorado Springs welcomed visitors to the "Switzerland of America," the spas at Manitou Springs became a North American Baden-Baden, and California boosters saw Naples in Monterey, Capri in Santa Barbara, and "Our Araby" in the southwestern deserts.

Railroads were essential in shaping playground landscapes in the West, determining opportunities for travel and opening corridors to remote loca-

FIG. 8.6 Camping, Yellowstone National Park, 1909. These women are enjoying their leisure time at Swan Lake Campground, not far from Mammoth Hot Springs. Tent camps were an affordable alternative to the park's highbrow hotels. *Courtesy of Utah State Archives, image no. 29860.*

FIG. 8.7 Automobile tourism, *Sunset* magazine cover, June 1933. With their dog and a mermaid hood ornament, this Depression-era family heads for the hills along with other auto campers of the period. *Source: author's collection.*

tions. Their promotional efforts shaped an entire generation of playground mythology and aesthetics (fig. I.21). In Colorado, California, the Southwest, and across the northern Rockies, a blizzard of evocative advertisements and package tours enticed thousands of people to tour western playgrounds to see Native Americans and spectacular scenery and to enjoy an escape from the modern world.

After 1910, railroad promotional bureaus were advocates of the "See America First" campaign, an initiative intended to keep wealthy Americans from traveling to Europe. World War I from 1914 to 1918 and the creation of the National Park Service in 1916 encouraged similar stay-at-home inclinations. Railroad companies directly financed hotels such as Glacier Park Lodge (Great Northern), the Del Monte Hotel (Southern Pacific), and El Tovar Lodge (Santa Fe) and indirectly spurred the growth of other resorts across the West.

Ordinary Americans also found diversion in simple outdoor activities such as hunting, fishing, and hiking. People took day trips into the foothills to pick berries, climb mountains, soak in hot springs, and relax by the shore. By 1900, for many of more modest means, western camping trips were popular as a way to spend longer periods of time away from home and work (fig. 8.6).

After 1920, increased automobile ownership, improved highways, and middle-class affluence rewrote the geography of western playgrounds. People took advantage of improved long-distance highways: an average daily road trip in 1920 was 170 miles; by 1936, it had more than doubled. Lower-cost lodging was built near grand resorts; municipal campgrounds, cottage camps, and motor courts paved the way for inexpensive motels; and tire shops, garages, gas stations, and diners redefined what it meant to "see the West" (fig. 8.7).

Economic growth following World War II created even more demand for leisure-time activities. Reflecting a more pluralistic, segmented society, playgrounds in the West allowed people to explore personal identity and lay claim

FIG. 8.8 Boondock living north of Quartzsite, Arizona. These RVs and fifth-wheel trailers are casually scattered across a convenient dry wash on public land. The ephemeral settlement takes shape for a few months each winter as snowbirds visit from colder lands.

to higher social status. Whether it was a whitewater raft ride down the Colorado, a ski trip to Aspen, or a retirement home in Arizona, people purchased an experience that had value in a postindustrial society increasingly oriented toward leisure and mass consumption.

More campgrounds, motel chains, and theme parks appeared (Disneyland opened in 1955), and people bought mountain bikes, Jet Skis, fifth-wheel trailers, golf carts, snowboards, scuba gear, and more. Weekend condos and seasonal time-shares began to thicken the slopes, line coastlines and lakeshores, and catch the desert sun. At the same time, the wealthy demanded boutique experiences and social exclusivity, prompting the exponential growth of bed-and-breakfast operations, art colonies, high-end golf resorts, huge vacation homes, and gated communities.

While continuing to cultivate their regional character, western playgrounds changed. Older resort communities such as Estes Park, Colorado; Sedona, Arizona; and La Jolla, California, ballooned in size. Blue-collar towns like Red Lodge, Montana; Madrid, New Mexico; Bend, Oregon; and Coeur d'Alene, Idaho, shed some of their working-class grit to attract second-home buyers and retirees. Entirely new settlements were built, including Vail, a planned ski resort in the Colorado high country, and seasonal snowbird communities proliferated across the southwestern deserts (fig. 8.8).

Playgrounds have not escaped conflict. Communities debate the costs and benefits of growth, and they sometimes struggle over competing land uses with mining, energy, and lumber companies. Playgrounds also have diverse participants. Backcountry skiers and snowmobilers have different priorities. Ultimately, these places remain firmly, problematically embedded in the culture that created them, and their landscapes offer great opportunities to explore ways that these tensions have unfolded in diverse regional settings. ❋

DUDE RANCHES

FIG. 8.9 Horse corrals, Tanque Verde Ranch, Tucson, Arizona. Especially popular in winter, this 640-acre getaway is one of southern Arizona's best-known dude ranches. Since the 1920s, it has offered visitors a mix of "Tall in the Saddle" experiences and comfortable ranch amenities.

LEATHERY-FACED CURLY (PLAYED BY JACK PALANCE) WAS "LIKE A SADdlebag with eyes" as the crusty trail boss in *City Slickers,* a 1991 film that told the story of three eastern rubes who head west for a dude ranch adventure. Their misdeeds drew little sympathy from Curly, whose cowboy poise and cowhand irascibility contrasted with the ignorance of the threesome from the effete East. The trio's search for identity in the visceral world of horses and cattle reflected a well-trodden pilgrimage to western dude ranches, an early type of playground and an opportunity, to use the slogan from the Dude Ranchers' Association (established in 1926), to embrace the virtues of "Horses, Hats, History, and Hospitality" (figs. 8.9 and 8.10).

Late-nineteenth-century guests at western ranches were often casual drop-ins who wanted to spend a week or two on a working ranch (**20**), an institution that symbolized the simple virtues of frontier life. Theodore Roosevelt's experiences on a Dakota ranch in the 1880s were an early example of the tradition and he helped popularize the practice in later years. Between 1900 and 1930, hundreds of dude ranches became tourist businesses, often promoted by railroads to entice people to the West. Dude ranches flowered again following World War II, but additional vacation spots and resorts have challenged the industry since 1980.

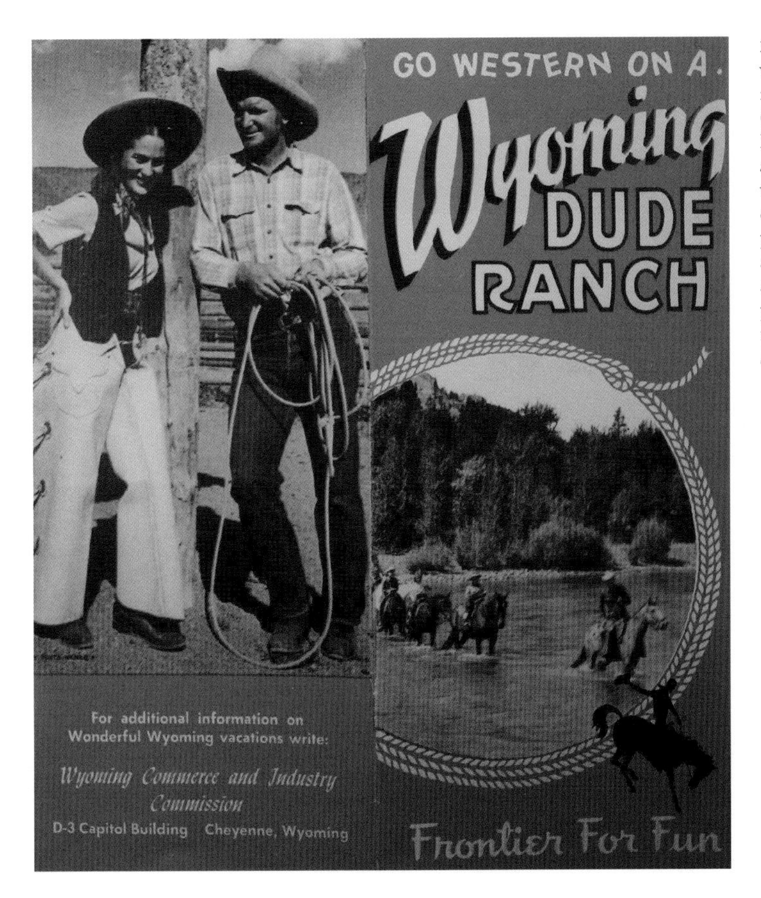

FIG. 8.10 Dude ranch vacation, Wyoming. This 1950s promotional brochure (Wyoming Commerce and Industry Commission) advises visitors to "go western on a Wyoming dude ranch." Thousands of postwar tourists followed the advice. Trail trips, cookouts, and socializing with cowboys proved powerful draws for newly arrived tenderfeet. *Source: author's collection.*

Most of the ranches owned by the one hundred members of the association are located in traditional clusters established decades ago. The northern dude ranching zone focuses both on working cattle ranches and summertime mountain guest lodges in western Montana (especially near Yellowstone), western Wyoming (near Cody, Sheridan, and Jackson), and the Colorado high country (fig. 8.10). The southern dude ranching zone stretches across the Southwest, especially near Tucson and Wickenburg, where sun and desert vistas attract wintertime visitors (fig. 8.9).

While dude ranch landscapes have changed with consumer tastes, horses, corrals, and trail rides remain de rigueur elements of any dude ranch experience. Associated cookouts, petting zoos, hay rides, and rodeos (**91**) are often ranch rituals, with hiking, mountain biking, swimming, hunting, fishing, and whitewater rafting connected to settings nearby (**89** and **90**). Other dude ranch experiences have diversified with consumer tastes. Upscale, resort-style variants include hot-stone massages, tennis courts, yoga clinics, and art classes (no chaps required). Accommodations range from tents and bunkhouses to log castles, but traditional dude ranches emphasize rustic, comfortable digs with pine furniture and wall art celebrating horses, cowboys, and cattle.

Even if you can't spare the dime for a dude ranch retreat, note the powerful echoes of this playground landscape by visiting any western wear store, sampling the West's equestrian culture, or simply by watching a late-night rerun of *City Slickers*. ❋

86 HOT-SPRINGS RESORTS

FIG. 8.11 Mount Princeton Hot Springs, Colorado. Since 1879, this central Colorado resort has offered upscale food and lodging, hot soaking and exercise pools, spa treatments, and nearby outdoor activities.

GEOLOGY CREATED HOT SPRINGS, BUT PEOPLE TRANSFORMED THESE curious natural features into hot-springs resorts. The chemical and thermal properties of hot springs have long been valued for their medicinal and restorative qualities. Traditions of "taking the waters" among Native Americans (for example, at Colorado's Pagosa Springs), European spa enthusiasts, and eastern health-seekers (at Saratoga Springs, New York, and White Sulphur Springs, West Virginia, for example) paved the way after 1860 for the development of health-related hot-springs resorts in the West.

In the late nineteenth-century, the best destinations included Manitou Springs and Glenwood Springs, Colorado; Calistoga and Paso Robles Hot Springs, California; and watering holes in southwestern Montana, Wyoming, and the Southwest (fig. 8.4).

As you dip a toe into the warm water, explore the setting by asking a few basic questions: What are the chemical and thermal properties of the water and the larger geological setting of the springs? What is the human history of the site? Is the springs a Victorian-era resort or a newer place to take the plunge? What is the scale of the operation? Is it an industrial-size giant like Glenwood Springs or a local family-run resort? Who is the clientele? Whom does it appeal to, for what activities, and at what cost?

FIG. 8.12 Thermopolis Hot Springs, Wyoming. Waterslides and pool play often dominate the scene at family-oriented hot springs.

Many older resorts have closed, such as those at Verde Hot Springs, Arizona, and Hunter's Hot Springs, Montana. At Manitou Springs, Colorado, while the large old resorts shut their doors decades ago, revived interest in restoring the area's spa culture has resulted in new public access to some of the area's famous mineral waters. Some survivors have built on tradition and maximized economies of scale (fig. 8.4). At Glenwood Springs, Calistoga, and Desert Hot Springs (near Palm Springs), for example, entire communities are built around their thermal treasures. Smaller operations survive as destination resorts, often in scenic settings like Mount Princeton Hot Springs, Colorado (fig. 8.11), and Chico Hot Springs, Montana, which offer lodges, horseback riding, and spa services. There are also family-centered hot-springs resorts, where the main attraction is a swimming pool with a waterslide (fig. 8.12). Clothing-optional springs have their own rules and local practices. Perhaps most memorably, primitive hot springs are worth exploring, in places like Yellowstone's Boiling River near Mammoth Hot Springs, Black Rock Hot Springs near Taos, and Granite Hot Springs near Jackson, Wyoming, where the small pool and simple changing room were built by the CCC (**63**).

Modern-day spa resorts can also be seen in relation to their roots in the hot-springs tradition. Their upscale aromatherapies and cleansing mud baths have much in common with the healing soaks and hot baths our ancestors found so appealing. ❋

87

COASTAL PLAYGROUNDS

FIG. 8.13 Long Beach Pike, circa 1915. This popular beach and amusement zone emerged after 1902 at the terminus of the interurban Red Car line from downtown Los Angeles. *Source: author's collection.*

SUN, SAND, AND SURF ARE THE THREE ESSENTIAL QUALITIES OF ANY coastal playground. In California, such settings extend from Imperial Beach near San Diego to the beaches of Los Angeles, Ventura, and Santa Barbara counties, and they also include quieter spots such as Big Sur, Point Reyes, and Mendocino. Northward, enthusiasts flock to Cannon Beach, Oregon, Washington's Olympic Peninsula, and the San Juan Islands (**10** and **57**). These West Coast playgrounds are rooted in older shoregoing habits developed in the East in such locations as Bar Harbor, Maine; Newport, Rhode Island; Martha's Vineyard, Massachusetts; and Atlantic City, New Jersey. For decades, promoters have urged people to spend time at the beach to improve health (sea air, sea breezes, sea bathing), to relax (lolling on the beach), and to have fun (beach diversions, fishing, sailing).

By 1915, Southern California coastal retreats included Coronado Beach and La Jolla near San Diego, Long Beach and the Pike amusement zone (fig. 8.13), Catalina Island, Bruce's Beach (one of the few Southern California beaches open to African Americans), Venice Beach (complete with gondolas), and Santa Monica, known as the Long Branch of the Pacific. To the north, near Monterey, seasonal playgrounds included Carmel, Del Monte, Pacific Grove, and Santa Cruz. Bay Area residents went to Cliff House, Playland amusement park, and Sausalito. In the Pacific Northwest, there were Newport, Seaside, and Clatsop beaches in Oregon and Long Beach and the San Juan Islands in Washington.

Later twentieth-century developments added variety. Some locations on the coast, such as Malibu, California, became essentially privatized resort and real estate ventures, while other beaches were developed or protected by states, municipal authorities, and the federal government. The Coastal Zone Management Act of 1972 encouraged states to create shoreline land-use plans, an initiative unevenly embraced by California, Oregon, and Washington.

Evolving cultural tastes have also shaped coastal playgrounds. Some people argue that contemporary American notions of the body developed on the beach—tanning and revealing swimsuits gained popularity by the 1920s, with bikinis coming into fashion after 1946—and after World War II Southern California's surfer culture (**89**) flowered, with mass production of surfboards in the 1950s and the Beach Boys' *Surfin' Safari* album appearing in 1962.

In the 1970s and 1980s, upscale condo complexes and luxury resorts were built along many stretches of open coastline. Beach and coastal towns such as Del Mar, Cardiff-by-the-Sea, Avalon, Venice Beach (fig. 8.14), Malibu, Morro Bay, Half Moon Bay, Cannon Beach, and Friday Harbor became destinations with cultural and economic importance. Particularly for westerners, scenic coast drives—along the central California coast from Morro Bay to Big Sur, Monterey's 17-Mile Drive, California Route 1 from the Golden Gate to Mendocino, and U.S. 101 from Crescent City into Oregon—became as much a part of the region's modern character as earlier notions of gold camps and cowboys. Shoreline arts communities in La Jolla, Laguna Beach, Carmel, and Mendocino, California, and La Conner, Washington (**97**), contributed to the connection between place identity and coastal landscapes.

Coastal playgrounds can usually be divided into public or private space, a distinction that is useful in understanding why landscapes look the way they do. As public space, popular local beaches and state parks attract large crowds who support nearby services (fig. 1.37). You can enjoy the cultural variety

FIG. 8.14 Ocean Echo Surf Shop, Venice Beach, California. A local institution since the late 1980s, this Venice retailer has outfitted generations of beachgoers and surfboard enthusiasts. Venice, complete with canals, was founded in 1905 and remains a center of California's "surfurbia" culture.

FIG. 8.15 Beachside living, La Jolla, California. Large windows and decks are oriented toward the sea in one of Southern California's prestigious coastal playgrounds near San Diego. Glass-enclosed properties protect owners from high surf, spray, and wind.

at Santa Monica Pier on a busy weekend or seek out San Francisco's Ocean Beach on a balmy, fog-free October afternoon.

Other public beaches, however, are quieter, landscapes where habitat is preserved and legal protections and regulations limit activities. Some of the coastline, such as Point Reyes National Seashore, is designated as protected land. Elsewhere, national forests, wildlife refuges, and state preserves also protect coastal habitat. There are coastal military lands—Camp Pendleton in Southern California, for example—and Indian reservations such as the Makah Reservation at Neah Bay in Washington.

Many privately owned coves and beaches remain exclusive playgrounds for the wealthy. Those less well off vacation in modest beachfront condos, in coastal time-shares, or at resorts (fig. 8.15). Marinascapes have also multiplied. Between 1885 and 1900, yacht clubs were created near major West Coast cities from Seattle to San Diego, where sailing and fishing enthusiasts created a prototype coastal landscape that is now mass-produced in marinas from San Diego Bay and Marina del Rey to Puget Sound. In some locations, condominiums and oceanside hotels have been built near docks and boat slips (fig. 8.16).

Private ownership also led in other directions. California's innovative Sea Ranch subdivision (1964) provided for the common ownership of the beach by residents. Subsequent political action—the Dunlap Act in 1968 and the establishment of the California Coastal Commission in 1972—guaranteed

public access to the beach, not only at Sea Ranch but on other privatized segments of the California coast. In other settings, private land trusts—for example, the Mendocino Land Trust, the Sonoma Land Trust, and the Land Trust for Santa Barbara County—purchased shoreline acreage, securing a part of the coast for future generations. ❋

FIG. 8.16 Marina del Rey, California. This placid marinascape includes boats, docks, and nearby condominiums and resort hotels only a few minutes from the Los Angeles International Airport. Much of the marina was built between 1953 and 1965.

88

LAKESIDE LANDSCAPES

FIG. 8.17 Dillon Reservoir, Colorado. South-facing vacation homes above the lake are oriented to catch the afternoon sun and views across the water. Boaters enjoy a perfect late-summer day.

FOR MANY WESTERNERS, A PERFECT SUMMER DAY IS SPENT RELAXING on a lake (fig. 8.17). Lakeside landscapes offer a place to stay in a weekend cabin, put a boat in the water, climb aboard a Jet Ski, or reel in a trout for dinner. These lakeside pleasures are part of traditions developed at eastern retreats such as Lake George, New York; Lake Winnipesaukee, New Hampshire; and Moosehead Lake, Maine, where people escaped during hot summer months. By 1900, resorts, seasonal cabins, and recreational activities had been built at Lake Tahoe, Yellowstone Lake, Lake McDonald in Montana, and Palmer Lake in Colorado.

None of these developments, however, presaged what the twentieth century would create. Thanks to seven decades of dam building (**64**)—much of it focused on irrigation, flood control, and hydroelectricity—the miles of lakeshore in the West multiplied, as did the opportunities to rent a houseboat or put a line in the water. Unlike hot springs (**86**) or coastlines (**87**), much of the modern lakeshore landscape is a peculiarly human creation, crafted for multiple purposes. Consider such cultural transformations along the Colorado River: reservoirs like lakes Granby, Powell, Mead, and Havasu (fig. 8.18) are utterly artificial creations.

Lake types vary widely, depending on topography and development. Alpine lakes are often less accessible, with fewer improvements, while canyon lakes are

typically artificial, the creation of high dams. Valley reservoirs, which are at lower elevation, have broad dams and good fishing. In the arid Southwest, winter retreats are lakes in desert settings primed for snowbirds (**99**) and winter sunbathers (fig. 8.18). Measured by investment dollars, seasonal cabin and fishing resorts at places like June Lake, California; Seeley Lake, Montana; or Lake Chelan, Washington, are much different from larger-scale developments at Lake Tahoe, Lake Coeur d'Alene, or Lake Havasu (figs. 8.18 and 8.19).

Western water is in short supply, and the region's freshwater lakes—both natural and contrived—are destined to be ground zero for future battles over that resource. Changing climate—especially the potential for a drier Southwest—will add uncertainty, particularly if populations continue to grow in that region of the country. Farmers, ranchers, industrial users, suburban residents, and amenity seekers are all staking their claims to a finite, potentially shrinking resource. Expect more bathtub rings around the shoreline. ❋

FIG. 8.18 Lake Havasu City, Arizona. Quiet waters near the London Bridge attract winter-sun enthusiasts to this master-planned community. January temperatures can reach into the seventies. Note the faux English-style half-timbered building on the opposite shoreline.

FIG. 8.19 Coeur d'Alene Resort, Idaho. Inspired by developer Duane Hagadone, the luxurious Coeur d'Alene Resort offers a high-rise lakeside hotel, extensive marina, and world-class golfing, complete with floating green.

89 ADVENTURE PLAY

FIG. 8.20 Rock climbing, Santa Catalina Mountains, Arizona. The West's varied terrain offers extraordinary opportunities for adventure play. This climber uses a top rope system to tackle the ascent of a fractured sandstone pinnacle above Tucson.

WHEN JOHN MUIR CLIMBED CALIFORNIA'S MOUNT RITTER IN 1872, he "was suddenly brought to a dead stop, with arms outspread, clinging close to the face of the rock, unable to move hand or foot either up or down. My doom appeared fixed. I must fall. . . . But . . . life blazed forth again with preternatural clearness. I seemed suddenly to become possessed of a new sense. . . . Then my trembling muscles became firm again, every rift and flaw in the rock was seen as through a microscope, and my limbs moved with positiveness and precision with which I seemed to have nothing at all to do. . . . I found a way without effort, and soon stood upon the topmost crag in the blessed light" (*The Mountains of California*, 64–65).

While Muir's high-country experiences have inspired generations of preservationists, he was also an early practitioner of adventure play, activities that have become a big business as thrill seekers reach for what Muir experienced almost 150 years ago. Whether you are a peak bagger (like Muir), a river rat (at home in whitewater), a tube rider (finding that perfect wave), a snowboarder (searching for big air; **92**), or a storm chaser (ferreting out funnel clouds), the West offers a landscape rich in opportunities (fig. 8.20). The activities appeal especially to wealthy urbanites who can afford to invest in the equipment, technology, and training often required to participate in such adventures.

Historians suggest that our love of outdoor challenges has roots in a masculine, Romantic, and Rousseauian search for the sublime. That may be true, but the current fascination with extreme outdoor avocations dates from the 1970s. REI (Recreational Equipment, Inc.) began expanding in 1975, with retail stores in Seattle and Berkeley, and *Outside* magazine published its first issue in 1977. Many developments stimulated the rush outdoors, including rising disposable incomes, baby boomers' fascination with recreational activities and the environment, increasing western populations, and improving technologies for everything from kayaking to wind surfing.

Each outdoor subculture, from ice climbing to shortboard surfing, has its own jargon, competitive events, champion practitioners, and prime destinations. Each activity also has economic impacts through clothing and equipment sales, schools and training programs, guided trips, and related tourist facilities; visual landscapes in the form of equipment stores, focal points of activity, and periodic events; environmental consequences from mountain-bike trails to off-road-vehicle tracks; and biophysical, spiritual, and social meaning for practitioners.

Geographies of adventure play are strongly shaped by physical settings, accessibility, and distribution and income levels of potential practitioners. Some activities can be enjoyed almost anywhere, while others are highly localized. For example, hang gliding enthusiasts must find appropriate terrain, available takeoff and landing spots, and wind patterns to catch the right thermals at places like Torrey Pines and Perris, California, and Draper, Utah.

Many land-based activities can take place near most western towns, especially on public lands, but some localities develop special reputations. Utah's slick-rock country around Moab is famous among mountain bikers nation-

FIG. 8.21 Old West and New: Saint Elmo, Colorado. This central Colorado mining town has both vintage architecture and off-road vehicle rentals. Numerous high-country roads promise nearby alpine adventures.

FIG. 8.22 Catching the breeze, Hood River, Oregon. Frequent west winds blow through the Columbia Gorge, making it some of the West's best windsurfing country.

wide, and the Nevada and California deserts with their BLM lands (**68**) are immensely attractive to off-road-vehicle (ORV) enthusiasts who explore the region with their dune bashers, sand rails, dirt bikes, and rock crawlers. Arizona's red-rock country near Sedona and Colorado's ear-popping passes also draw thousands of people each year (fig. 8.21). Rock climbers have their own promised lands: Joshua Tree, California; Devils Tower, Wyoming; the Santa Catalina Mountains in Arizona (fig. 8.20); the City of Rocks in Idaho; and Smith Rock State Park in Oregon. Mountaineers climb thousands of western peaks (**6**), but many focus on Colorado's famed fourteeners, California's High Sierra (Mount Whitney remains a popular trek), and the Cascade volcanoes (especially Rainier, Hood, and Shasta).

Water sports have shaped their own recreational geography across the West. Whitewater rafting and kayaking are tied to particular rivers and seasonal flows (**7**). Because of their proximity to population centers, the Sierra, Cascade, and Colorado rivers attract large numbers of people, but some of the West's best runs are in the more remote interior, including the Middle Fork of the Salmon in Idaho and segments of the Green-Colorado drainage (the classic Grand Canyon trip). Windsurfers have fewer options, but they have utterly transformed the economic geography of the zephyr-prone Columbia Gorge, where the epicenter of the sport is Hood River (fig. 8.22). Traditional surfers play along the northern Pacific coast or at Mavericks, a legendary stretch near Half Moon Bay, but most participants choose to surf in relatively warmer southern waters at Rincon, Malibu, Manhattan Beach, and La Jolla (fig. 8.23). ❋

FIG. 8.23 Surfer, La Jolla, California. A wetsuit tempers winter chill as this ace surfer prepares to enjoy the ride on a perfect January day in Southern California.

90 HUNTING AND FISHING

FIG. 8.24 Fly-fishing the Shields River, Montana. This woman in waders is enjoying the good life as she lets her fly float through a quiet stretch of water.

THE DEVELOPMENT OF HUNTING AND FISHING PLAYGROUNDS IN THE West was shaped both by local practitioners and by exotic visitors. Native peoples hunted game and fished streams and rivers in the West for millennia, primarily for sustenance and clothing. Early settlers threw a line in the water or hunted local game for both food and pleasure. Between 1880 and 1920, visitors such as Theodore Roosevelt, George Bird Grinnell, and Charles F. Holder experienced firsthand the rewards of "life in the open," and their publications and political efforts also helped secure an enduring place for outdoor pursuits in the West.

Consider the enduring success of the Boone and Crockett Club (founded in 1887), *Field and Stream* magazine (1895), and *Outdoor Life* magazine (1898), as well as the more recent appearance of the Rocky Mountain Elk Foundation and Trout Unlimited. You can also appreciate the impact of Norman Maclean's short story "A River Runs through It," made into a 1992 film honoring the sport of fly-fishing and set in Montana. Five years after the film's appearance, national rates of fly-fishing had doubled (fig. 8.24; **7**). While such evidence reveals broader national passions for hunting and fishing, the West has played a particularly formative role in shaping appetites for everything from native trout to elk pepperoni.

FIG. 8.25 Duran's Lodge and Outfitters, Ocate, New Mexico. Many professional outfitters across the West offer visitors guided deer, elk, and antelope hunts. In some settings, black bear, bighorn sheep, and mountain lion hunts are also available.

It is hard to overstate the economic and cultural significance of hunting and fishing in the West (fig. 8.25). Certainly there is variation. While passions for fly tying and bow hunting may be muted in suburban San Diego or on the Vegas Strip—California and Nevada have the lowest per-capita participation rates for hunting and fishing in the West—localities such as Stanley, Idaho; Ennis, Montana; and Newport, Oregon, live and breathe the joys of pursuing fish and game (fig. 8.26). One simple way to gauge the local passion for these activities is to count the wildlife trophies and mounted fish in local bars and cafés: Berkeley differs from Butte!

Residents of Montana (31 percent), Wyoming (28 percent), and Idaho (24 percent) are most likely to hunt or fish in any particular year, well above the national average (15 percent), but nonresidents are vital to the recreational economy in some states: among anglers, about 40 percent are visitors in Wyoming, Idaho, and Montana, and visiting hunters constitute at least one-third of participants in Wyoming, Colorado, Idaho, and New Mexico. You can tell how significant hunting and fishing are to a particular place by considering how its economy is shaped by these activities, how these pursuits shape the cultural landscape, and the extent to which these activities define an area's place identity.

Aggregate statistics compiled by the U.S. Fish and Wildlife Service sum up the annual value of purchased equipment (guns, ammo, rods, reels, etc.); tags, licenses, and stamps; trip-related expenses (hunting lodges, fishing

FIG. 8.26 Fly-fishing statue, Ennis, Montana. One of many pieces of public art in this southwestern Montana town, this metallic fly fisherman has hooked a leaping trout on Main Street. Ennis, a sports enthusiast's paradise, is home to the annual Madison Fly Fishing Festival and is close to several blue-ribbon trout streams.

guides, fuel, etc.); land owned and leased for sport; and membership dues and subscriptions to related organizations and publications. Across the West, hunters spend more than $3.5 billion annually, while fishing generates about $7 billion. This economy is especially robust in Wyoming, Montana, and Idaho, where small resident populations are joined by big-spending visitors to boost the impact.

Looking beyond dollars, you can see the imprint of hunting and fishing on the landscape. As any careful angler or hunter will tell you, the very presence of wild fish and game (**18**) is related to their conservation efforts. Revenues from annual tags and licenses are used to purchase land and fisheries and to maintain healthy wildlife populations and a host of federal and state regulations are in place to guarantee that these activities thrive. Land devoted to hunting and fishing includes private parcels and specially maintained spring creeks where visitors pay a fee for access. Large tracts of Forest Service (**67**), BLM (**68**), and state trust and wildlife management lands are also prime destinations, although there are few visual cues suggesting precise uses.

You can also track specialized businesses and activities such as fly shops, fish hatcheries, gun stores, outfitters, and guide services (fig. 8.25). And you see anglers in waders hip-deep in a mountain river, gun racks in pickup trucks, orange vests in the woods, hunter check stations along the highway, and odd-looking pieces of luggage—rods, guns, and harvested souvenirs—at an airport.

Place identities and cultural meanings are also shaped by hunting and fishing. Longtime anglers and hunters develop personal relationships with place as they get to know a stream or an area where game animals are likely to congregate. They covet hot spots, a favorite pool to float a line (fig. 8.24) or the perfect place up the valley where elk often graze.

Most important, particularly in smaller communities, the annual round of activities associated with hunting and fishing—maintaining equipment, planning trips, hunter education classes, summer fishing expeditions, fall hunting seasons, or winter ice-fishing—are key elements of everyday life. They define community identity, friendships, and family relationships. Deeper cultural values can be tied to the meditative virtues of a day on the river or time spent bird hunting with a daughter. These rituals often speak to basic, place-based life lessons about nature, ethical behavior, life and death, and the centrality of friendships and family that westerners seldom articulate.

Lastly, watch for examples of how western hunting and fishing are steeped in vernacular humor and tall-tale telling. You can enjoy stuffed jackalopes (mythical jack rabbits with antlers), singing wall-mounted bass, comical art, and storytelling that celebrates bagging the prize (fig. 8.27) or describing the one that got away. ❋

FIG. 8.27 Hunting and fishing postcard humor. Vernacular humor and storytelling are time-honored traditions that help confer larger cultural meanings on hunting and fishing activities. *Source: author's collection.*

91

RODEOS AND ROUNDUPS

FIG. 8.28 Livingston Roundup, Livingston, Montana. The action heats up as a bucking bronc tries to unseat his rider. In the background, the announcer narrates the action from his vantage point opposite the grandstand. Signs advertise local businesses and organizations.

YOU CAN LISTEN FOR THE LOUDSPEAKER AS YOU AMBLE TOWARD THE rodeo. In a sharp, staccato style, the announcer barks out events, competitors' names, and hometowns, a liturgy delivered in a bourbon-kissed twang originating somewhere between West Texas and eastern Montana. Rodeos and roundups are rooted in the dusty, gritty work of the range cattle industry (**20**) and in the showmanship of the Wild West shows, popularized between the 1880s and the 1920s by promoters such as Buffalo Bill Cody (figs. 8.28 and 8.29). Modern-day rodeos and roundups are landscapes where you can soak up the pageantry and sheer rowdiness of the rural West.

Rodeo history runs deep in the West. Southwestern vaqueros competed in the early nineteenth century, and similar celebrations marked seasonal roundups in the mid and late nineteenth century. Early claimants for the more formal beginnings of competitive events include Deer Trail, Colorado (1869); Cheyenne, Wyoming (1872); and Prescott, Arizona (1888). By the 1920s, rodeos were staple fare in western towns. They included traveling professional circuits with national celebrities such as Will Rogers and local events organized to coincide with county fairs or the Fourth of July.

Twenty-first-century rodeos and roundups have retained some of the traditions. In the county-fair setting, there will be a Main Street parade, the

grand entry, the crowning of a rodeo queen, pancake breakfasts, and diversions such as evening concerts and no doubt a bar fight or two. The main events—including steer wrestling, bronc and bull riding, and barrel racing for women—take place inside the arena, a large open area surrounded by grandstands and gated chutes for releasing animals (figs. 8.28 and 8.30). Rodeo clowns entertain crowds and protect the riders by diverting the attention of threatening bulls or horses. You can also wander and explore nearby parking lots, corrals, and stock pens to gather in the full ambiance of sequined vests, horse trailers, hay bales, and water troughs.

FIG. 8.29 Decal of the Wyoming bucking horse. Celebrating rodeo and patterned after Steamboat, a legendary Wyoming bucking horse, this state symbol appears on license plates, decals, T-shirts, belt buckles, road signs, and more. *Source: author's collection.*

FIG. 8.30 Pendleton Round-Up arena, Pendleton, Oregon. The annual Pendleton Round-Up is one of the West's largest rodeos. For one week every September, grandstands fill with appreciative fans as bulls and broncs buck to cast competitors into the dust.

The Professional Rodeo Cowboys Association (PRCA) is the nation's largest sanctioning organization, and its seven thousand members are involved with more than six hundred events a year that attract more spectators than professional golf. Rodeos have a huge economic and cultural impact on sponsoring communities. Big-time rodeo competitions include the Pendleton Round-Up in Oregon (since 1910), Wyoming's Frontier Days in Cheyenne (since 1897), the Grand National Rodeo in San Francisco (first at the Cow Palace in 1941), and the National Finals Rodeo. The elaborate ProRodeo Hall of Fame is located in Colorado Springs. There are also gay rodeos (**47**); cowboy poetry readings in Elko, Nevada, and other locations; and Agmes de Mille's delightful ballet, set to an Aaron Copland score, simply titled *Rodeo* (1942). In addition, traditional Mexican-style rodeos known as *charreadas* feature similar daredevil equestrian tricks and precision riding, and these events enjoy growing popularity, especially among the West's Latino population (**41**).

Still, countercurrents exist. Animal-rights advocates have criticized both rodeos and *charreadas* because they argue these activities cause animal suffering and potential injury. Future practices may be shaped by changing state legislation and evolving standards adopted by the PRCA. ❋

92 SKI TOWNS

FIG. 8.31 Vail Village, Colorado. Opened in the 1960s, Vail remains one of the West's most popular ski destinations. Downtown architecture reflects "European-inspired" alpine flourishes that differ from Aspen's earlier Victorian blocks (see fig. 8.2).

WESTERN SKI TOWNS COME IN TWO FLAVORS: THOSE WITH A VINtage patina that have their roots in older, mostly mining-related settlements (Breckenridge, Telluride, Park City), and newer localities developed specifically for sport (Big Sky, Squaw Valley). You can see the differences in these towns in a comparison of Aspen (fig. 8.2) and Vail (fig. 8.31), Colorado. Either way, both towns represent concentrated investments in housing (resorts, condos, second homes), services (restaurants, retailing, entertainment, amenities), and infrastructure (roads, utilities, chairlifts, groomed terrain) in spectacular, fragile settings (fig. 8.32).

While ski towns have characteristic landscape features, they have much in common with other regional settlements. Like mining towns (**31**), ski towns transform the physical fundament into a marketable commodity. As with farm towns (**26**), the fortunes of ski towns are determined by the weather: droughts can have similar impacts on the pocketbook whether you are in Montana's wheat country or Colorado's high country. And as with urban areas more generally, many ski towns boast mega consumer landscapes (**73**), where après-ski shopping, eating, and drinking help support the local economy—The Village at Mammoth in California and Vail Village in Colorado are examples (fig. 8.31). Larger settlements also feature what are essentially suburban

FIG. 8.32 Breckenridge, Colorado. Once a sleepy mining town, "Breck" became the site of large-scale ski-inspired development after 1961 and is one of North America's most visited winter resorts. Condos crowd the foreground, while distant slopes have been cropped for winter's return.

master-planned communities (**79**), clusters of residences where landscapes are manicured and permissible land uses controlled—Sun Valley's Elkhorn Resort and Montana's Big Sky Resort, for example (fig. 8.33). Ski-town equivalents of the "city invisible" (**74**) can be found, too—ordinary, less-legible places where visitors park, wastewater is treated (often a major issue), and low-paid service workers live, creating the "down-valley syndrome" in which employees live miles away from the ski area in cheaper housing (fig. 8.34).

Ski-town history is wedded to the broader evolution of the sport. The recreational use of skis—either cross-country (Nordic style) or downhill (alpine style)—flowered after 1900. Steamboat Springs, Colorado, is one of the earliest ski towns; in 1913, Norwegian Carl Howelsen organized the first winter festival and ski-jumping contest there. Sun Valley, Idaho, the West's first modern, integrated ski resort, was created by the Union Pacific Railroad in 1936. The town's developers borrowed directly from European Alpine resort traditions and offered a model for later development after World War II. Sun Valley's star-studded clientele also solidified connections between the elite and winter sports.

Better economic times and a growing appreciation for winter sports—much of it cultivated by veterans of World War II's Tenth Mountain Division of alpine skiers—encouraged later developments. Aspen flowered after the war, the brainchild of businessman and intellectual Walter Paepcke. His vision brought together a heady mixture of summertime educational insti-

FIG. 8.33 Big Sky Mountain Village, Montana. The Arrowhead Chalets offer handy access to ski runs on nearby Andesite Mountain at this popular southwestern Montana resort.

tutes and music—Aspen was known as the Athens of the Rockies—with the more physical challenges of winter sport.

After 1960, the scale of ski-town development mushroomed, reflecting growing national demands for outdoor recreation and an increasingly corporatized business model (Twentieth Century Fox bought the Aspen Skiing Company in 1978) that mass-produced second homes—50 percent to 70 percent of housing in many ski towns—and packaged the whole experience in the cultural cachet of deep powder. Squaw Valley hosted the 1960 Olympics, the planned community of Vail dated from 1962, and Jackson Hole was developed as a resort after 1965.

Many variables shape the distribution of ski towns in the West. Physical setting is critical—enough land for settlement, the potential for downhill runs, and dependable winter snow. They also require proximity to airports, interstate highways, and large cities. Add to this a need for entrepreneurial expertise, financing for construction budgets, cheap labor, and a supportive political climate for expansion.

In the twenty-first century, Colorado has the largest accumulation of ski towns, which can be found in many of the more mountainous, well-watered parts of the state. California's Tahoe Basin claims the densest concentration of resorts, but no single one dominates. Elsewhere, a line of settlements follows the Sierra Nevada and the Cascades. Other landscapes—including Alta and Park City in Utah, Big Mountain and Big Sky in Montana, Sun Valley in Idaho, and Jackson Hole in Wyoming—have sparked development across the Wasatch Mountains and northern Rockies. To the south, Big Bear, California; Flagstaff, Arizona; and Cloudcroft, New Mexico, also offer wintertime activities.

The full range of settlement features in a typical ski town includes in-town hotels and resorts, exurban trophy homes (**95**), and worker housing (dormitories, mobile-home parks, low-cost apartments). Building sites for ski towns are typically constrained by steep slopes and narrow valleys, and adjacent public lands, usually national forests (**67**), limit private development while

FIG. 8.34 Down-valley housing near Edwards, Colorado. Not far from Vail's Tyrolean touches, Edwards (just off Interstate 70) offers mobile home parks and rental units for service workers who cannot afford Vail's lofty cost of living.

they guarantee both open space and high housing costs. You can also examine how mountain environments have been modified and how slopes bear the mark of ski runs and altered vegetation (fig. 8.32).

Ask yourself, What is the ski town's place identity? Is the town low-key and locally oriented or faster-paced, home to the glitterati? In some towns, Tyrolean touches dress up the main street (fig. 8.31), while down-valley workers live in duplexes and mobile homes (fig. 8.34). In many settings, the past is evident (figs. 3.11 and 8.2), while others have a postmodern mix of landscape elements (figs. 8.32 and 8.33). Many localities, building on the Aspen model, market their four-season appeal with golf courses (fig. 1.27), arts and music festivals (**97**), and opportunities for adventure play (**89**).

The slopes above the ski town are divided into different varieties of social terrain. You can spot family-centered bunny hills and the expert turf of double black diamond slopes. Note the snowboarders (participants in the Olympics since 1998), distant cousins to surfers and skateboarders, mingling with the skiers. On some ski hills, you can see their signature half pipes, ramps, and boxes. Whatever their expertise or preferences, every snow enthusiast shares one bond, an unrequited love for the next dump that delivers them to heaven. ❋

95

GOLF COURSES

FIG. 8.35 Practice green, Sonoma Ranch Golf Course, Las Cruces, New Mexico. New homes line these desert links, a reminder of the close relations between the sport of golf and the business of real estate development.

THERE ARE MORE THAN THREE THOUSAND GOLF COURSES IN THE West, more than 30 percent of them in California. From above, especially in the arid West, you can spot them at thirty thousand feet, long narrow strips of green set within some of the country's most desolate environments. The average course consumes about one hundred to two hundred acres of land, and each course creates its own cultural landscape, a circumscribed world of tee boxes, fairways, water hazards, sand traps, and putting greens. It is an intense and intricate geography, contoured for both sport and torture.

With roots in the Northeast and Europe, golf became popular in the West in the 1890s, as courses were established at places like Riverside, Pasadena, and Redlands in Southern California and in San Francisco, Monterey, and Vallejo farther north. Between 1900 and 1930, hundreds of courses both public and private were built, most of them in the suburbs. Major western cities created the greatest demand.

Eventually, resort settings in localities such as Palm Springs, California, accounted for much of the industry's western expansion. In particular, the notion of integrating a golf course with other real estate and resort developments, often as components of master-planned or retirement communities

(**79** and **98**), transformed landscapes from Bandon, Oregon, to Scottsdale, Arizona. Homes and condominiums line the fairways, and nearby hotels and resorts host visitors (fig. 8.35). These real estate ventures have served as the nucleus for mega clusters of golf courses in Southern California, southern Nevada, and southern Arizona, and they serve as destinations in amenity-filled localities in areas such as Colorado, central Oregon, and western Montana (fig. I.27).

Consider western golf courses from several viewpoints. As a real estate development, every golf course demands huge, often speculative capital investments. Land values change enormously as residential lots along the fairways are subdivided and integrated with other commercial shopping and entertainment opportunities nearby. As a cultural experience, golf courses create a local sense of place. Pebble Beach, with its oceanfront vistas; Montana's Old Works course in Anaconda, built on colorful copper-smelting slag; and the legendary third hole on the Mountain Course at the Ventana Canyon Golf and Racquet Club in Tucson are western landscapes that create rich memories for those who visit them (fig. 8.36). Finally, golf courses can be seen through the lens of ecological transformation. They create new landscapes of native and exotic species (**17**; figs. 8.3, 8.35, and 8.36). In the clear-cut corridors of pine and fir along Pacific Northwest links and in the southwestern deserts, thousands of acres of Bermuda grass, ryegrass, and bluegrass have replaced native species. Critics argue that golf courses use far too much water and alter sensitive natural settings for plants and animals. Enthusiasts see opportunities among the links for creating open space and for using more local plant species, producing settings where golf carts and duffers can coexist with native birds and wildlife. ❋

FIG. 8.36 Mountain Course, Ventana Canyon Golf and Racquet Club, Tucson, Arizona. Choose your club wisely: the legendary third hole is unforgiving at this posh golf resort in the Santa Catalina foothills. Here, tiny greens set amid rocks and cacti challenge golfers as upscale homeowners enjoy the view.

94 SEXUAL COMMERCE

FIG. 8.37 Exotic Kitty's Gentlemen's Club, Bakersfield, California. This central California strip club is on a quiet street near the Kern River in a nonresidential area of Bakersfield's north side. Industrial developments and several motels are nearby.

FROM LIGHT-BEER COMMERCIALS TO BROTHELS, COMMODIFIED SEX sells. Sexual commerce on the western landscape is both revealed and concealed. Billboards, store windows, and the sexualized fashion industry—even for children—exemplify how sex and money mingle. In its more explicit forms, however, community standards and legal restrictions often segregate or hide sexual commerce—massage parlors, strip clubs, adult book and lingerie stores, adult film studios, legal prostitution—in ways that minimize its visual presence or restrict it to areas zoned for its use, likely a part of the city invisible (**74**).

Western states and communities have adopted ordinances to regulate sexually oriented business, including zoning restrictions on where businesses such as adult bookstores and strip clubs can operate and licensing procedures that define when certain activities can occur. Conflicts develop when First Amendment rights are pitted against the proximity of sexual commerce to schools, churches, and residential neighborhoods. While patterns vary, a common result for strip-club geography, for example, locates many of these businesses in industrial areas, along commercial thoroughfares beyond city limits (the same is true of fireworks stands), in older downtown neighborhoods, or near military bases (**62**; fig. 8.37). Adult bookstores, sex shops, and

FIG. 8.38 Sagebrush Ranch, Mound House, Nevada. This legal Nevada brothel, located near Carson City, is on a quiet cul-de-sac (with similar businesses). The main building (with its long, narrow, multiroom layout) is fenced and gated for privacy.

massage parlors also locate in a complex landscape that reflects local ordinances, land costs, consumer demand, and a nuanced web of community standards and law enforcement practices.

Particular regional expressions of sexual commerce merit attention. Geographer Darrick Danta found that the San Fernando Valley in the Los Angeles metro area has long been one of the world's leading producers of pornographic films and videos. Porn Valley's more than two hundred production studios account for much of the industry's global revenues. Thousands are employed in a business that is attracted to the valley's cheap rental space, the availability of low-cost labor, and flexible regulations and law enforcement. In valley communities such as Van Nuys, Canoga Park, and Chatsworth, most porn studios are housed in older industrial or warehouse facilities or in nondescript office buildings.

While there are many forms of illegal prostitution, only in the West—in Nevada—can you find legal prostitution. About two dozen brothels operate in the state, typically in rural counties (fig. 8.38). Most brothels have small welcoming signs, isolated fenced properties, and a horizontally configured array of rooms oriented around a bar and meeting area or pool. In towns such as Elko and Pahrump, brothels cluster in particular districts or along streets where such activities are sanctioned. ❋

95

AMENITY EXURBS

FIG. 8.39 Amenity exurb, Bitterroot Valley, Montana. This scattered, low-density pattern of valley farmland (foreground), rural nonfarmers on scenic hillside tracts, and nearby national forests (distant slopes) characterizes many western settings. Some residents live here and commute to nearby Missoula, about thirty minutes away. Ski runs (left) represent part of a failed resort subdivision (a "zombie development").

SINCE 1970, A QUIET REVOLUTION HAS RECONFIGURED THE RURAL West. Massive tracts of land beyond the edge of cities and suburbs (**83**) have become amenity exurbs, places where people who are not farmers or ranchers settle in rural areas. They are drawn to the attractions of privacy, open space (**1**), outdoor recreation, low-density living, and social homogeneity while also being able to enjoy the conveniences of the modern world (figs. 6.10, 8.1, and 8.39).

In the nineteenth century, many elite Americans pined for a country estate where they could leave the city behind and cultivate rural pleasures. Later on, suburbanization and the flowering of western resort communities (**86–89**, **92–93**) also pointed the way. It wasn't until after 1970, however, that economic prosperity, improvements in rural transportation and communications, and cultural tastes combined to create a demand for amenity exurbs.

By definition, amenity exurbs lie beyond the metropolitan fringe, in areas once dependent on agriculture, particularly ranching. As agricultural profits shrank and farm and ranch owners aged, nonfarmers bought up the land in these areas. The new population includes long-distance commuters, blue-collar families looking for an inexpensive lifestyle, wealthy families in search of privacy and exclusivity (**96**), aging hippies, upscale retirees (**98**), outdoor

FIG. 8.40 Low-density amenity exurb north of Boise, Idaho. These "lone-eagle" residents each claim sizable turf in an upscale retreat just off busy State Highway 55.

enthusiasts, and gentleman farmers. In fact, these eclectic exurban neighborhoods are often settings for conflict as people create different visions of the good life: one resident's notion of acceptable behavior—burning stubble in the fields, spraying herbicide, parking an RV in the front yard—may not be another's.

Every city has exurbs, some of them seventy miles outside the suburban fringe—the islands of Puget Sound, the hills of western Oregon, and the forested peripheries beyond Bend and Spokane are examples. Transect the slopes of the Sierra Nevada foothills from Mariposa to Grass Valley and consider the nooks and crannies of the California Coast Ranges, one of exurbia's pioneering promised lands. Boise, Salt Lake City, and Denver also have extensive amenity exurbs in the nearby foothills and mountains beyond the suburban fringe, and you can find similar places in western Montana valleys, northern New Mexico highlands, and picturesque desert retreats in the Southwest.

The cultural landscapes of these exurbs are as diverse as their inhabitants and the natural settings they occupy. Ask the following questions to make sense of the patterns you encounter.

How large are the land parcels? While one version of amenity exurbs emphasizes isolated hilltop structures linked by wandering access roads (fig. 8.40), other developments may have houses tucked around cul-de-sacs, all surrounded by extensive tracts of open land (fig. 8.41). Local and state subdivision laws also determine the pattern. For years, Colorado's law exempting parcels greater than thirty-five acres from subdivision review, for example, produced a landscape of thirty-six-acre ranchettes. Sometimes a few tracts are hived off from a large rural parcel (fig. 8.1), while nearby a neighborhood is created by subdividing large ranches (fig. 8.39). Some exurbs grow organically, irregularly expanding along highways. Others are master-planned communities (**79**) that may be so-called conservation developments designed to preserve open space and wildlife habitat. Some may be speculative real estate ventures, evident in regularly sized parcels, elaborate road networks, and consistent signage.

FIG. 8.41 High-density amenity exurb near Genoa, Nevada. This cluster of newer homes (foreground) features small lots but sweeping views. Lower tracts in the nearby Carson Valley remain working ranches, while rural nonfarmers seek higher land. The towns of Minden and Gardnerville are in the distance.

What are the houses like? Size does seem to matter in these landscapes. Count picture windows, fireplaces, outdoor decks, garage spaces, and recreational toys, but be wary of stereotypes. While much has been written about the "gentrified range" and the upscale nature of amenity exurbs (figs. 8.40 and 8.41), the reality is more complex. Manufactured houses and double-wide mobile homes far outnumber trophy retreats featured in slick real estate brochures (fig. 8.42). This vernacular landscape of lower- and middle-class housing suggests that many amenity exurbs are attractive simply for their lower living costs and casual lifestyles.

How close is the exurb to a major city, high-speed highway, or airport? These variables measure accessibility and help determine who lives in the exurbs. The blossoming of exurbs far beyond Los Angeles, Denver, and Phoenix, for example, is closely linked to the interstate highways that make long-distance commutes possible. Many exurban residents economically tied to these cities are willing to drive sixty to ninety minutes to enjoy open space and rural life (figs. 8.39 and 8.40). In other cases, isolation may limit options, suggesting that people who live in these places may be retired or otherwise untethered from cities.

What are key local amenities attracting exurban residents? In addition to the general attractions of exurban life, particular outdoor amenities also have drawn people to these areas (**87–90, 92–93**). Many of the West's most rapidly growing amenity-oriented exurbs are found near national parks and forests,

BLM land, or wilderness areas. Precise physical setting can be essential. Exurbanites especially fancy foothill properties positioned between open valleys and forested mountains (often public lands), although these landscapes are particularly vulnerable to wildfire hazards (**16**; figs. 8.39 and 8.41). Other desirable destinations include riparian tracts (**7**; fig. 8.1), scenic desert real estate (fig. 6.10), and wide-open spaces. ❋

FIG. 8.42 Home in an amenity exurb north of Las Vegas, New Mexico. Far from log castles and posh subdivisions, most of the West's amenity exurbs display the benefits of casual living and low-cost (often manufactured) housing.

96

GATED COMMUNITIES

FIG. 8.43 Falcon View Estates, Orange County, California. An exclusive neighborhood near Irvine, this gated community offers five-thousand-square-foot (minimum) homes and a setting that is "close to nature, but not too far from civilization."

AS CULTURAL LANDSCAPES, GATED COMMUNITIES PRODUCE HARDened, privatized terrain across the West, and they suggest how physical and social spaces intertwine. The message of encapsulation is clear: perimeter walls, gates, guard shacks, private security patrols, and electronic surveillance separate a safe and orderly world within from the heterogeneous, unpredictable, potentially dangerous world beyond. Led by California and Arizona, the West claims the nation's largest number of gated communities. Elite nineteenth-century suburbs, mostly in the Northeast, pioneered the practice of "forting up" in the United States. In the West, beginning in the 1950s and 1960s, western land developers recognized the cachet of designing master-planned, gated retirement communities (**98**) and resorts in places such as Palm Springs, Phoenix, and Los Angeles. In the late 1980s, the idea of these developments spread to more of the region's major cities and many nonmetropolitan places. Since 2000, much of the new home construction in California and Arizona has been in gated developments.

You will find gated communities in two predictable settings. In upscale suburbs in places like Irvine, California; Highlands Ranch, Colorado; and North Scottsdale, Arizona, gated neighborhoods are often incorporated into master-planned communities (**79**; fig. 8.43) that advertise safety, shared com-

FIG. 8.44 Stock Farm Club, Bitterroot Valley, Montana. Carved from the estate of a copper baron (Marcus Daly), the Stock Farm Club gated community offers a mix of ranch homes (shown), mountain homes, and cabins, plus amenities such as golf, tennis, and swimming.

munity standards (often meaning racial homogeneity), higher home resale values, and common-interest developments such as parks, playgrounds, swimming pools, and tennis courts. As you look into the interior, assess the development's name and signage, the size and design of gates, methods of gaining entry, and the amenities within (often visible on the Internet via Google Earth).

Gated communities also appear in amenity exurbs (**95**) that lie beyond the metropolitan periphery, where most "residents" may visit only occasionally. Many of these settlements add to their exclusivity by enclosing homesites, open space, and amenities with walls and fences. Private golf clubs (**93**), lakeshore and oceanfront resorts (**87** and **88**), and retirement communities (**98**) have incorporated these security elements at places such as Indian Wells, California; Entrada at Snow Canyon, near Saint George, Utah; and the ultra-exclusive Stock Farm Club near Hamilton, Montana (fig. 8.44).

Gated communities will remain a fixture on the western American landscape. In fact, the appeal of gated communities has spread into the ranks of middle-class westerners, and a growing number of walled retreats offer smaller lots, manufactured homes, and scaled-down frills to meet more modest budgets. ❋

FIG. 8.45 Santa Fe arts community, New Mexico. As its name suggests, the Kokochile de Santa Fe sells a diverse assortment of local and regionally inspired arts, crafts, and souvenirs.

97

REGIONAL ARTS COMMUNITIES

CHANCE OFTEN SHAPES PEOPLE AND PLACES. IN 1898, ARTISTS ERNEST Blumenschein and Bert Phillips were touring the Southwest when their wagon broke down on the rough roads of northern New Mexico. Waiting for repairs, they dallied near Taos. The locality caught their eye: sparkling blue skies, mountain backdrops, and high-desert hues, a landscape steeped in Native and Hispanic history. They remained in town for several months, initiating a geographic courtship that connected both men to New Mexico for decades. That by-chance rendezvous started Taos down the road to becoming one of the West's foremost regional arts communities.

The West's first-generation art colonies had antecedents in Europe, where nineteenth- and early twentieth-century artists (both European and American) were attracted to the camaraderie and creativity found in towns such as Volendam (Netherlands), Barbizon (France), Newlyn (Great Britain), and Worpswede (Germany). In the West, while cities such as San Francisco and Denver had active arts communities, more remote colonies evolved in favored settings. Booster Edgar Hewett and the advertising weight of the Santa Fe

Railroad (**50**), for example, propelled Santa Fe, New Mexico, into the creative limelight between 1915 and 1925 (fig. 8.45). The New Mexico Museum of Art, the Santa Fe Art Museum, and the annual Santa Fe Festival solidified the city's role as a creative community. By 1920, Carmel, Laguna Beach, and La Jolla in California had also emerged as centers for artists, artists' associations, art museums, and galleries (fig. 8.46).

Dozens of western localities have reinvented themselves as arts communities (fig. 8.47). Laguna Beach, Carmel, and Mendocino in California and Cannon Beach in Oregon remain very active among the coastal communities. Desert, canyon, and Native American art is prominent in Arizona and New Mexico (fig. 8.48). In the mountainous interior, you can find galleries in localities such as Aspen, Colorado; Sun Valley and Sandpoint, Idaho; and Kalispell and Livingston, Montana.

The setting obviously matters for regional arts communities. Whether it be scenic Sedona, Arizona, or Victorian Port Townsend, Washington, evocative environments energize artistic imaginations. Creative people in these communities have often come from distant places. New Mexico's Georgia O'Keeffe (Wisconsin), John Marin (New Jersey), and Maynard Dixon (California) all became ardent neonatives. Key promoters of regional art also include founding visionaries (Santa Fe's Hewett), local artists, and business interests (railroads, real estate agents, chambers of commerce). A thriving arts community also depends on the presence of people willing to spend money to harvest the fruits of cultural capital they wish to call their own.

These playgrounds deliberately market their place-based identities, often in ways that parallel initiatives directed toward urban consumers (**73**). Clusters of art galleries and studios may highlight local artists, and collectively

FIG. 8.46 La Jolla, California. This Southern California arts community is anchored by grand La Valencia Hotel (opened in 1926), which celebrates the fine points of Spanish Colonial Revival architecture (left) and sits adjacent to galleries that sell work by local artists.

FIG. 8.47 Selected regional arts communities. The map suggests the attractions of scenic coastlines, rugged mountain settings, and colorful southwestern deserts. An arts community's potential popularity is also elevated by its proximity to large urban areas and other amenity activities.

owned artist cooperatives provide shared venues for showcasing members' work. Art supply stores, framing shops, museums, schools, and foundations also provide key support. Examine the larger cultural footprint, the architectural ambiance (figs. 8.45–8.46; 8.48–8.49), which may include theaters, live music venues, and film festivals (Park City's Sundance began in 1978). In a related fashion, you can also see how place meanings in these localities are emphasized by independent bookstores and restaurants where the area's setting and culture are promoted and often interwoven with examples of local art.

Once-dying mining towns like Madrid, New Mexico, and Jerome and Bisbee, Arizona, are now vibrant arts communities, home to creative people attracted by scenery, local character, and inexpensive housing (figs. 3.27 and 8.49). Places like Aspen, Whitefish, and Jackson have integrated creative

FIG. 8.48 Tubac Festival of the Arts, Arizona. Since 1959, this small Arizona town has hosted dozens of local artists and visiting arts vendors (see tent stalls, right). Set amid the town's pueblo-style buildings, the festival celebrates the iconic cultural symbols of the Desert Southwest.

FIG. 8.49 Channeled Gallery, Madrid, New Mexico. Once a dying coal town, this northern New Mexico community redefined itself as an arts and cultural center. This gallery sells imaginative works by mentally ill and schizophrenic artists.

amenities with recreation (**86–93**). Also note cyclical pulses: crowds arrive for scheduled art walks or annual arts festivals, but vanish during off-season lulls (fig. 8.48). The art scene in some communities is especially ephemeral. For one breathless, incandescent week each year, the Burning Man art event in Nevada's Black Rock Desert creates an instant city of more than fifty thousand people, and then it all vanishes as quickly as a dust devil on a summer afternoon. ❋

98

RETIREMENT COMMUNITIES

FIG. 8.50 Green Valley, Arizona. Quiet streets, well-maintained homes and yards, local security, and nearby golf courses, shops, and medical care prove powerful draws for residents age fifty-five and older in search of a planned retirement community.

IN GREEN VALLEY, ARIZONA, ONE OF THE WEST'S LARGEST RETIREMENT communities, founded in 1963, the streets and sidewalks are well maintained. Upscale homes and condos feature Southwest-style architecture and desert-themed front yards (fig 8.50). Health-care services are readily available for the more than twenty thousand people who live in the town, nearby fairways invite residents to play a round, and shopping centers offer parking spaces for golf carts. Green Valley's success is part of a reinvention of aging that has marked fundamental economic, social, political, and spatial shifts since 1960. The gerontocracy's ascent reflects the affluence of baby boomers, improved health care for seniors, and a voting bloc no politician can ignore.

The growing power of seniors is reflected on the landscape, as master-planned retirement communities have blossomed across the West (**79**), especially in Arizona, California, Colorado, and Nevada. Many communities stress active retirement lifestyles, centered on golf courses (**93**); environmental, recreational, and cultural amenities; and comfortable exurban living (**87–88**, **90**, **95**, **97**). Del Webb's Sun City development (1960) in Arizona pioneered this approach (fig. 7.48). Elsewhere, gated variations (**96**) at places like Leisure World in Seal Beach, California, suggest an emphasis on security and social homogeneity (fig. 8.5).

Some of these retirement settlements emphasize luxurious single-family homes, while others offer single-story condominium living. Smaller retirement "villages" with modular housing and mobile homes cater to modest budgets (fig. 8.51). Other emergent alternatives appeal to particular preferences, including retirees who are gay (**47**), Mormon (**42**), and Asian (**46**). Levels of care also vary. Larger facilities stress "aging in place," where a person can sample the virtues of independent living, assisted living, skilled nursing care, and Alzheimer's care as their needs evolve. In more diffuse fashion, entire towns have acquired reputations as best places to retire. Bend, Oregon; Cedar City and Saint George, Utah (fig. 8.52); Prescott, Arizona; and Las Cruces, New Mexico (fig. 7.1), have grown rapidly since 1980. While neighborhoods, businesses, and amenities cater to seniors in these communities, the resulting largely unplanned growth produces sprawling settlements radically different from the Green Valley model. ❋

FIG. 8.51 Monte del Lago community near Castroville, California. Seniors on modest budgets are attracted to the affordable living options in many retirement-oriented villages that emphasize manufactured housing and mobile homes.

FIG. 8.52 Saint George, Utah. Since 1980, much of Saint George's growth has been generated by retirees who enjoy its desert climate and small-town amenities. Diverse subdivisions offer many choices, but overall planning initiatives have been modest. This view, taken just northeast of town, shows how new residents seek out attractive view lots on nearby mesas.

99

SNOWBIRD SETTLEMENTS

FIG. 8.53 Roosting snowbirds, Quartzsite, Arizona. During the winter, Quartzsite's population expands from 3,600 to more than 80,000. Distant settlements on the desert are groups of boondockers, RV enthusiasts who are untethered from support services and utility hookups.

EVERY WINTER, PEOPLE FLOCK TO PALM SPRINGS, CALIFORNIA; Quartzsite, Arizona; and Deming and Alamogordo, New Mexico. The seasonal migration from the northern tier of states to areas known as snowbird settlements balloons Quartzsite's population, for example, from 3,600 to more than 80,000 (fig. 8.53). Some stop briefly for the classic car or gem show, while others "boondock"—that is, stay in unimproved campgrounds—for several months in nearby dry washes (fig. 8.8).

Some snowbirds are full-timers who live year-round in trailers (fifth-wheelers) and recreational vehicles (RVs), returning every winter to the same camp. The Good Sam Club, the world's largest RV club, and the Wally Byam Caravan Club (for people with Airstream trailers) strengthen these bonds and help members organize annual activities. Other retirees maintain homes in the north and then lose latitude in winter, either living on the road or finding seasonal condos or long-term rentals (**98**) in places such as Lake Havasu (fig. 8.18), Mesa, or Tucson, Arizona.

The roots of snowbirding reach into the nineteenth century, when wealthy northeasterners wintered in places like Pasadena, California. Middle-class participation blossomed after World War II with mass-produced trailers (Bing Crosby opened Palm Springs' Blue Skies Trailer Park in 1952), but the real surge came

FIG. 8.54 RV and mobile-home park, Quartzsite, Arizona. The paved streets and landscaped yards of this seasonal community stand in contrast to more casual settlement patterns in nearby dry washes (compare to fig. 8.8).

FIG. 8.55 Retail district, Quartzsite, Arizona. Snowbird necessities are all within strolling distance along the main road through town. The local McDonald's, grocery store, and public library also do a booming business in January but are all but deserted in July.

between 1970 and 1990 when seniors began to travel south in large numbers to established campgrounds in southeastern California and southern Arizona.

For some itinerant snowbirds, upscale RV parks offer paved grids, landscaping, streetlights, and full hookups (fig. 8.54), while more casual arrangements prevail in Long-Term Visitor Areas managed by the BLM, or in the dry-wash boondocks (fig. 8.8). Iowans may cluster in one setting, while people from the West Coast surround campfires nearby. Sprawling residential land uses predominate. Watch for "white cities"—clusters of fiberglass rooftops and carports (fig. 8.53)—and notice their own version of commercial roadside strips (**80**) where seasonal businesses meet visitors' needs (fig. 8.55).

Also note the seasonal flip side: every summer, thousands of *zonies* (out-migrating Arizonans) descend upon San Diego's mild coast when desert temperatures soar. ❋

100 LAS VEGAS

FIG. 8.56 Luxor, Las Vegas, Nevada. Completed in 1993, the Luxor is an early example of the Strip's theme-based resort complex. The thirty-story, 365-foot-high pyramid houses more than four thousand rooms, all accessed by "inclinators" that ride along the interior at a 39-degree angle.

LAS VEGAS IS UNLIKE ANY OTHER CITY IN THE WEST. WITH MORE THAN 150,000 hotel rooms and 200,000 slot machines at their disposal, the city's forty million annual visitors find plenty of ways to play. How many places offer a landscape oriented toward an *indelicato* conjoining of eroticism and family fun? Only within the Strip's malleable social space can three generations of Midwesterners mingle effortlessly with strolling Guatemalans who peddle blond escorts a phone call away (**94**). It is a place where volcanoes erupt, pirate ships sally forth, and pyramids pierce the desert sky (fig. 8.56). Architect Robert Venturi and historian Hal Rothman have suggested that Las Vegas is a prototype for tomorrow's West, with its heady mix of postindustrial mass consumption, automobility, speculative real estate, decentralized living, cultural pluralism, and visual theater.

The city's story parallels larger regional narratives. Built as a railroad town in 1905, Las Vegas (named for nearby meadows) later blossomed thanks to federal spending. The construction of Boulder (now Hoover) Dam from 1928 to 1935, the New Deal, World War II air bases and defense factories, and the atomic age spurred the economy, and Nevada's relegalization of gambling in 1931 fueled tourist-related growth. Frontier-themed casinos and motels proved popular with early visitors (especially Californians), and by the 1950s and 1960s, mob-financed operations (Bugsy Siegel's Flamingo opened in 1947) gradually combined older western themes with the hip, raffish appeal of the Rat Pack, led by Frank Sinatra, Dean Martin, and Sammy Davis Jr., to create the Sin City mystique and the early Strip landscape.

Nevada's approval of the Corporate Gaming Act in 1969 sparked a new scale of entertainment tourism, opening the way for the construction of the destination resorts that dominate Las Vegas today. The boom transformed the landscape, as the Old West style was cast aside in favor of themed resorts. Developer Steve Wynn's Mirage (1989) set the trend, inviting visitors into a fanciful French Polynesian landscape, complete with a dinner buffet, air-conditioning, and nonstop gaming. The Luxor, Treasure Island, and New York–New York properties were followed by the even grander Venetian, Bellagio, and Wynn resorts (fig. 8.57). In fact, American urban planners and developers cite the city's formative role in shaping larger national landscapes of commercial strips (**80**) and urban mass consumption (**73**).

In the early twenty-first century, growth in Las Vegas outpaced every other western city. The gaming economy, retirees (**98**), and a real estate and construction boom increased the population from 850,000 in 1990 to about 2,000,000 in 2010. Employment opportunities attracted an increasingly diverse population, and Las Vegas became more than 30 percent Latino, 11 percent African American, and 6 percent Asian American.

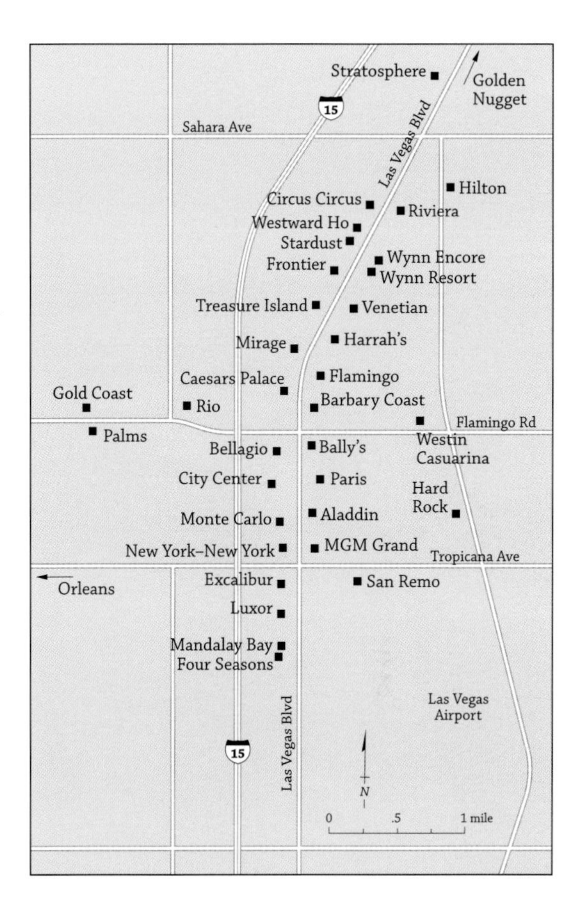

FIG. 8.57 Major casino resorts, Las Vegas Strip, Las Vegas, Nevada. For several miles, high-rise hotels, fantasy landscapes, and opulent casinos line Las Vegas Boulevard, making the Strip one of the West's most recognizable commercial landscapes.

Las Vegas also became a world city, hosting more than seven million international visitors and twenty thousand convention events a year. A global clearinghouse for the next generation of manufactured products, tech toys, and consumer goods, the city reflects and shapes popular culture in ways that may exceed New York, London, or Beijing.

Still, as is true for other western boomtowns, the Las Vegas economy has periodically contracted. In downturns, unemployment and foreclosures soar, gambling revenues sag, and construction declines. East of town (in Henderson), a luxurious master-planned community (**79**), the sprawling 3,600-acre Lake Las Vegas (fig. 8.58), included golf courses, hotels, and Tuscan-style villas, but the fantasies and financing soured and the development went bankrupt, leaving an unfinished landscape of browning fairways and half-occupied neighborhoods.

It is hard to be neutral about Las Vegas: people either embrace its whims or decry its excess. Whatever one's visceral take, we can all learn from Las Vegas by seeing how the Strip conceals reality behind the lights. There are two important stories to keep in mind. First, Las Vegas is a thirsty city surrounded by desert. Much of the city's water (suburbs like Henderson and Summerlin

FIG. 8.58 Lake Las Vegas near Las Vegas, Nevada. This subdivision (in Henderson) included golf courses, a Ritz-Carlton hotel, and hundreds of well-appointed Tuscan-style homes and condominiums. When developers later declared bankruptcy, it was a reminder that the West's playgrounds are never immune to economic downturns.

FIG. 8.59 Las Vegas Strip, Las Vegas, Nevada. Multiple fantasies mingle to create the Strip's diverse cultural landscape.

FIG. 8.60 Wynn's Encore, Las Vegas, Nevada. The Encore is part of Steve Wynn's enduring signature on the Las Vegas landscape.

use far more than Strip hotels and fountains) comes from nearby Lake Mead. By 2035, the area's four million residents will need 985,000 acre-feet of water annually, 50 percent more than in 2010. Even plans to tap underground aquifers in northern Nevada will fail to close the gap. Second, despite its well-earned reputation as a playground, the city is all about work. More than eight hundred thousand people work in Las Vegas, 70 percent of the state's total. Unions are relatively strong, and many jobs pay well. Twenty-four hours a day, people cook in restaurants, change sheets, and drive cabs. Daily routines revolve around shift changes, days off, and the search for better jobs. Much of this laboring landscape is all but invisible, a seamless part of an experience designed to lighten your pocketbook (fig. 8.59).

So how will Las Vegas look in 2080? Despite looming hazards from global warming and other threats, Las Vegas appears positioned to outlive us all (fig. 8.60). Like the larger West, however, left unceremoniously to the next generation, its precise future landscape remains undecipherable, full of yet-to-be-imagined fantasies. ❋

CHRONOLOGY

~1.7 billion B.P.	(before present) Creation of Vishnu Schist, visible at bottom of Grand Canyon
~60 million B.P.	Laramide orogeny creates modern Rocky Mountains
~2.5 million–12,000 B.P.	Pleistocene epoch: repeated glaciations shape landscape
35,000–15,000 B.P.	Ancestors of modern Native Americans migrate from Asia
15,000–12,500 B.P.	Glacial Lake Missoula floods, reworks Pacific Northwest landscape
12,000–9000 B.P.	Clear archaeological record of Native American life in multiple settings around the West (Clovis and Folsom cultures)
5000–4000 B.P.	Domesticated varieties of maize arrive in Southwest from Mesoamerica
1000–800 B.P.	Pueblo villages at Oraibi and Acoma are settled in Southwest
900–600 B.P.	Ancestors of modern Navajo (Athapaskans) arrive in Southwest
A.D. 1492	Columbus arrives in Bahamas, initiates Columbian Exchange
1540–42	Coronado explores area from Arizona to Kansas for Spain
1598	Juan de Oñate and colonists settle northern New Mexico for Spain
1610	Santa Fe founded
1680–1750	Widespread diffusion of horses in the West via Spaniards
1692	Spanish mission at San Xavier del Bac (southern Arizona) founded
1706	Albuquerque founded
1769	First Spanish mission founded in Alta California at San Diego de Acalá
1776–77	Domínguez-Escalante expedition across interior Southwest
1778	Captain James Cook explores Pacific Coast for British
1785	Federal Land Ordinance of 1785, beginning of public lands survey system
1792	Mouth of Columbia River "discovered" by Captain Robert Gray
1803	Louisiana Purchase from France costs the United States $15 million
1804–6	Lewis and Clark expedition
1812–41	Russians settle Fort Ross along Northern California coast
1820–35	Height of fur-trade era in the interior West
1821	Mexico achieves independence from Spain
1837	Smallpox epidemic decimates Hidatsa, Mandan, and Arikara tribes
1842–49	Frémont leads four expeditions in interior West
1846	Treaty with Great Britain secures Oregon Country for United States
1846–47	Ill-fated Donner Party resorts to cannibalism on its way to California
1847	Mormon settlers arrive in Utah
1848	Treaty of Guadalupe Hidalgo ends Mexican War, adds California and much of the Southwest to United States
1848	Gold discovered in California at Sutter's Mill
1849	California gold rush begins
1849	U.S. Department of the Interior established
1850s	Federally sponsored Pacific Railroad surveys explore the West
1854	Gadsden Purchase secures from Mexico a strip of southern Arizona and New Mexico
1859	Mining rushes in Nevada and Colorado
1860–61	Pony Express speeds east-west mail delivery
1862	Homestead Act
1864	Sand Creek Massacre in southeastern Colorado

Tourists at Mesa Verde National Park, Colorado

1869	Transcontinental railroad completed at Promontory Point, UT
1869	Wyoming Territory grants right to vote to women
1870–1910	Commercial lumber industry expands in Pacific Northwest
1872	Creation of Yellowstone National Park
1872	General Mining Act passed; authorizes and governs private mining activity on public lands
1873	Timber Culture Act encourages tree planting on enlarged homesteads
1876	Battle of the Little Bighorn in Montana Territory
1877	Desert Land Act allows settlers to apply for desert land in return for irrigating the land
1878	Bland-Allison Act stimulates western silver mining and production
1879	Creation of United States Geological Survey
1880s	Plains bison herds almost exterminated by commercial hunters
1882	Chinese Exclusion Act limits Asian immigration
1882	Butte, Montana, sees rapid expansion of copper mining at Anaconda Mine
1883	Southern Pacific railroad completed to Los Angeles
1883	Northern Pacific railroad reaches Tacoma, Washington
1885	Atchison, Topeka, and Santa Fe Railroad reaches Southern California
1886–87	End of open-range cattle era as blizzards decimate western herds
1887	Dawes Act initiates breakup and sale of Indian reservation lands
1887	Wright Act encourages creation of irrigation districts in California
1891	Forest Reserve Act passes, creates presidential authority to reserve western timber lands
1892	John Muir helps to found the Sierra Club
1893	Great Northern railroad reaches Seattle
1894	Carey Act encourages western states to develop irrigation projects
1897	Organic Act grants federal government authority over western forest reserves
1902	Newlands Reclamation Act expands federal role in promoting irrigation
1905	U.S. Forest Service created within Department of Agriculture
1906	San Francisco earthquake and fire
1906	Antiquities Act passes, granting expanded presidential power to protect public lands and create national monuments
1907	First Dry-Farming Congress, Denver, Colorado
1909	Enlarged Homestead Act recognizes needs of western ranchers
1910	Huge wildfires engulf northern Rockies
1910	First Pendleton Round-Up in Oregon
1914	Ludlow Massacre ends a strike by United Mine Workers in southern Colorado
1915–17	Panama-California Exposition in San Diego
1916	Organic Act creates National Park Service
1916	Federal Aid Road Act expands federal role in road construction
1917–18	U.S. involvement in World War I
1921	Signal Hill (Long Beach) oil boom begins in Southern California
1922	Colorado River Compact initiates regional water planning in Southwest
1924	National Origins Act further limits immigration
1924	Gila Wilderness created, first in national forests
1926	Dude Rancher's Association forms
1926	U.S. federal numbered highway system established
1929–41	Great Depression hits much of the West
1930s	Periodic Dust Bowl conditions strike Great Plains
1931	Nevada legalizes gambling
1933	New Deal initiatives begin under President Franklin D. Roosevelt
1934	Taylor Grazing Act regulates livestock use on western public lands
1934	Indian Reorganization Act grants greater tribal sovereignty

1935	Hoover Dam completed
1936	Opening of Sun Valley ski resort in Idaho by Union Pacific railroad
1940	Fort Peck Dam completed
1941–45	U.S. involvement in World War II, massive investments in the West
1942	Grand Coulee Dam completed
1942–45	Internment of Japanese Americans in western interior
1945	Atomic bomb successfully tested at Trinity Site, New Mexico
1946	Bureau of Land Management is created from the merging of the Government Land Office and U.S. Grazing Service
1950	Planned suburb of Lakewood, California, opens
1951	Stanford Industrial Park expands in Palo Alto, California
1954	Federal tax laws modified to accelerate depreciation of investments in commercial development; stimulates construction of strips and malls
1955	Disneyland opens in Anaheim, California
1955	*Gunsmoke* television series premieres
1956	Federal-Aid Highway Act initiates interstate system
1958	Boeing 707 enters into commercial air service
1958	Brooklyn Dodgers and New York Giants move west to California
1960	Sun City retirement community opens in southern Arizona
1962	*Surfin' Safari* album by the Beach Boys hits charts
1962	Vail ski resort opens in Colorado
1963	Planned suburb of Westlake Village opens in Southern California
1964	Federal Wilderness Act passes
1965	Immigration and Nationality Act liberalizes immigration laws
1965	Watts Riots in South Los Angeles
1966	Black Panther Party created in California
1968	César Chávez organizes farmworker strikes against grape growers
1968	Edward Abbey publishes *Desert Solitaire*
1970	National Environmental Policy Act passed by Congress; mandates environmental impact assessments
1972	Coastal Zone Management Act passes
1973	Endangered Species Act passes
1973	Oil crisis stimulates energy conservation and regional energy boom
1976	Apple Computer founded in California
1976	National Forest Management Act passes; shapes ecosystem management policies of Forest Service
1977	*Outside* magazine begins publication
1980	Mount Saint Helens erupts in southwestern Washington
1988	Indian Gaming Regulatory Act allows regulated gambling on reservations
1989	Steve Wynn's Mirage resort opens on Las Vegas Strip
1991	First authorized Burning Man event in Nevada's Black Rock Desert
1992	Los Angeles race riots
1992	Film adaptation of Norman Maclean's "A River Runs through It" is released
1994	North American Free Trade Agreement signed by United States, Canada, and Mexico
1994	Operation Gatekeeper allocates increased federal resources to protect international borders
2006	Secure Fence Act allocates more federal resources to harden the border, especially with Mexico
2007–10	Real estate slump crushes housing market in many western settings

FURTHER READING

NAVIGATING WESTERN LANDSCAPES: AN INTRODUCTION

For digging into western history generally, I recommend seven books: Patricia Limerick, *The Legacy of Conquest: The Unbroken Past of the American West* (New York: W. W. Norton, 1987); Richard White, *"It's Your Misfortune and None of My Own": A New History of the American West* (Norman: University of Oklahoma Press, 1991); D. W. Meinig, *The Shaping of America: A Geographical Perspective on 500 Years of History*, vol. 3: *Transcontinental America, 1850–1915* (New Haven, CT: Yale University Press, 1998), 29–183; Clyde A. Milner II, Carol A. O'Connor, and Martha A. Sandweiss, eds., *The Oxford History of the American West* (New York: Oxford University Press, 1994); Robert M. Utley, ed., *The Story of the West: A History of the American West and Its People* (New York: DK Publishing, 2003); Howard R. Lamar, ed., *The New Encyclopedia of the American West* (New Haven, CT: Yale University Press, 1998); and Derek Hayes, *Historical Atlas of the American West* (Berkeley: University of California Press, 2009).

For more on the complex notion of "landscape," read the thoughtful essays in D. W. Meinig, ed., *The Interpretation of Ordinary Landscapes* (New York: Oxford University Press, 1979); Michael P. Conzen, ed., *The Making of the American Landscape* (New York: Routledge, 2010); and Barry Lopez, ed., *Home Ground: Language for an American Landscape* (San Antonio: Trinity University Press, 2006). The contributions of J. B. Jackson can be sampled in *A Sense of Place, A Sense of Time* (New Haven, CT: Yale University Press, 1994), and *Landscape in Sight: Looking at America* (New Haven, CT: Yale University Press, 1997).

On the idea of landscape, see Denis Cosgrove and S. Daniels, eds., *The Iconography of Landscape: Essays on the Symbolic Representation, Design, and Use of Past Environments* (Cambridge: Cambridge University Press, 1988); Kent C. Ryden, *Mapping the Invisible Landscape: Folklore, Writing, and the Sense of Place* (Iowa City: University of Iowa Press, 1993); Don Mitchell, "New Axioms for Reading the Landscape: Paying Attention to Political Economy and Social Justice," in *Political Economies of Landscape Change: Places of Integrative Power*, ed. J. Wescoat and D. Johnston (Dordrecht: Springer, 2007), 29–50; and Richard H. Schein, "A Methodological Framework for Interpreting Ordinary Landscapes: Lexington, Kentucky's Courthouse Square," *Geographical Review* 99 (2009): 377–402.

On the role played by myths, literary works, art, and photography, consult Henry Nash Smith, *Virgin Land: The American West as Symbol and Myth* (Cambridge, MA: Harvard University Press, 1950); William H. Truettner, ed., *The West as America: Reinterpreting Images of the Frontier: 1820–1920* (Washington, D.C.: Smithsonian Institution, 1991); Richard Etulain, *Re-Imagining the Modern American West: A Century of Fiction, History, and Art* (Tucson: University of Arizona Press, 1996); and Martha A. Sandweiss, *Print the Legend: Photography and the American West* (New Haven, CT: Yale University Press, 2002).

On ordinary landscapes, ponder D. W. Meinig, "Environmental Appreciation: Localities as a Humane Art," *Western Humanities Review* 25 (1971): 1–11; Charles F. Wilkinson, "Toward an Ethic of Place," in *Beyond the Mythic West*, ed. Stewart L. Udall et al. (Salt Lake City, UT: Peregrine Smith Books, 1990), 71–104; Grady Clay, *Real Places: An Unconventional Guide to America's Generic Landscape* (Chicago, IL: University of Chicago Press, 1994); and William Cronon, "The Trouble with Wilderness; or, Getting Back to the Wrong Nature," in *Uncommon Ground: Rethinking the Human Place in Nature*, ed. William Cronon (New York: W. W. Norton, 1996), 69–90.

A sampling of studies that highlight slices of the western landscape include William deBuys, *Enchantment and Exploitation: The Life and Hard Times of a New Mexico Mountain Range* (Albuquerque: University of New Mexico Press, 1985); Thomas R. Vale and Geraldine R. Vale, *Western Images, Western Landscapes: Travels along U.S. 89* (Tucson: University of Arizona Press, 1989); Thomas T. Veblen and Diane C. Lorenz, *The Colorado Front Range: A Century of Ecological Change* (Salt Lake City: University of Utah Press, 1990); Kenneth I. Helphand, *Colorado: Visions of an American Landscape* (Niwot, CO: Roberts Rinehart, 1991); Carlos A. Schwantes, *So Incredibly Idaho! Seven Landscapes that Define the Gem State* (Moscow: University of Idaho Press, 1996); Judith L.

Castle Valley, southeast Utah

Meyer, *The Spirit of Yellowstone: The Cultural Evolution of a National Park* (Lanham, MD: Rowman and Littlefield, 1996); John B. Wright, *Montana Ghost Dance: Essays on Land and Life* (Austin: University of Texas Press, 1998); William Wyckoff, *Creating Colorado: The Making of a Western American Landscape, 1860–1940* (New Haven, CT: Yale University Press, 1999); Lary M. Dilsaver, William Wyckoff, and William Preston, "Fifteen Events that Have Shaped California's Human Landscape," *The California Geographer* 40 (2000): 3–78; Peter Goin and Paul Starrs, *Black Rock* (Reno: University of Nevada Press, 2005); and Nancy Langston, *Where Land and Water Meet: A Western Landscape Transformed* (Seattle: University of Washington Press, 2006).

Another source of local history and geography is the American Guide Series, originally published in the New Deal era. Produced by the Federal Writers' Project, these state guidebooks include highway tours that describe what travelers can see along major routes. Most have also been reissued and are in print. Also worth reading are back issues of *High Country News* (www.hcn.org) and the daily newspaper synopses provided by mountainwestnews.org, which scours the pages of major western newspapers and provides daily updates to subscribers.

Lastly, my nominees of some western classics that you can still feast on are Edward Abbey, *Desert Solitaire: A Season in the Wilderness* (New York: McGraw-Hill, 1968); Mary Austin, *The Land of Little Rain* (Boston, MA: Houghton Mifflin, 1903); Bernard DeVoto, ed., *The Journals of Lewis and Clark* (Boston, MA: Houghton Mifflin, 1953); Ivan Doig, *This House of Sky: Landscapes of a Western Mind* (New York: Harcourt Brace Jovanovich, 1978); Dan Flores, *Caprock Canyonlands: Journeys into the Heart of the Southern Plains* (Austin: University of Texas Press, 1990); Joseph Wood Krutch, *The Desert Year* (New York: Viking Press, 1951); Carey McWilliams, *Southern California Country: An Island on the Land* (New York: Duell, Sloane, and Pearce, 1946); John Muir, *The Mountains of California* (New York: Century Company, 1921); Rebecca Solnit, *Infinite City: A San Francisco Atlas* (Berkeley: University of California Press, 2010); and Wallace Stegner, *Wolf Willow: A History, a Story, and a Memory of the Last Plains Frontier* (New York: Viking, 1962).

1. NATURE'S FUNDAMENT

Frémont's narrative can be found in his *Report of the Exploring Expedition to the Rocky Mountains in the Year 1842* (Washington, D.C.: Gales and Seaton, 1845). The significance of the West's wide-open spaces is considered in Gretel Ehrlich, *The Solace of Open Spaces* (New York: Penguin Books, 1985); Dayton Duncan, *Miles from Nowhere: Tales from America's Contemporary Frontier* (New York: Viking, 1993); and Leonard Engel, ed., *The Big Empty: Essays on Western Landscapes as Narrative* (Albuquerque: University of New Mexico Press, 1994).

For understanding the geology and terrain in the West, begin with classics such as Nevin M. Fenneman, *Physiography of Western United States* (New York: McGraw-Hill, 1931); John S. Shelton, *Geology Illustrated* (San Francisco, CA: W. H. Freeman, 1966); and Charles B. Hunt, *Physiography of the United States* (San Francisco, CA: W. H. Freeman, 1967). The Roadside Geology series is a great companion on a trip. See, for example, David R. Lageson and Darwin R. Spearing, *Roadside Geology of Wyoming* (Missoula, MT: Mountain Press, 1988). In California, you can follow the San Andreas Fault with Robert Iacopi, *Earthquake Country: A Sunset Book* (Menlo Park, CA: Lane Books, 1971). In national parks, feast on Robert J. Lillie, *Parks and Plates: The Geology of Our National Parks* (New York: W. W. Norton, 2005). Keith Meldahl's splendid mix of geology and history can be sampled in *Hard Road West: History and Geology along the Gold Rush Trail* (Chicago, IL: University of Chicago Press, 2007), and *Rough-Hewn Land: A Geologic Journey from California to the Rocky Mountains* (Berkeley: University of California Press, 2011). Also consult the United States Geological Survey Web site, www.usgs.gov.

For more on the symbolic importance of western mountains, sample Kevin S. Blake's fine work in "Peaks of Identity in Colorado's San Juan Mountains," *Journal of Cultural Geography* 18, no. 2 (Spring–Summer 1999): 29–56, "Sacred and Secular Landscape Symbolism at Mount Taylor, New Mexico," *Journal of the Southwest* 41 (1999): 487–509, and "Imagining Heaven and Earth at Mount of the Holy Cross, Colorado," *Journal of Cultural Geography* 25:1 (February 2008): 1–30.

For an excellent book on key issues confronting western river systems, see Char Miller, ed., *River Basins of the American West* (Corvallis: Oregon State University Press, 2009). For specific settings, wade into Richard White, *The Organic Machine: The Remaking of the Columbia River* (New York: Hill and Wang, 1995); Blake Gumprecht, *The Los Angeles River: Its Life, Death, and Possible Rebirth* (Baltimore, MD: Johns Hopkins Press, 1999); William L. Lang and Robert C. Carricker, eds., *Great River of the West: Essays on the Columbia River* (Seattle: University of Washington Press, 1999); Ellen E. Wohl, *Virtual Rivers: Lessons from the Mountain Rivers of the Colorado Front Range* (New Haven, CT: Yale University Press, 2001); and April R. Summitt, *Contested Waters: An Environmental History of the Colorado River* (Boulder: University Press of Colorado, 2013). Online, explore www.americanrivers.org.

The Peterson Field Guide series has books on western birds, flowers, insects, animal tracks, and more. See, for example, Theodore E. Niehaus and Charles L. Ripper, *A Field Guide to Pacific States Wildflowers* (Boston, MA: Houghton Mifflin Harcourt, 1998). One of the best National Audubon Society Nature Guides is Stephen Whitney, *Western Forests* (New York: Alfred A. Knopf, 1997). For wild creatures, it is hard to beat Paul Schullery, *The Bears of Yellowstone* (Worland, WY: High Plains Publishing, 1992).

On western clouds and weather, the best regional overview is Bette Roda Anderson, *Weather in the West: From the Midcontinent to the Pacific* (Palo Alto, CA: American West Publishing, 1975). Also see John A. Day, *The Book of Clouds* (New York: Sterling Publishing, 2006). For sheer meteorological drama, read George R. Stewart's novel, *Storm* (New York: Random House, 1941).

Finally, there is a large scholarly literature in the fields of ecology, geography, and environmental history. Among many fine examples are Conrad Joseph Bahre, *A Legacy of Change: Historic Human Impact on Vegetation of the Arizona Borderlands* (Tucson: University of Arizona Press, 1991); Timothy Egan, *The Worst Hard Time: The Untold Story of Those Who Survived the Great American Dust Bowl* (Boston, MA: Houghton Mifflin, 2006); Mark Klett et al., *Third Views, Second Sights: A Rephotographic Survey of the American West* (Santa Fe: Museum of New Mexico Press, 2004); Laura Cunningham, *A State of Change: Forgotten Landscapes of California* (Berkeley, CA: Heyday Books, 2010); Robert M. Wilson, *Seeking Refuge: Birds and Landscapes of the Pacific Flyway* (Seattle: University of Washington Press, 2010); and Robin Grossinger, *Napa Valley Historical Ecology Atlas: Exploring a Hidden Landscape of Transformation and Resilience* (Berkeley: University of California Press, 2012). An excellent field guide to exotic and invasive plants is Sylvan R. Kaufman and Wallace Kaufman, *Invasive Plants: A Guide to Identification, Impacts, and Control of Common North American Species* (Mechanicsburg, PA: Stackpole Books, 2007).

On the role of fire, see Stephen J. Pyne, *Year of the Fires: The Story of the Great Fires of 1910* (New York: Penguin, 2001); Thomas R. Vale, ed., *Fire, Native Peoples, and the Natural Landscape* (Washington, D.C.: Island Press, 2002); and Mark Hudson, *Fire Management in the American West: Forest Politics and the Rise of Megafires* (Boulder: University Press of Colorado, 2011).

2. FARMS AND RANCHES

Excellent overviews of the rural landscape include John Fraser Hart's *The Rural Landscape* (Baltimore, MD: Johns Hopkins University Press, 1998), and John C. Hudson's *Across This Land: A Regional Geography of the United States and Canada* (Baltimore, MD: Johns Hopkins University Press, 2002). Although he has more of an eastern North American bias, don't neglect Eric Sloane and his amazing bookshelf of contributions on the rural landscape; see, for example, *Our Vanishing Landscape* (New York: Ballantine Books, 1974).

On western ranching, see Terry G. Jordan, *North American Cattle-Ranching Frontiers: Origins, Diffusion, and Differentiation* (Albuquerque: University of New Mexico Press, 1993); Terry G. Jordan, Jon T. Kilpinen, and Charles F. Gritzner, *The Mountain West: Interpreting the Folk Landscape* (Baltimore, MD: Johns Hopkins University Press, 1997); and Paul F. Starrs, *Let the Cowboy Ride: Cattle Ranching in the American West* (Baltimore, MD: Johns Hopkins University Press, 1998).

On the Great Plains, harvest copies of John C. Hudson, *Plains Country Towns* (Minneapolis: University of Minnesota Press, 1985); Mary W. M. Hargreaves, *Dry Farming in the Northern Great Plains: Years of Readjustment, 1920–1990* (Lawrence: University Press of Kansas, 1993); and R. Douglas Hurt, *The Big Empty: The Great Plains in the Twentieth Century* (Tucson: University of Arizona Press, 2011). A hugely useful reference is David Wishart, ed., *Encyclopedia of the Great Plains* (Lincoln: University of Nebraska Press, 2004). For those captivated by the aesthetics of grain elevators, see Bruce Selyem and Barbara Krupp Selyem, *Old Time Grain Elevators: Stories and Photography of a Vanishing Way of Life* (Bozeman, MT: Headhouse Books, 2007).

Selected intermountain locales have also attracted the attention of historians and historical geographers. Among the best and most detailed studies remain D. W. Meinig, *The Great Columbia Plain: A Historical Geography, 1805–1910* (Seattle: University of Washington Press, 1968); Robert A. Sauder, *The Lost Frontier: Water Diversion in the Growth and Destruction of Owens Valley Agriculture* (Tucson: University of Arizona Press, 1994); Marshall Bowen, *Utah People in the Nevada Desert: Homestead and Community on a Twentieth-Century Farmers' Frontier* (Logan: Utah State University Press, 1994); Mark Fiege, *Irrigated Eden: The Making of an Agricultural Landscape in the American West* (Seattle: University of Washington Press, 2000); and Robert A. Sauder, *The Yuma Reclamation Project: Irrigation, Indian Allotment, and Settlement along the Lower Colorado River* (Reno: University of Nevada Press, 2009).

California's agricultural landscape can also be explored. Begin with the visual feast within Paul F. Starrs and Peter Goin, *Field Guide to California Agriculture* (Berkeley: University of California Press, 2010). The citrus industry is the focus of Douglas C. Sackman, *Orange Empire: California and the Fruits of Eden* (Berkeley: University of California Press, 2007). The same industry is assessed from labor's perspective in

Gilbert G. González, *Labor and Community: Mexican Citrus Worker Villages in a Southern California County, 1900–1950* (Urbana: University of Illinois Press, 1994). The long, close, often violent relationship between California farmworkers and agribusiness has been traced by Carey McWilliams in *Factories in the Field: The Story of Migratory Farm Labor in California* (Boston, MA: Little, Brown, 1939) and *Ill Fares the Land: Migrants and Migratory Labor in the United States* (Boston, MA: Little, Brown, 1942). Also see Richard Walker, *The Conquest of Bread: 150 Years of Agribusiness in California* (New York: New Press, 2004), and Don Mitchell, *They Saved the Crops: Labor, Landscape, and the Struggle over Industrial Farming in Bracero-Era California* (Athens: University of Georgia Press, 2012).

For more on the story of organic gardening, community-supported agriculture, and the location of farmers' markets in the West, see the LocalHarvest Web site, the nation's leading online source of information on organic and locally produced food (visit www.localharvest.org/).

3. LANDSCAPES OF EXTRACTION

Any digging into the West's extractive landscape requires a larger economic and political context. See William G. Robbins, *Colony and Empire: The Capitalist Transformation of the American West* (Lawrence: University Press of Kansas, 1994), and D. W. Meinig, *The Shaping of America: A Geographical Perspective on 500 Years of History*, vol. 3: *Transcontinental America, 1850–1915* (New Haven, CT: Yale University Press, 1998).

On mining in the West, there are several useful overviews that offer historical context. For example, see Rodman Paul, *Mining Frontiers of the Far West, 1848–1880* (New York: Holt, Rinehart and Winston, 1963); William S. Greever, *The Bonanza West: The Story of the Western Mining Rushes, 1848–1900* (Norman: University of Oklahoma Press, 1963); and, for Montana, Kent Curtis, *Gambling on Ore: The Nature of Metal Mining in the United States, 1860–1910* (Boulder: University Press of Colorado, 2013).

When it comes to mining landscapes and interpreting what you see, consult Richard V. Francaviglia, *Hard Places: Reading the Landscape of America's Historic Mining Districts* (Iowa City: University of Iowa Press, 1991). For creative visual explorations of mining landscapes, see Peter Goin and C. Elizabeth Raymond, *Changing Mines in America* (Santa Fe, NM: Center for American Places, 2004). For the nitty-gritty details of poking around old mines and mining equipment, bring along a copy of Beth Sagstetter and Bill Sagstetter, *The Mining Camps Speak: A New Way to Explore the Ghost Towns of the American West* (Denver, CO: Benchmark, 1998).

Mining towns remain fascinating places. Colorado scholar Duane A. Smith has penned many detailed portraits of mining towns. Also see his useful overview, *Rocky Mountain Mining Camps: The Urban Frontier* (Bloomington: Indiana University Press, 1967). The special character of company mining towns is portrayed in James B. Allen, *The Company Town in the American West* (Norman: University of Oklahoma Press, 1966), and in Monica Perales, *Smeltertown: Making and Remembering a Southwest Border Community* (Chapel Hill: University of North Carolina Press, 2010). The West's roller-coaster uranium mining economy is recalled in Raye C. Ringholz, *Uranium Frenzy: Boom and Bust on the Colorado Plateau* (New York: W. W. Norton, 1989).

Fine portraits of particular mining communities can be explored in Carlos A. Schwantes, ed., *Bisbee: Urban Outpost on the Frontier* (Tucson: University of Arizona Press, 1992); Mary Murphy, *Mining Cultures: Men, Women, and Leisure in Butte, 1914–1941* (Urbana: University of Illinois Press, 1997); Michael A. Amundson, *Yellowcake Towns: Uranium Mining Communities in the American West* (Boulder: University Press of Colorado, 2002); and David Robertson, *Hard as the Rock Itself: Place and Identity in the American Mining Town* (Boulder: University Press of Colorado, 2006).

The large popular literature on ghost towns is fun to peruse and take along to your favorite mining district. Vintage examples include Muriel Sibell Wolle, *Stampede to Timberline: The Ghost Towns and Mining Camps of Colorado* (Denver, CO: Muriel S. Wolle, 1949); Lambert Florin, *Western Ghost Towns* (Seattle, WA: Superior Publishing, 1961); and William Carter, *Ghost Towns of the West* (Menlo Park, CA: Lane Magazine and Book Company, Sunset Books, 1971). Probing the deeper cultural meanings of such places (in Bodie, California) is Dydia DeLyser, "Authenticity on the Ground: Engaging the Past in a California Ghost Town," *Annals of the Association of American Geographers* 89 (1999): 602–32.

A growing literature has revealed how mining has affected western environments. An excellent sampling includes Duane A. Smith, *Mining America: The Industry and the Environment, 1800–1980* (Lawrence: University Press of Kansas, 1987); David Stiller, *Wounding the West: Montana, Mining, and the Environment* (Lincoln: University of Nebraska Press, 2000); Andrew C. Isenberg, *Mining California: An Ecological History* (New York: Hill and Wang, 2005); and Timothy J. LeCain, *Mass Destruction: The Men and Giant Mines that Wired America and Scarred the Planet* (New Brunswick, NJ: Rutgers University Press, 2009).

On the West's logging industry and its impact on regional landscapes, see Robert Ficken, *The Forested*

Land: History of Lumbering in Western Washington (Seattle: University of Washington Press, 1987); Michael Williams, *Americans and Their Forests: A Historical Geography* (Cambridge: Cambridge University Press, 1989); Nancy Langston, *Forest Dreams, Forest Nightmares: The Paradox of Old Growth in the Inland West* (Seattle: University of Washington Press, 1996); and William G. Robbins, *Hard Times in Paradise: Coos Bay, Oregon* (Seattle: University of Washington Press, 2006).

Western energy is a topic of enormous interest, but no one has written a comprehensive regional assessment. Parts of the story are nicely told in volumes such as Lee Scamehorn, *High Altitude Energy: A History of Fossil Fuels in Colorado* (Boulder: University Press of Colorado, 2002); Paul Sabin, *Crude Politics: The California Oil Market, 1900–1940* (Berkeley: University of California Press, 2004); and Robert Righter, *Windfall: Wind Energy in America Today* (Norman: University of Oklahoma Press, 2011). Solar enthusiasts may want to ponder Judith Lewis Mernit, "Sacrificial Land: Will Renewable Energy Devour the Mojave Desert?" *High Country News,* April 15, 2013.

Some of the best ways to follow recent developments in topics such as western coal mining, the Bakken oil boom, fracking, oil shale development, coal-bed methane drilling, and alternative energy sources are through research reports (often available online) from organizations such as Boulder's Center of the American West (visit www.centerwest.org), Tucson's Sonoran Institute (visit www.sonoraninstitute.org), or Bozeman's Headwaters Economics (visit www.headwaterseconomics.org). Also consult *High Country News* (visit www.hcn.org).

4. PLACES OF SPECIAL CULTURAL IDENTITY

Two useful textbooks to peruse in studying the concept of culture and how it is applied to landscape are Bret Wallach, *Understanding the Cultural Landscape* (New York: Guilford Press, 2005), and Paul L. Knox and Sallie A. Marston, *Human Geography: Places and Regions in Global Context* (Upper Saddle River, NJ: Pearson Prentice Hall, 2010).

The literature on the West's remarkable cultural diversity has flowered since 1970, and several useful collections examine the topic of place-based ethnic identity in North America. For a sampling of material that includes the West, see Allen G. Noble, ed., *To Build in a New Land: Ethnic Landscapes in North America* (Baltimore, MD: Johns Hopkins University Press, 1992); Richard L. Nostrand and Lawrence E. Estaville, eds., *Homelands: A Geography of Culture and Place across America* (Baltimore, MD: Johns Hopkins University Press, 2001); Ines M. Miyares and Christopher A. Airriess, eds., *Contemporary Ethnic Geographies in America* (Lanham, MD: Rowman and Littlefield, 2006); Richard H. Schein, ed., *Landscape and Race in the United States* (New York: Routledge, 2006); and Wilbur Zelinsky, *Not Yet a Placeless Land: Tracking an Evolving American Geography* (Amherst: University of Massachusetts Press, 2011). For a careful look at one of the West's most ethnically diverse settings, see James P. Allen and Eugene Turner, *Changing Faces, Changing Places: Mapping Southern Californians* (Northridge, CA: Center for Geographical Studies, 2002).

For the Native American experience in the West, begin with four splendid overviews. For the pre-1800 era, see Colin G. Calloway, *One Vast Winter Count: The Native American West before Lewis and Clark* (Lincoln: University of Nebraska Press, 2003). For the tumultuous and violent encounter with Euro-Americans, Dee Brown's classic, *Bury My Heart at Wounded Knee: An Indian History of the American West* (New York: Holt, Rinehart, and Winston, 1971), remains essential reading. The more recent Native American experience in North America is assessed in Ian Frazier, *On the Rez* (New York: Farrar, Straus and Giroux, 2000), and in Charles Wilkinson, *Blood Struggle: The Rise of Modern Indian Nations* (New York: W. W. Norton, 2005).

Detailed scholarly studies of the historical Native American experience in different western settings have multiplied greatly. An excellent regional sampling that emphasizes varied themes should include Keith H. Basso, *Wisdom Sits in Places: Landscape and Language among the Western Apache* (Albuquerque: University of New Mexico Press, 1996); Peter Nabokov and Lawrence Loendorf, *American Indians and Yellowstone Park: A Documentary Overview* (Yellowstone Park, WY: National Park Service, 2002); Steven W. Hackel, *Children of Coyote, Missionaries of Saint Francis: Indian-Spanish Relations in Colonial California, 1769–1850* (Chapel Hill: University of North Carolina Press, 2005); Ned Blackhawk, *Violence over the Land: Indians and Empires in the Early American West* (Cambridge, MA: Harvard University Press, 2008); Marsha Weisiger, *Dreaming of Sheep in Navajo Country* (Seattle: University of Washington Press, 2009); and Pekka Hämäläinen, *The Comanche Empire* (New Haven, CT: Yale University Press, 2009).

For the Hispano homeland in New Mexico and its distinctive imprint on the regional landscape, see Alvar Carlson, *The Spanish-American Homeland: Four Centuries in New Mexico's Rio Arriba* (Baltimore, MD: Johns Hopkins University Press, 1990); Richard L. Nostrand, *The Hispano Homeland* (Norman: University of Oklahoma Press, 1992); and John B. Wright and Carol L. Campbell, "Landscape Change in Hispano and

Chicano Villages of New Mexico," *Geographical Review* 98 (2008): 551–65.

The contemporary latinization of the United States is examined in Eliot Tiegel, *Latinization of America: How Hispanics Are Changing the Nation's Sights and Sounds* (Los Angeles, CA: Phoenix Books, 2007). Landscape expressions of that process are detailed in Lawrence A. Herzog, *From Aztec to High Tech: Architecture and Landscape across the Mexico–United States Border* (Baltimore, MD: Johns Hopkins University Press, 1999), and in Daniel D. Arreola, ed., *Hispanic Spaces, Latino Places: Community and Cultural Diversity in Contemporary America* (Austin: University of Texas Press, 2004). Two evocative case studies are James Rojas, "The Enacted Environment: Examining the Streets and Yards of East Los Angeles," in *Everyday America: Cultural Landscape Studies after* J. B. *Jackson,* ed. Chris Wilson and Paul Groth (Berkeley: University of California Press, 2003), 275–92, and Alex P. Oberle and Daniel D. Arreola, "Resurgent Mexican Phoenix," *Geographical Review* 98 (2008): 171–96.

Mormon regional history and cultural landscapes have been carefully documented. Begin with a classic and read Wallace Stegner's *Mormon Country* (New York: Duell, Sloan, and Pearce, 1942). For Mormon country as a distinctive culture region, see D. W. Meinig, "The Mormon Culture Region: Strategies and Patterns in the Geography of the American West, 1847–1964," *Annals of the Association of American Geographers* 55 (1965): 191–219, and Richard H. Jackson, "Mormon Wests: The Creation and Evolution of an American Region," in *Western Places, American Myths: How We Think about the West,* ed. Gary J. Hausladen (Reno: University of Nevada Press, 2003), 135–65. For thoughtful snapshots of the Mormon landscape, see Richard V. Francaviglia, *The Mormon Landscape* (New York: AMS Press, 1979); Gary B. Peterson and Lowell C. Bennion, *Sanpete Scenes: A Guide to Utah's Heart* (Eureka, UT: Basin/Plateau Press, 1987); and Paul F. Starrs, "Meetinghouses in the Mormon Mind: Ideology, Architecture, and Turbulent Streams of an Expanding Church," *Geographical Review* 99 (2009): 323–55.

African American communities in the West have provided settings for a growing academic literature. Valuable regional overviews are Quintard Taylor, *In Search of the Racial Frontier: African Americans in the West, 1528–1990* (New York: W. W. Norton, 1999), and Douglas Flamming, *African Americans in the West* (Santa Barbara, CA: ABC-Clio, 2009). Three excellent case studies are Shirley Moore, *To Place Our Deeds: The African American Community in Richmond, California, 1910–1963* (Berkeley: University of California Press, 2000); Josh Sides, *L.A. City Limits: African American Los Angeles from the Great Depression to the Present* (Berkeley: University of California Press, 2003); and Richard Morrill, "The Seattle Central District (CD) over Eighty Years," *Geographical Review* 103 (2013): 315–35.

Asian American cultural imprints across the West are an integral part of the regional landscape. On the internment of Japanese Americans during World War II, see the excellent summary provided by Jeffrey F. Burton, Mary M. Farrell, Florence B. Lord, and Richard W. Lord in *Confinement and Ethnicity: An Overview of World War II Japanese American Relocation Sites* (Seattle: University of Washington Press, 2002). For a fascinating tale of how the camps evolved after the war see Robert Wilson, "Landscapes of Promise and Betrayal: Reclamation, Homesteading, and Japanese American Incarceration," *Annals of the Association of American Geographers* 101 (2011): 424–44. For broader assessments of contemporary Asian American cultural geographies, see Shilpa Davé, LeiLani Nishime, and Tasha G. Oren, eds., *East Main Street: Asian American Popular Culture* (New York: New York University Press, 2005), and Wei Li, *Ethnoburb: The New Ethnic Community in Urban America* (Honolulu: University of Hawai'i Press, 2009). A good case study of the Vietnamese experience in Orange County, California, is offered by Karin Aguilar–San Juan, *Little Saigons: Staying Vietnamese in America* (Minneapolis: University of Minnesota Press, 2009). An excellent exploration of South Asian immigrants in the Pacific Northwest (via oral histories) is Amy Bhatt and Nalini Iyer's *Roots and Reflections: South Asians in the Pacific Northwest* (Seattle: University of Washington Press, 2013).

Both gender and sexual orientation have become major themes in regional cultural interpretations of the West. For example, see Karen M. Morin, *Frontiers of Femininity: A New Historical Geography of the Nineteenth-Century American West* (Syracuse, NY: Syracuse University Press, 2008), and Virginia Scharff and Carolyn Brucken, *Home Lands: How Women Made the West* (Berkeley: University of California Press, 2010). Gay and lesbian communities in the West can be explored in Gary J. Gates and Jason Ost, *The Gay and Lesbian Atlas* (Washington, D.C.: Urban Institute Press, 2004). A growing body of regional scholarship on the West can be sampled in Peter Boag, *Same-Sex Affairs: Constructing and Controlling Homosexuality in the Pacific Northwest* (Berkeley: University of California Press, 2003); Gary Atkins, *Gay Seattle: Stories of Exile and Belonging* (Seattle: University of Washington Press, 2003); Nan A. Boyd, *Wide-Open Town: A History of Queer San Francisco to 1965* (Berkeley: University of California Press, 2005); and Lillian Faderman and Stuart Timmons, *Gay L.A.: A History of Sexual Outlaws, Power Politics, and Lipstick Lesbians* (New York: Basic Books, 2006).

5. CONNECTIONS

Want to pick up the trail, follow the rails, or hitch a ride on the open road? Both popular and academic writers have paved the way. For a slower pace, seek out some of the detailed guidebooks that allow modern travelers to revisit many of the West's classic long-distance trails. Historian William E. Hill has followed just about all of them: see, for example, *The Santa Fe Trail: Yesterday and Today* (Caldwell, ID: Caxton Printers, 1992). Journey on the Oregon Trail with Julie Fanselow, *Traveling the Oregon Trail* (Guilford, CT: Falcon, 2001). For Lewis and Clark buffs, you cannot beat the three-volume compilation by Martin Plamondon II titled *Lewis and Clark Trail Maps: A Cartographic Reconstruction* (Pullman, WA: Washington State University Press, 2000–2004). This extraordinary set of resources takes you on a mile-by-mile sojourn and overlays modern maps with the historic route. For a great bird's-eye view, see Joseph A. Mussulman and Jim Wark's beautiful *Discovering Lewis and Clark from the Air* (Missoula, MT: Mountain Press, 2004).

The West's close connections with the railroad have received well-deserved attention. For understanding the larger national significance of railroad technology in shaping economic systems, landscapes, and modernity, read John Stilgoe, *Metropolitan Corridor: Railroads and the American Scene* (New Haven, CT: Yale University Press, 1983); James E. Vance Jr., *The North American Railroad: Its Origin, Evolution, and Geography* (Baltimore, MD: Johns Hopkins University Press, 1995); and Richard White, *Railroaded: The Transcontinentals and the Making of Modern America* (New York: W. W. Norton, 2011). The environmental impact of the railroads is creatively assessed by Mark Fiege in *The Republic of Nature: An Environmental History of the United States* (Seattle: University of Washington Press, 2012), 228–64.

Fine regional assessments of the impact of the railroads are Carlos A.Schwantes, *Railroad Signatures across the Pacific Northwest* (Seattle: University of Washington Press, 1996), and Carlos A. Schwantes and James P. Ronda, *The West the Railroads Made* (Seattle: University of Washington Press, 2008). Railroad buffs will also find detailed studies on individual lines and corporations. As examples, see Lucius Beebe, *Mixed Train Daily: A Book of Short-Line Railroads* (Berkeley, CA: Howell-North, 1961); Robert G. Athearn, *Rebel of the Rockies: The Denver and Rio Grande Western Railroad* (New Haven, CT: Yale University Press, 1962), and Ted Wurm and H. W. Demoro, *The Silver Short Line: A History of the Virginia and Truckee Railroad* (Glendale, CA: Trans-Anglo Books, 1983).

In the twentieth century, highways and the automobile transformed life across the West. The most comprehensive and well-written regional overview of twentieth-century connections is Carlos A. Schwantes, *Going Places: Transportation Redefines the Twentieth-Century West* (Bloomington: Indiana University Press, 2003). In addition, for broader national perspective, consult John A. Jakle and Keith A. Sculle, *Motoring: The Highway Experience in America* (Athens: University of Georgia Press, 2008), and Christopher W. Wells, *Car Country: An Environmental History* (Seattle: University of Washington Press, 2013). For a regional assessment, see William Wyckoff, *On the Road Again: Montana's Changing Landscape* (Seattle: University of Washington Press, 2006). The origin and evolution of the interstate highway system can be followed in Tom Lewis, *Divided Highways: Building the Interstate Highways, Transforming American Life* (New York: Penguin, 1997), and in D. W. Meinig, *The Shaping of America: A Geographical Perspective on 500 Years of History*, vol. 4: *Global America, 1915–2000* (New Haven, CT: Yale University Press, 2004).

The rich cultural landscapes of commercial strips and roadside architecture are celebrated by Chester H. Liebs in *Main Street to Miracle Mile: American Roadside Architecture* (Baltimore, MD: John Hopkins University Press, 1995); and by a collection of fine volumes by John A. Jakle and Keith A. Sculle, including *The Gas Station in America* (Baltimore, MD: Johns Hopkins University Press, 1994), and (with coauthor Jefferson S. Rogers) *The Motel in America* (Baltimore, MD: Johns Hopkins University Press, 1996). Also consult Warren Belasco, *Americans on the Road: From Autocamp to Motel, 1910–1945* (Cambridge, MA: MIT Press, 1979). For related cartographic explorations into the stylistic charms of old road maps, open a colorful copy of Douglas A. Yorke Jr., John Margolies, and Eric Baker, *Hitting the Road: The Art of the American Road Map* (San Francisco, CA: Chronicle Books, 1996).

Getting off the interstate to explore the old road? You will have some fine company if you throw a few more books into the backseat. In addition to the state guides produced by the Federal Writers' Project, include a copy of Jamie Jensen, *Road Trip USA: Cross-Country Adventures on America's Two-Lane Highways* (Chico, CA: Moon Publications, 1996). Particular highways have attracted their own enthusiasts, and none more so than Route 66 between Chicago and Los Angeles. You can follow the old road with strip maps in Tom Snyder, *The Route 66 Traveler's Guide and Roadside Companion* (New York: St. Martin's Press, 1990). Even better, Arthur Krim's *Route 66: Iconography of the American Highway* (Santa Fe, CA: Center for American Places, 2005) delves into the historical origins and cultural gravitas associated with the Mother Road.

Other western highways have also attracted attention. For a trip down the memory lane of two of the region's early named highways, enjoy Harold A. Meeks, *On the Road to Yellowstone: The Yellowstone Trail and American Highways 1900–1930* (Missoula, MT: Pictorial Histories, 2000), and Brian Butko, *Greetings from the Lincoln Highway: A Road Trip Celebration of America's First Coast-to-Coast Highway*, Centennial edition (Mechanicsburg, PA: Stackpole Books, 2013). For east-west U.S. 40, see the classic volume by George R. Stewart, *U.S. 40: Cross Section of the United States of America* (Boston, MA: Houghton Mifflin, 1953), and the follow-up volume, Thomas R. Vale and Geraldine R. Vale, *U.S. 40 Today: Thirty Years of Landscape Change in America* (Madison: University of Wisconsin Press, 1983). Elsewhere, enjoy getting lost with Joseph V. Tingley and Kris Ann Pizarro in *Traveling America's Loneliest Road: A Geologic and Natural History Tour through Nevada along U.S. Highway 50*, Nevada Bureau of Mines and Geology, Special Publication 26 (Reno: University of Nevada, 2000), or with Earl Thollander in his ageless *Back Roads of California* (Menlo Park, CA: Lane Magazine and Book Company, 1971).

Other western connections enjoy more uneven coverage. On electricity, see David E. Nye, *Electrifying America: Social Meanings of a New Technology, 1880–1940* (Cambridge, MA: MIT Press, 1990); James C. Williams, *Energy and the Making of Modern California* (Akron, OH: University of Akron Press, 1997); and Jay L. Brigham, *Empowering the West: Electrical Politics before FDR* (Lawrence: University Press of Kansas, 1998). The best regional history is Paul W. Hirt, *The Wired Northwest: The History of Electric Power, 1870s-1970s* (Lawrence: University Press of Kansas, 2012).

No single volume synthesizes western coastal connections, but you can explore both John R. Stilgoe's *Alongshore* (New Haven, CT: Yale University Press, 1994) and Elinor DeWire's *The DeWire Guide to Lighthouses of the Pacific Coast: California, Oregon and Washington* (Arcata, CA: Paradise Cay, 2010). For those with two-wheeled ambitions, bring along Vicky Spring and Tom Kirkendall, *Bicycling the Pacific Coast: A Complete Route Guide, Canada to Mexico* (Seattle: Mountaineers Books, 2005).

6. LANDSCAPES OF FEDERAL LARGESSE

The federal role in the West emerges as a key theme in major historical interpretations of the region, but additional studies provide added detail. On the role of federal scientists and military expeditions in exploring and opening up the West, see William H. Goetzmann, *Exploration and Empire: The Explorer and the Scientist in the Winning of the American West* (New York: Alfred A. Knopf, 1966). The historical evolution of public-lands policy in the West and the influence of the conservation ethic in shaping the management of public lands can be followed in Samuel P. Hays, *Conservation and the Gospel of Efficiency: The Progressive Conservation Movement, 1890–1920* (Cambridge, MA: Harvard University Press, 1959), and in Dyan Zaslowsky, *These American Lands: Parks, Wilderness, and the Public Lands* (New York: Henry Holt, 1986).

The twentieth-century role of federal agencies in western land management is reviewed in Richard H. Jackson, "Federal Lands in the Mountainous West," in *The Mountainous West: Explorations in Historical Geography*, ed. William Wyckoff and Lary M. Dilsaver (Lincoln: University of Nebraska Press, 1995), 253–77, and in John B. Wright, "Land Tenure: The Spatial Musculature of the American West," in *Western Places, American Myths: How We Think about the West*, ed. Gary J. Hausladen (Reno: University of Nevada Press, 2003), 85–110.

Making sense of the West's federally imposed survey system and the evolution of its political geography requires historical context. On the township-and-range surveys, trace the grid's history in William D. Pattison, *Beginnings of the American Rectangular Land Survey System, 1784–1800*, University of Chicago Department of Geography Research Paper 50 (Chicago, IL: University of Chicago Press, 1964), and in Hildegard Binder Johnson, *Order upon the Land: The U.S. Rectangular Land Survey and the Upper Mississippi Country* (New York: Oxford University Press, 1976). For the evolution of national and state boundaries across the West, see the story displayed in Charles O. Paullin and John K. Wright, *Atlas of the Historical Geography of the United States* (Washington, D.C.: Carnegie Institution of Washington, 1932), plates 57–65, and in Derek Hayes, *Historical Atlas of the American West* (Berkeley: University of California Press, 2009), 180–89. The hardening of the Mexico-U.S. border is assessed by Maurice Sherif, *The American Wall: From the Pacific Ocean to the Gulf of Mexico* (Austin: University of Texas Press, 2011).

Various elements of the military presence in the West have been assessed. The historical saga can be gleaned through William H. Goetzmann, *Army Exploration in the American West, 1803–1863* (New Haven, CT: Yale University Press, 1959), and Robert Wooster, *The American Military Frontiers: The United States Army in the West, 1783–1900* (Albuquerque: University of New Mexico Press, 2012). For the impacts of World War II and the Cold War, see Gerald D. Nash, *The American West Transformed: The Impact of the Second World War* (Bloomington: Indiana University Press, 1985), and Peter J. Westwick, ed., *Blue Sky Metropolis: The Aerospace Century in Southern California* (Berkeley: University of California Press, 2012).

The impact of the New Deal on the western landscape is examined in Richard Lowitt, *The New Deal and the West* (Bloomington: University of Indiana Press, 1984), and in Michael P. Malone and Richard W. Etulain, *The American West: A Twentieth-Century History* (Lincoln: University of Nebraska Press, 1989), 94–107. Specific New Deal programs have also been studied with impressive results. See, for example, Brian Q. Cannon, *Remaking the Agrarian Dream: New Deal Rural Resettlement in the Mountain West* (Albuquerque: University of New Mexico Press, 1996), and Neil Maher, *Nature's New Deal: The Civilian Conservation Corps and the Roots of the American Environmental Movement* (New York: Oxford University Press, 2009). Also admire the evocative images in Mary Murphy, *Hope in Hard Times: New Deal Photographs in Montana, 1936–1942* (Helena: Montana Historical Society Press, 2003).

The federal role in reengineering western water is masterfully detailed in Donald Worster, *Rivers of Empire: Water, Aridity, and the Growth of the American West* (New York: Oxford University Press, 1985), and in Marc Reisner, *Cadillac Desert: The American West and Its Disappearing Water* (New York: Penguin Books, 1986). Also read a pair of works of enduring and meticulous scholarship by Donald J. Pisani: *To Reclaim a Divided West: Water, Law, and Public Policy, 1848–1902* (Albuquerque: University of New Mexico Press, 1992) and *Water and American Government: The Reclamation Bureau, National Water Policy, and the West, 1902–1935* (Berkeley: University of California Press, 2002). For the West's mega-dam projects, see David P. Billington, Donald C. Jackson, and Martin V. Melosi, *The History of Large Federal Dams: Planning, Design, and Construction* (Denver, CO: U.S. Department of the Interior, Bureau of Reclamation, 2005).

A growing literature is tracing the evolution of the atomic West. To follow the story, consult two coauthored books by a productive pair of historians: Bruce Hevly and John M. Findlay, eds., *The Atomic West* (Seattle: University of Washington Press, 1998), and John M. Findlay and Bruce Hevly, *Atomic Frontier Days: Hanford and the American West* (Seattle: University of Washington Press, 2011). For the wide-ranging cultural consequences of the atomic age, see Scott C. Zeman and Michael A. Amundson, eds., *Atomic Culture: How We Learned to Stop Worrying and Love the Bomb* (Boulder: University Press of Colorado, 2004). Effective visual essays on the regional detritus (environmental and human) of the era are Richard Misrach and Myriam W. Misrach, *Bravo 20: The Bombing of the American West* (Baltimore, MD: Johns Hopkins University Press, 1990); Peter Goin, *Nuclear Landscapes* (Baltimore, MD: Johns Hopkins University Press, 1991); and Carole Gallagher, *American Ground Zero: The Secret Nuclear War* (New York: Random House, 1994).

The management of the West's federal lands is dominated by several agencies. The Web sites of the National Park Service (visit www.nps.gov), the U.S. Forest Service (visit www.fs.fed.gov), the Bureau of Land Management (visit www.blm.gov), and the U.S. Fish and Wildlife Service (visit www.fws.gov) are treasure troves of historical and current information on western lands.

Of these four agencies, the National Park Service has attracted the greatest scholarly interest. The genesis and evolution of the system are described in Alfred Runte, *National Parks: The American Experience* (Lincoln: University of Nebraska Press, 1979), and in Dayton Duncan and Ken Burns, *The National Parks: America's Best Idea* (New York: Knopf, 2009). A great summary of key federal legislation is Lary M. Dilsaver, ed., *America's National Park System: The Critical Documents* (Lanham, MD: Rowman and Littlefield, 1994). Key elements of the built environment are reconstructed in Linda Flint McClelland, *Building the National Parks: Historic Landscape Design and Construction* (Baltimore, MD: Johns Hopkins University Press, 1998), and in Ethan Carr, *Mission 66: Modernism and the National Park Dilemma* (Amherst: University of Massachusetts Press, 2007). Two fine examples of interpretive park histories are Lary M. Dilsaver and William C. Tweed, *Challenge of the Big Trees: A Resource History of Sequoia and Kings Canyon National Parks* (Three Rivers, CA: Sequoia Natural History Association, 1990) and Paul Schullery, *Searching for Yellowstone: Ecology and Wonder in the Last Wilderness* (Boston, MA: Houghton Mifflin, 1997).

On the role of the U.S. Forest Service in the West, see Harold K. Steen, *The United States Forest Service: A History* (Seattle: University of Washington Press, 1976), and Samuel P. Hays, *The American People and the National Forests: The First Century of the U.S. Forest Service* (Pittsburgh: University of Pittsburgh Press, 2009). For a thorough examination of the Bureau of Land Management, consult James R. Skillen, *The Nation's Largest Landlord: The Bureau of Land Management in the American West* (Lawrence: University Press of Kansas, 2009). A descriptive guide to the nation's wildlife refuges is Russell D. Butcher, *America's National Wildlife Refuges: A Complete Guide* (Lanham, MD: Roberts Rinehart, 2003).

On the evolution of federal wilderness in the West, see Craig Allin's excellent legislative summary in *The Politics of Wilderness Preservation* (Fairbanks: University of Alaska Press, 2008). Regional statistics are summarized in Ross W. Gorte, *Wilderness: Overview and Statistics* (Washington, D.C.: Congressional Research Office, 2010). On the cultural significance of wilderness, ponder the essays in William Cronon, ed., *Uncommon Ground: Rethinking the Human Place in Nature* (New York: W. W. Norton, 1995), and the various meanings of wilderness explored by Thomas R. Vale in *The American Wilderness: Reflections on Nature Protection in the United States* (Charlottesville: University of Virginia Press, 2005). The pioneering intellectual history on wilderness remains Roderick Nash, *Wilderness and the American Mind* (New Haven, CT: Yale University Press, 1967). Another excellent narrative is Donald Worster, *A Passion for Nature: The Life of John Muir* (New York: Oxford University Press, 2008).

Other outstanding studies have examined the causes and consequences of the wilderness movement in the West. Four of the best are Mark W. T. Harvey, *A Symbol of Wilderness: Echo Park and the American Conservation Movement* (Seattle: University of Washington Press, 1994); Paul S. Sutter, *Driven Wild: How the Fight against Automobiles Launched the Modern Wilderness Movement* (Seattle: University of Washington Press, 2002); David Louter, *Windshield Wilderness: Cars, Roads, and Nature in Washington's National Parks* (Seattle: University of Washington Press, 2006); and Kevin R. Marsh, *Drawing Lines in the Forest: Creating Wilderness Areas in the Pacific Northwest* (Seattle: University of Washington Press, 2007). For current wilderness issues and updates, visit www.wilderness.net.

7. CITIES AND SUBURBS

Many excellent studies open doors into western cities and suburbs. Important historical aspects of urbanization in the West are covered in all of the major interpretive works on the region. For more on the early evolution of the urban grid in the West and variations on that pattern, see John Reps, *The Forgotten Frontier: Urban Planning in the American West before 1890* (Columbia: University of Missouri Press, 1981). Regional overviews of twentieth-century trends can be followed in John Findlay, *Magic Lands: Western Cityscapes and American Culture after 1940* (Berkeley: University of California Press, 1992); William Wyckoff, "Cities of the Mountain West: Twentieth-Century Transformations," *Journal of the West* 41 (2002): 30–40; and Carl Abbott, *How Cities Won the West: Four Centuries of Urban Change in Western North America* (Albuquerque: University of New Mexico Press, 2008).

Urban landscapes have been studied from varied angles. The creation of mega civic and consumer landscapes is nicely summarized in Edward K. Muller, "Building American Cityscapes," in *The Making of the American Landscape,* ed. Michael P. Conzen (New York: Routledge, 2010), 303–28, and in Michael P. Conzen, "Developing Large-Scale Consumer Landscapes," in *Making of the American Landscape,* 423–50. The key role played by the City Beautiful design impulse is examined in William H. Wilson, *The City Beautiful Movement* (Baltimore, MD: Johns Hopkins University Press, 1989). The Denver story is nicely told in Thomas J. Noel and Barbara S. Norgren, *Denver: The City Beautiful and Its Architects, 1893–1941* (Denver, CO: Historic Denver, 1987).

Enduring guideposts to the everyday vernacular scene (including "city invisible") are Grady Clay's *Close-Up: How to Read the American City* (Chicago, IL: University of Chicago Press, 1973), and his *Real Places: An Unconventional Guide to America's Generic Landscape* (Chicago, IL: University of Chicago Press, 1994); and read many of J. B. Jackson's evocative essays on urban America in *A Sense of Place, A Sense of Time* (New Haven, CT: Yale University Press, 1994) and *Landscape in Sight: Looking at America* (New Haven, CT: Yale University Press, 1997). Larry Ford's excellent eye has also closely examined North America's urban landscapes: see, for example, *Cities and Buildings: Skyscrapers, Skid Rows, and Suburbs* (Baltimore, MD: Johns Hopkins University Press, 1994), *The Spaces between Buildings* (Baltimore, MD: John Hopkins University Press, 2000), and *America's New Downtowns: Revitalization or Reinvention?* (Baltimore, MD: Johns Hopkins University Press, 2003). For additional context on the recent forces reshaping the North American urban landscape, see Edward Relph, *The Modern Urban Landscape* (Baltimore, MD: Johns Hopkins University Press, 1987). Questions about preserving the urban landscape are addressed in Judy Mattivi Morley, *Historic Preservation and the Imagined West: Albuquerque, Denver, and Seattle* (Lawrence: University Press of Kansas, 2006).

For more on urban and suburban architecture in Southern California, enjoy Reyner Banham's engaging descriptions of Southern California in *Los Angeles: The Architecture of Four Ecologies* (Berkeley: University of California Press, 1971), and Barbara Rubin's "A Chronology of Architecture in Los Angeles," *Annals of the Association of American Geographers* 67 (1977): 521–37. Also see the colorful examples provided in Charles Phoenix, *Southern California in the '50s* (Santa Monica, CA: Angel City Press, 2001).

The popularity of the bungalow is explored in Robert Winter and Alexander Vertikoff, *American Bungalow Style* (New York: Simon and Schuster, 1996), and in Janet Ore, *The Seattle Bungalow: People and Houses, 1900–1940* (Seattle: University of Washington Press, 2006). John M. Faragher assesses the evolution of both the bungalow and the ranch house in "Bungalow and Ranch House: The Architectural Backwash of California," *Western Historical Quarterly* 32 (2001): 149–73. For the classic popular treatise on designing the ranch house, see Cliff May, *Western Ranch Houses* (Menlo Park, CA: Lane Publishing, 1946). Hunting for dingbats? You will need a copy of the "Field Guide to Dingbats," available from the Los Angeles Forum for Architecture and Urban Design (visit http://laforum.org/content/competitions/field-guide-to-dingbats). On the integration of nature and suburban life, find a spot on the patio to peruse Elizabeth Carney, "Suburbanizing Nature and Naturalizing Suburbanites: Outdoor-Living Culture and Landscapes of Growth," *Western Historical Quarterly* 38 (2007): 477–500.

In the suburbs and edge cities, retail and commercial settings have been studied by both Grady Clay (see his *Close-Up* and *Real Places*) and J. B. Jackson (see *A Sense of Place, A Sense of Time* and *Landscape in Sight: Looking at America*). Also consult Chester H. Liebs, *Main Street to Miracle Mile: American Roadside Architecture* (Baltimore, MD: Johns Hopkins University Press, 1995), and John A. Jakle and Keith A. Sculle, *Motoring: The Highway Experience in America* (Athens: University of Georgia Press, 2008) for their assessments of commercial strip landscapes. In addition, examine Joel Garreau, *Edge City: Life on the New Frontier* (New York: Doubleday, 1991); William Wyckoff, "Denver's Aging Commercial Strip," *Geographical Review* 82 (1992): 282–94; Richard Longstreth, *City Center to Regional Mall: Architecture, the Automobile, and Retailing in Los Angeles, 1920–1950* (Cambridge, MA: MIT Press, 1997); and Richard Longstreth, *The Drive-In, the Supermarket, and the Transformation of Commercial Space in Los Angeles, 1914–1941* (Cambridge, MA: MIT Press, 1999).

Suburbanization (including master-planned communities) has also been examined. For the larger historical picture, see Dolores Hayden, *Building Suburbia: Green Fields and Urban Growth, 1820–2000* (New York: Vintage Books, 2003). Hayden then examines some of the modern expressions of these processes in *A Field Guide to Sprawl* (New York: W. W. Norton, 2004). Different models of suburban growth in Southern California are evaluated in Greg Hise, *Magnetic Los Angeles: Planning the Twentieth-Century Metropolis* (Baltimore, MD: Johns Hopkins University Press, 1999). The larger social and political significance of suburbanization is also probed. Two valuable studies, both with a Southern California setting, are Eric Avila, *Popular Culture in the Age of White Flight: Fear and Fantasy in Suburban Los Angeles* (Berkeley: University of California Press, 2006), and Laura R. Barraclough, *Making the San Fernando Valley: Rural Landscapes, Urban Development, and White Privilege* (Athens: University of Georgia Press, 2011).

On the urban periphery, explore hillside letters with James J. Parsons, "Hillside Letters in the Western Landscape," *Landscape* 30, no. 1 (1988): 15–23, and Evelyn Corning, *Hillside Letters A to Z:A Guide to Hometown Landmarks* (Missoula, MT: Mountain Press, 2007). While it has an eastern U.S. focus, Jim Sterba's *Nature Wars* (New York: Crown Books, 2012) is a useful reminder of the interesting ecological dynamics encountered in places where the suburbs meet wildlands or open countryside.

Specific cities around the West have garnered additional attention. For a hard-edged look at the social inequalities and environmental challenges facing Los Angeles, see Mike Davis, *City of Quartz: Excavating the Future in Los Angeles* (New York: Vintage Books, 1990), and his *Ecology of Fear: Los Angeles and the Imagination of Disaster* (New York Vintage Books, 1998). For those keen to venture out into the Southern California scene, take along a copy of Laura Pulido, Laura Barraclough, and Wendy Cheng, *A People's Guide to Los Angeles* (Berkeley: University of California Press, 2012). Also of interest to those bound for Southern Cal is Dolores Hayden's excellent work *The Power of Place: Urban Landscapes as Public History* (Cambridge, MA: MIT Press, 1995). For a fine collection of essays focused on the region's longer environmental history, see William Deverell and Greg Hise, eds., *Land of Sunshine: An Environmental History of Metropolitan Los Angeles* (Pittsburgh, PA: University of Pittsburgh Press, 2006).

San Francisco's urban origins in the mining era are reviewed in Gunther Barth, *Instant Cities: Urbanization and the Rise of San Francisco and Denver* (New York: Oxford University Press, 1975). In addition, Gray Brechin's *Imperial San Francisco: Urban Power, Earthly Ruin* (Berkeley: University of California Press, 1999) describes how that city extended its economic, political, and environmental reach across a broad portion of the Golden State. From different perspectives, San Francisco's parks and protected spaces are the subject of Terence Young, *Building San Francisco's Parks, 1850–1930* (Baltimore, MD: Johns Hopkins University Press, 2004), and Richard A. Walker, *The Country in the City: The Greening of the San Francisco Bay Area* (Seattle: University of Washington Press, 2007). Most evocatively, Rebecca Solnit's *Infinite City: A San Francisco Atlas* (Berkeley: University of California Press, 2010) manages to portray the complex social and environmental history of the city in ways that are both aesthetically pleasing and profoundly unsettling.

Denver's urban ascent is traced in Barth's *Instant Cities* and can be supplemented by Stephen J. Leonard and Thomas J. Noel's *Denver: Mining Camp to Metropolis* (Niwot, CO: University Press of Colorado, 1990), which takes the story through much of the twentieth century. In the Southwest, read Patricia Gober's *Metropolitan Phoenix: Place Making and Community Building in the Desert* (Philadelphia: University of Pennsylvania Press, 2005) and Michael F. Logan's *Desert Cities: The Environ-*

mental History of Phoenix and Tucson (Pittsburgh, PA: University of Pittsburgh Press, 2006). Both human-environment relationships and questions of social class figure prominently in Matt Klingle, *Emerald City: An Environmental History of Seattle* (New Haven, CT: Yale University Press, 2007).

8. PLAYGROUNDS

Vintage guidebooks are still great windows into the playgrounds of the past, both in the East and in the West. For a representative assortment, imagine yourself traveling with John B. Bachelder, *Popular Resorts and How to Reach Them* (Boston, MA: John B. Bachelder, 1875); Bushrod W. James, *American Resorts with Notes upon Their Climate* (Philadelphia, PA: F. A. Davis, 1889); D. Appleton, *Appleton's General Guide to the United States and Canada,* part 2: *Western and Southern States* (New York: D. Appleton, 1893); or Karl Baedeker, *The United States with an Excursion into Mexico: A Handbook for Travelers,* 3rd rev. ed. (Leipzig: Karl Baedeker, 1904). Travelers in the West might also have consulted a sampling of popular publications such as Frederick E. Shearer, *The Pacific Tourist: An Illustrated Guide to the Pacific Railroad, California, and Pleasure Resorts across the Continent* (New York: Adams and Bishop, 1884); George A. Crofutt, *Crofutt's Grip-Sack Guide of Colorado* (Omaha: Overland Publishing, 1885); and Stanley Wood, *Over the Range to the Golden Gate: A Complete Tourist's Guide* (Chicago: R. R. Donnelley and Sons, 1912).

The tourist experience in the West has been explored by a growing number of scholars. Important topical and regional overviews would include Earl Pomeroy, *In Search of the Golden West: The Tourist in Western America* (New York: Alfred A. Knopf, 1957); Valerie Fifer, *American Progress: The Growth of the Transport, Tourist, and Information Industries in the Nineteenth-Century West* (Chester, CT: Globe Pequot, 1988); John Sears, *Sacred Places: American Tourist Attractions in the Nineteenth Century* (New York: Oxford University Press, 1989); Hal K. Rothman, *Devil's Bargains: Tourism in the Twentieth-Century American West* (Lawrence: University Press of Kansas, 1998); Orvar Löfgren, *On Holiday: A History of Vacationing* (Berkeley: University of California Press, 1999); and David M. Wrobel and Patrick T. Long, eds., *Seeing and Being Seen: Tourism in the American West* (Lawrence: University Press of Kansas, 2001).

Some studies focus on how a place is culturally and economically reinvented by tourism and by reimagining the past in fresh ways. In addition to the works cited above, creative examples of this scholarship from western American settings can be explored in Bonnie Christensen, *Red Lodge and the Mythic West: Coal Miners to Cowboys* (Lawrence: University Press of Kansas, 2002); Liza Nicholas, Elaine M. Bapis, and Thomas J. Harvey, eds., *Imagining the Big Open: Nature, Identity, and Play in the New West* (Salt Lake City: University of Utah Press, 2003); Dydia DeLyser, *Ramona Memories: Tourism and the Shaping of Southern California* (Minneapolis: University of Minnesota Press, 2005); and William Philpott, *Vacationland: Tourism and Environment in the Colorado High Country* (Seattle: University of Washington Press, 2013).

Specific recreational activities have also attracted more extensive attention. On dude ranching, see Laurence R. Bourne, *Dude Ranching: A Complete History* (Albuquerque: University of New Mexico Press, 1983). Sidle up to the Web site of the Dude Ranching Association of America to explore the industry today (visit www.duderanch.org). For rodeos, see Pomeroy's overview, *In Search of the Golden West,* and explore the Professional Rodeo Cowboys Association (visit www.prorodeo.com) to get a feel for life in the saddle. The rodeo scenes in Ivan Doig's novel *English Creek* (New York: Atheneum, 1984) also offer a terrific feel for the Montana version of that western tradition.

In the realm of sport hunting and fishing, to get a flavor for elite historical traditions, trawl through the riches of vintage narratives such as those found in Alfred M. Mayer, ed., *Sport with Gun and Rod in American Woods and Waters* (New York: Century, 1893), and Charles F. Holder, *Life in the Open: Sport with Rod, Gun, Horse, and Hound in Southern California* (New York: G. P. Putnam's Sons, 1906). On the special attractions of fly-fishing, see Ken Owens, "Fishing the Hatch: New West Romanticism and Fly-Fishing in the High Country," in *Imagining the Big Open,* ed. Nicholas, Bapis, and Harvey, 111–21, and Paul Schullery, *Cowboy Trout: Western Fly Fishing as if It Matters* (Helena: Montana Historical Society Press, 2006).

Other western diversions have also attracted interest. The historical importance of hot springs in the West is covered by general historical narratives (see above) but also detailed in Marshall Sprague's *Newport in the Rockies: The Life and Good Times of Colorado Springs* (Chicago, IL: Sage Books, 1961). For modern-day soaking options, consult the Falcon Guide series, which covers the entire West. For example, dip into Matt C. Bischoff, *Touring Hot Springs California and Nevada: A Guide to the Best Hot Springs in the Far West* (Guilford, CT: Falcon Press, 2012). For duffers interested in the recent diffusion of golf in the United States, tee up Darrell E. Napton and Christopher R. Laingen, "Expansion of Golf Courses in the United States," *Geographical Review* 98 (2008): 24–41. For a more detailed assessment of the history of Palm Springs as a winter golf resort, see Lawrence Culver's

The Frontier of Leisure: Southern California and the Shaping of Modern America (New York: Oxford University Press, 2010), 139–97. On the development of western ski resorts, see Rothman's *Devil's Bargains*, 202–86. For an alpine Sierra Nevada case study, read Roland Loeffler and Ernst Steinicke, "Amenity Migration in the U.S. Sierra Nevada," *Geographical Review* 97 (2007): 67–88. In the realm of sexual commerce in the West, in addition to a plethora of evocative online resources, consult Darrick Danta, "Ambiguous Landscapes of the San Pornando Valley," *Yearbook of the Association of Pacific Coast Geographers* 71 (2009): 15–30.

Amenity-oriented communities across the West have generated a growing literature that examines their history, land-use patterns, environmental impacts, and special planning challenges. Though now dated, explore the diagnostic maps of "New West" activities displayed in the *Atlas of the New West*, ed. William E. Riebsame, Hannah Gosnell, and David Theobald (New York: W. W. Norton, 1997). Two of the best general assessments of amenity communities in the West are Philip L. Jackson and Robert Kuhlken, *A Rediscovered Frontier: Land Use and Resource Issues in the New West* (Lanham, MD: Rowman and Littlefield, 2006), and William R. Travis, *New Geographies of the American West: Land Use and the Changing Patterns of Place* (Washington, D.C.: Island Press, 2007). Also explore California's growing Sierra Nevada foothills zone with Timothy P. Duane, *Shaping the Sierra: Nature, Culture, and Conflict in the Changing West* (Berkeley: University of California Press, 1998).

The sprawling, low-density "ranchette landscape" of the West has been assessed from different angles. Rural Colorado offers textbook examples of the phenomenon. Three excellent case studies to consult are Jeremy D. Maestas, Richard L. Knight, and Wendell C. Gilgert, "Biodiversity and Land-Use Change in the American Mountain West," *Geographical Review* 91 (2001): 509–24; Steven M. Schnell, Curtis J. Sorenson, Soren Larsen, Matthew Dunbar, and Erin McGrogan, "Old West and New West in Garden Park, Colorado," *Montana: The Magazine of Western History* 54, no. 4 (2004): 32–47; and John Harner and Bradley Benz, "The Growth of Ranchettes in La Plata County, Colorado, 1988–2008," *The Professional Geographer* 65 (2013): 329–44.

Retirement communities offer other variations. See John Findlay's chapter, "Sun City, Arizona: New Town for Old Folks" in his *Magic Lands: Western Cityscapes and American Culture after 1940* (Berkeley: University of California Press, 1992), 160–213. Also settle into Phillip B. Stafford, *Elderburbia: Aging with a Sense of Place in America* (Santa Barbara, CA: ABC-Clio, 2009). For more itinerant types interested in snowbird settlements, see James J. Parsons, "Quartzsite, Arizona: A Woodstock for RV'ers," *Focus* 42, no. 3 (1992): 1–3, and Nate Berg, "Mobile Nation," *High Country News*, March 15, 2010. For those in search of a bit of culture in the mix, see John Villani, *The 100 Best Art Towns in America* (Woodstock, VT: Countryman Press, 2005). The complex reinvention of Santa Fe as a center of art and culture is further probed by Rothman in *Devil's Bargains*, 81–112. More interested in security? Safely ponder the causes and consequences of gated communities with Edward J. Blakely and Mary Gail Snyder, *Fortress America: Gated Communities in the United States* (Washington, D.C.: Brookings Institution Press, 1997); Florence Williams, "Behind the Gate: A Look into the Fortified Rural Retreats of the West's Moneyed Elite," *High Country News*, November 11, 2002; and Setha Low, *Behind the Gates: Life, Security, and the Pursuit of Happiness in Fortress America* (New York: Routledge, 2003).

On Las Vegas, the greatest amenity community of them all, you will hit the jackpot with Hal K. Rothman, *Playing the Odds: Las Vegas and the Modern West* (Albuquerque: University of New Mexico Press, 2007), and Rex J. Rowley, *Everyday Las Vegas: Local Life in a Tourist Town* (Reno: University of Nevada Press, 2013). A lively look at how Las Vegas has been woven into America's larger popular culture is Larry Gragg, *Bright Light City: Las Vegas in Popular Culture* (Lawrence: University Press of Kansas, 2013). For the classic exegesis of the earlier Vegas strip, cruise along with Robert Venturi, Denise Scott Brown, and Steven Izenour, *Learning from Las Vegas* (Cambridge, MA: MIT Press, 1972).

INDEX

Mission Mountains, western Montana

D

E

F

I

J

K

L

M

N

O

P

Q

R

S

T

X

Y

Z

WEYERHAEUSER ENVIRONMENTAL BOOKS

The Natural History of Puget Sound Country, by Arthur R. Kruckeberg

Forest Dreams, Forest Nightmares: The Paradox of Old Growth in the Inland West, by Nancy Langston

Landscapes of Promise: The Oregon Story, 1800–1940, by William G. Robbins

The Dawn of Conservation Diplomacy: U.S.-Canadian Wildlife Protection Treaties in the Progressive Era, by Kurkpatrick Dorsey

Irrigated Eden: The Making of an Agricultural Landscape in the American West, by Mark Fiege

Making Salmon: An Environmental History of the Northwest Fisheries Crisis, by Joseph E. Taylor III

George Perkins Marsh, Prophet of Conservation, by David Lowenthal

Driven Wild: How the Fight against Automobiles Launched the Modern Wilderness Movement, by Paul S. Sutter

The Rhine: An Eco-Biography, 1815–2000, by Mark Cioc

Where Land and Water Meet: A Western Landscape Transformed, by Nancy Langston

The Nature of Gold: An Environmental History of the Alaska/Yukon Gold Rush, by Kathryn Morse

Faith in Nature: Environmentalism as Religious Quest, by Thomas R. Dunlap

Landscapes of Conflict: The Oregon Story, 1940–2000, by William G. Robbins

The Lost Wolves of Japan, by Brett L. Walker

Wilderness Forever: Howard Zahniser and the Path to the Wilderness Act, by Mark Harvey

On the Road Again: Montana's Changing Landscape, by William Wyckoff

Public Power, Private Dams: The Hells Canyon High Dam Controversy, by Karl Boyd Brooks

Windshield Wilderness: Cars, Roads, and Nature in Washington's National Parks, by David Louter

Native Seattle: Histories from the Crossing-Over Place, by Coll Thrush

The Country in the City: The Greening of the San Francisco Bay Area, by Richard A. Walker

Drawing Lines in the Forest: Creating Wilderness Areas in the Pacific Northwest, by Kevin R. Marsh

Plowed Under: Agriculture and Environment in the Palouse, by Andrew P. Duffin

Making Mountains: New York City and the Catskills, by David Stradling

The Fishermen's Frontier: People and Salmon in Southeast Alaska, by David F. Arnold

Shaping the Shoreline: Fisheries and Tourism on the Monterey Coast, by Connie Y. Chiang

Dreaming of Sheep in Navajo Country, by Marsha Weisiger

The Toxic Archipelago: A History of Industrial Disease in Japan, by Brett L. Walker

Seeking Refuge: Birds and Landscapes of the Pacific Flyway, by Robert M. Wilson

Quagmire: Nation-Building and Nature in the Mekong Delta, by David Biggs

Iceland Imagined: Nature, Culture, and Storytelling in the North Atlantic, by Karen Oslund

A Storied Wilderness: Rewilding the Apostle Islands, by James W. Feldman

The Republic of Nature: An Environmental History of the United States, by Mark Fiege

The Promise of Wilderness: American Environmental Politics since 1964, by James Morton Turner

Nature Next Door: Cities and Their Forests in the Northeastern United States, by Ellen Stroud

Pumpkin: The Curious History of an American Icon, by Cindy Ott

Car Country: An Environmental History, by Christopher W. Wells

Vacationland: Tourism and Environment in the Colorado High Country, by William Philpott

Tangled Roots: The Appalachian Trail and American Environmental Politics, by Sarah L. Mittlefehldt

Loving Nature, Fearing the State: American Environmentalism and Antigovernment Politics before Reagan, by Brian Allen Drake

Whales and Nations: Environmental Diplomacy on the High Seas, by Kirkpatrick Dorsey

Pests in the City: Flies, Bedbugs, Cockroaches, and Rats, by Dawn Day Biehler

How to Read the American West: A Field Guide, by William Wyckoff

Behind the Curve: Science and the Politics of Global Warming, by Joshua P. Howe

WEYERHAEUSER ENVIRONMENTAL CLASSICS

The Great Columbia Plain: A Historical Geography, 1805–1910, by D. W. Meinig

Mountain Gloom and Mountain Glory: The Development of the Aesthetics of the Infinite, by Marjorie Hope Nicolson

Tutira: The Story of a New Zealand Sheep Station, by Herbert Guthrie-Smith

A Symbol of Wilderness: Echo Park and the American Conservation Movement, by Mark Harvey

Man and Nature: Or, Physical Geography as Modified by Human Action, by George Perkins Marsh; edited and annotated by David Lowenthal

Conservation in the Progressive Era: Classic Texts, edited by David Stradling

DDT, Silent Spring, *and the Rise of Environmentalism: Classic Texts,* edited by Thomas R. Dunlap

The Environmental Moment, 1968–1972, by David Stradling

The Wilderness Writings of Howard Zahniser, edited by Mark Harvey

CYCLE OF FIRE, BY STEPHEN J. PYNE

Fire: A Brief History

World Fire: The Culture of Fire on Earth

Vestal Fire: An Environmental History, Told through Fire, of Europe and Europe's Encounter with the World

Fire in America: A Cultural History of Wildland and Rural Fire

Burning Bush: A Fire History of Australia

The Ice: A Journey to Antarctica